TRANSMISSION LINES, WAVEGUIDES, AND SMITH CHARTS

TRANSMISSION LINES, WAVEGUIDES, AND SMITH CHARTS

RICHARD L. LIBOFF
G. CONRAD DALMAN
Cornell University

MACMILLAN PUBLISHING COMPANY
NEW YORK

Collier Macmillan Publishers
LONDON

Macmillan Publishing Company
866 Third Avenue, New York, NY 10022

Collier Macmillan Canada, Inc.

Printed in the United States of America

printing number

1 2 3 4 5 6 7 8 9 10

Library of Congress Cataloging in Publication Data

Liboff, Richard L., 1931–
Transmission lines, waveguides, and Smith charts.

Includes index.
1. Microwave transmission lines. 2. Wave guides.
3. Smith charts. I. Dalman, G. Conrad. II. Title.
TK7876.L4783 1985 621.381'32 84-28542
ISBN 0-02-949540-7

For

Lisa
David
Diana
Kristine
Karen
Conrad

CONTENTS

PREFACE

This text is designed for a course in electromagnetic theory which is usually offered during the junior year to students of electrical engineering. The syllabus includes some electromagnetic theory, a strong emphasis on transmission lines and waveguides, and instruction on the theory and application of the Smith chart.

The present description begins with a brief review of some AC circuit theory. The main subject matter starts in Chapter 2, where terminology important in the study of transmission lines is set forth. Chapter 3 concerns the theory of DC pulses. Emphasis here is placed on the understanding of "bounce diagrams." The chapter concludes with some important applications, including that of the Blumlein configuration.

Expressions for impedance along a transmission line with a generator at one end and a load at the other end are developed in Chapter 4. In this chapter notions of the standing-wave ratio and the quarter-wave transformer are introduced. The chapter concludes with a brief description of transport along a nerve cell. This material was included due to the recent growth in interest among electrical engineering students to topics in bioengineering.

The Smith chart is developed in the next chapter, and application is made to the single- and double-stub tuner.

Elements of statics and Maxwell's equations are introduced in Chapter 6. Attention is directed to plane-wave solutions in the following chapter. The notion of wave impedance emerges in the study of transmission of radiation through an interface.

The text continues with a study of the mode theory of rectangular waveguides. Distinction is drawn between TE and TM waves, and the significant notion of cutoff frequency for waveguides is developed. The important analogy between (V, I) for transmission lines and relevant components of $(\mathbf{E}, \mathbf{H})$ for waves is drawn, thereby allowing application of the Smith chart to waveguide problems.

The work concludes with a description of transmission lines important to present-day technology. Microstrips, strip lines, and other practical transmission lines are discussed.

Four appendices are included. Of these, Appendix D is most closely related to the text. This appendix includes a compilation of transmission lines, waveguides, and resonant cavities. Properties of junctions, directional couplers, and other components important to microwave transmission are also found in this Appendix.

A list of the symbols and abbreviations follows this Preface.

We have sought to keep the presentation clear, self-contained, and readable. The text should provide students with a basic understanding of transmission lines and waveguides. It will also prove valuable as a basic reference work fundamental to the study of analog and digital devices and systems.

We wish to express our deep appreciation to Allyson Yarbrough, Kenneth Gardner and Rebecca Greenberg for their careful reading of the manuscript of this work.

Richard L. Liboff and G. Conrad Dalman
Ithaca, New York

LIST OF SYMBOLS AND ABBREVIATIONS

$\mathbf{a}_x, \mathbf{a}_y, \mathbf{a}_z$	Unit vectors, Cartesian
a	Radius; width
B	Susceptance
b	Susceptance, normalized; width; separation; radius
B_C, B_L	Susceptance: capacitive, inductive
$\mathbf{B}$	Magnetic field
C	Capacitance; capacitance per unit length
c	Velocity of light
C_m	Capacitance, membrane
$\mathbf{D}$	Displacement field
D	Diameter, outer
d	Diameter, inner; length
$\mathbf{E}$	Electric field
f	Frequency
f_c	Cutoff frequency
$\mathbf{F}$	Vector field, arbitrary
F	Fraction of transmitted power
F	Farad
G	Conductance
G_m	Conductance, membrane
g	Conductance, normalized
h	Thickness; separation
$\mathbf{H}$	Magnetic field intensity
H	Henry

$\mathbf{I}$	Current, complex
I	Current amplitude
I_S	Current, magnitude of source
I_+, I_-	Current wave: forward, reverse
I_D	Current, displacement
$\mathbf{J}$	Current density
J	joule
$J_0(x)$	Bessel function, first kind, zero order, argument x
$J_1(x)$	Bessel function, first kind, first order, argument x
k	Wavenumber
k_c	Wavenumber, cutoff
K	$1/4\pi\epsilon_0$; current, surface (A/m)
L	Inductance; inductance per unit length; length
l	Length
P	Power
P_{loss}	Power lost from system
$\mathbf{P}$	Poynting vector
Q	Energy parameter of system, dimensionless
Q_C, Q_L	Q: capacitive, inductive
q	Charge
R	Resistance
r	Resistance, normalized; radius
R_C, R_L, R_S	Resistance: capacitive, inductive, source
res	Resonant
rms	Root-mean-square
S	Standing-wave ratio; center-strip width
s	seconds
t	Time; thickness
U	Energy, electromagnetic
u	Energy density; reflection coefficient, real part
v	Velocity wave; reflection coefficient, imaginary part
v_p	Velocity, phase
v_G	Velocity, group
$\mathbf{V}$	Voltage, complex
V	Voltage, magnitude
V_S	Voltage, magnitude of source
V_+, V_-	Voltage wave: forward, reverse
W	Watt
w	Width
W_C, W_L	Energy: capacitive, inductive
X	Reactance
X_C, X_L	Reactance: capacitive, inductive
x	Reactance, normalized; distance, horizontal; argument of the Bessel function

(x, y, z)	Displacement, Cartesian
$\mathbf{Y}$	Admittance, complex
$\mathbf{Y}_C, \mathbf{Y}_L, \mathbf{Y}_R$	Admittance, complex: capacitive, inductive, resistive
y	Admittance, normalized; distance, vertical
Z	Distance
$\mathbf{Z}$	Impedance, complex
$\mathbf{Z}_C, \mathbf{Z}_L, \mathbf{Z}_R$	Impedance, complex: capacitive, inductive, resistive
Z_0	Impedance, characteristic
z	Impedance, normalized
α	Real part of ν. Real part of λ
$\boldsymbol{\beta}$	Wavenumber, vector
β	Wavenumber, scalar
γ	Propagation constant
δ	Penetration depth
ϵ	Permittivity (dielectric constant)
ϵ_r	Dielectric constant, relative
ϵ_{re}	Dielectric constant, relative effective
ϵ', ϵ''	Dielectric constant: real part, imaginary part
θ	Angle
κ	Decay constant
λ	Wavelength
λ_g	Wavelength, guide
ν	Frequency, complex
ρ	Reflection coefficient; charge density
$\rho_{sc}, \rho_{oc}, \rho_L$	Reflection coefficient: short-circuit, open-circuit, load
σ	Charge density, surface; conductivity (mho/m)
τ	Transmission coefficient
ϕ	Phase
Φ	Scalar function, arbitrary
Φ_B	Flux, magnetic field
ω	Frequency, angular
ω_c	Frequency, cutoff
$\bar{\omega}$	Imaginary part of ν

PART ONE

VOLTAGE AND CURRENT WAVES

CHAPTER

ONE

ELEMENTS OF CIRCUIT THEORY

In this introductory chapter we review some elementary notions and terminology of circuit theory which are important in the theory of waveguides and transmission lines.

1.1 THE SERIES *LRC* CIRCUIT. IMPEDANCE

Consider that a series *LRC* circuit oscillates in response to an harmonic voltage supply as shown in Fig. 1.1. We are concerned with the steady-state properties of the circuit. Voltage drops across the various elements are written in terms of complex impedance **Z** and complex current **I**. In our notation boldface symbols represent complex values.[1] Complex component impedances are given by

$$\mathbf{Z}_L = j\omega L \equiv jX_L$$

$$\mathbf{Z}_C = \frac{1}{j\omega C} = -jX_C$$

$$\mathbf{Z}_R = R \tag{1.1}$$

Equating the source voltage to the sum of the voltage drops across the three elements gives

$$\mathbf{V}_S = V_S e^{j\omega t} = \mathbf{I}(\mathbf{Z}_L + \mathbf{Z}_C + \mathbf{Z}_R) \tag{1.2}$$

[1]A brief review of complex variables is given in Appendix C.

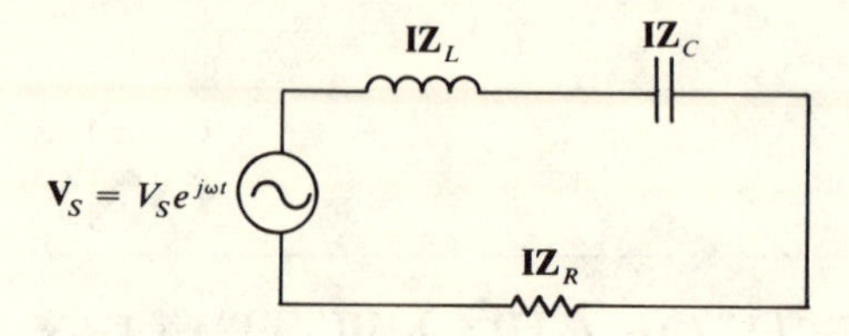

Figure 1.1 The series *LRC* circuit.

The total complex impedance may be rewritten

$$\mathbf{Z}_L + \mathbf{Z}_C + \mathbf{Z}_R = Ze^{j\phi} \tag{1.3}$$

where

$$Z = |R + j(X_L - X_C)|$$
$$= \sqrt{R^2 + (X_L - X_C)^2} = \sqrt{R^2 + \left(\omega L - \frac{1}{\omega C}\right)^2} \tag{1.4}$$

and

$$\tan\phi = \frac{X_L - X_C}{R} \tag{1.5}$$

(See Fig. 1.2.) Combining (1.2) and (1.3) gives the steady-state current

$$\mathbf{I}(t) = \frac{V_S e^{j(\omega t - \phi)}}{Z} = \frac{\mathbf{V}_S e^{-j\phi}}{Z} \tag{1.6}$$

Note that we have set $\mathbf{V}_S \equiv V_S e^{j\omega t}$. The current lags behind the source voltage by ϕ radians. On the complex $\mathbf{V}, \mathbf{I}$ plane this phase lag appears as shown in Fig. 1.3. Taking the real part of (1.2), we find

$$I = \operatorname{Re}\mathbf{I} = \frac{V_S}{Z}\cos(\omega t - \phi)$$
$$= I_S \cos(\omega t - \phi) \tag{1.7}$$

where I_S, the amplitude of the current, is related to V_S, the amplitude of the

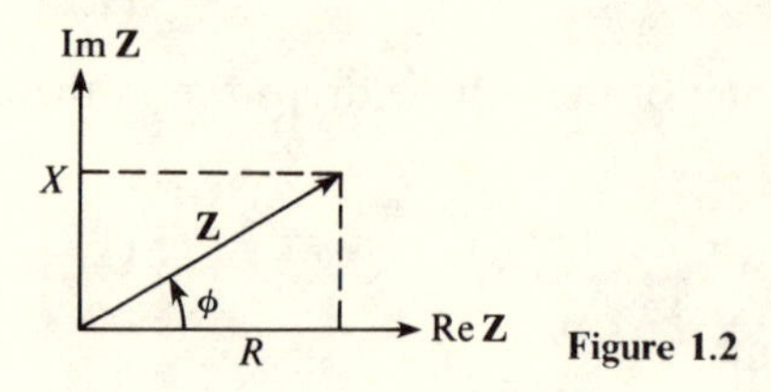

Figure 1.2

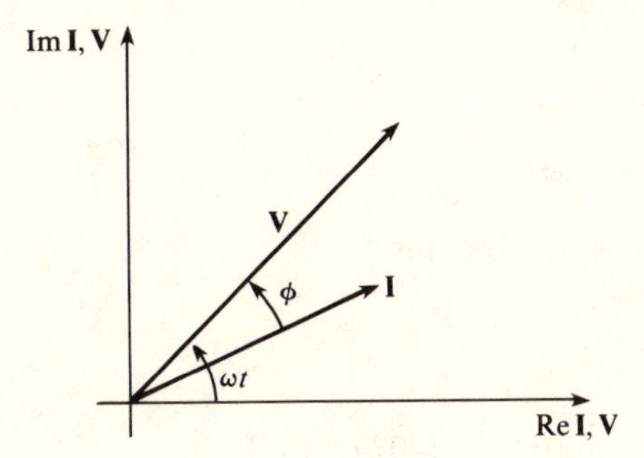

Figure 1.3 The voltage leads the current by ϕ radians in the series *LRC* circuit.

source voltage, through the magnitude of impedance:

$$I_S = \frac{V_S}{Z} = \frac{V_S}{\sqrt{R^2 + (X_L - X_C)^2}} = \frac{V_S}{\sqrt{R^2 + \left(\omega L - \frac{1}{\omega C}\right)^2}} \tag{1.8}$$

At resonance $X_L = X_C$, and the network is purely resistive. At the corresponding frequency, given by

$$\omega^2 = \frac{1}{LC}$$

the amplitude of current is *maximum*:

$$I_{S,\,\mathrm{res}} = \frac{V_0}{R}$$

From (1.5) we see that at resonance the current is in phase with the source voltage:

$$\tan\phi_{\mathrm{res}} = 0, \qquad \phi_{\mathrm{res}} = 0$$

Example 1.1

A voltage supply oscillates at 10 MHz and has peak voltage 15.0 volts. This source voltage is connected to a series *LRC* circuit with component values $L = 3.0\ \mu\text{H}$, $C = 10.0$ pF, and $R = 100$ ohms.

a. What is the complex impedance of the network?
b. What is the peak current through the circuit?
c. What is the phase difference between the current and applied voltage?
d. What is the rms voltage across the capacitor?

Ans.

a.

$$\mathbf{Z} = R + j\left(\omega L - \frac{1}{\omega C}\right)$$

$$\mathbf{Z} = 100 - j1403 \text{ ohms}$$

b.

$$I_S = V_S/Z$$

$$Z = \sqrt{(100)^2 + (1403)^2} = 1407 \text{ ohms}$$

$$I_S = 10.7 \text{ mA}$$

c.

$$\tan\phi = \frac{X_L - X_C}{R} = -14.03$$

$$\phi = -1.50 \text{ radians}$$

The current *leads* the applied voltage by 1.50 radians.

d.

$$\mathbf{V}_C = \mathbf{Z}_C \mathbf{I}_C$$

$$= \frac{\mathbf{I}_C}{j\omega C}$$

With (1.6) we obtain

$$\mathbf{V}_C = \frac{I_S e^{j(\omega t - \phi)}}{j\omega C} = \frac{I_S}{\omega C} e^{-j\pi/2} e^{j(\omega t - \phi)}$$

Forming the real part of this expression gives

$$V_C = \operatorname{Re} \mathbf{V}_C = \frac{I_S}{\omega C} \cos\left(\omega t - \phi - \frac{\pi}{2}\right)$$

$$= \frac{I_S}{\omega C} \sin(\omega t - \phi)$$

To obtain the rms voltage we must calculate the integral

$$V_{C,\text{rms}}^2 = \langle V_C^2 \rangle = \frac{\omega}{2\pi} \int_0^{2\pi/\omega} V_C^2(t)\, dt$$

which gives

$$V_{C,\text{rms}} = \frac{1}{\sqrt{2}} \left(\frac{I_S}{\omega C}\right)$$

For the example at hand we obtain

$$V_{C,\text{rms}} = 11.9 \text{ volts}$$

1.2 THE PARALLEL *LRC* CIRCUIT. ADMITTANCE

Now we consider that a parallel *LRC* circuit oscillates in response to an oscillating voltage supply. The circuit is shown in Fig. 1.4.

The current through the generator may be written in terms of the admittance of the network **Y** (see Fig. 1.5):

$$\mathbf{I}(t) = \mathbf{Y}\mathbf{V}_S \tag{1.9}$$

We see that admittance is merely the inverse of the impedance:

$$\mathbf{Y} = \frac{1}{\mathbf{Z}} \tag{1.10}$$

This current through the generator is equal, by Kirchhoff's law, to the sum of the individual currents through the parallel circuit elements:

$$\mathbf{I}(t) = \mathbf{I}_R + \mathbf{I}_C + \mathbf{I}_L$$

$$\mathbf{I}(t) = \mathbf{V}_S(\mathbf{Y}_R + \mathbf{Y}_C + \mathbf{Y}_L) \tag{1.11}$$

Comparison of this last expression with (1.9) indicates that for the parallel *LRC* circuit we may write

$$\mathbf{Y} = \mathbf{Y}_R + \mathbf{Y}_C + \mathbf{Y}_L \tag{1.12}$$

(See Fig. 1.6.) We may conclude that *admittances of parallel circuit elements add*. On the other hand, referring to (1.3) we see that for the *LRC series* circuit

$$\mathbf{Z} = \mathbf{Z}_R + \mathbf{Z}_C + \mathbf{Z}_L \tag{1.13}$$

(See Fig. 1.7.) It follows that *impedances of series-circuit elements add.*

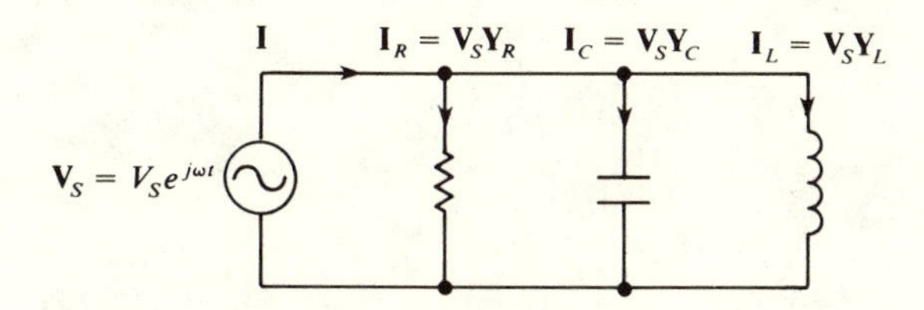

Figure 1.4 The parallel *LRC* circuit.

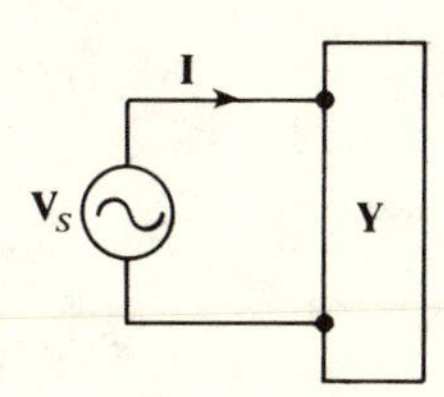

Figure 1.5

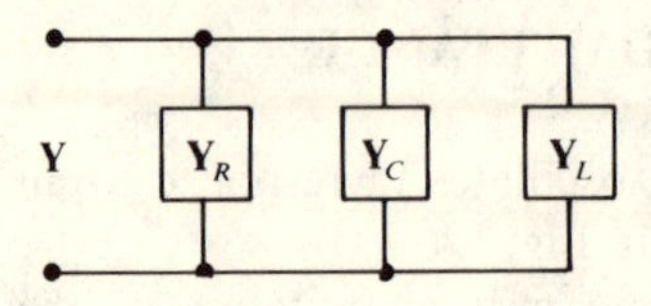

Figure 1.6 Admittances of parallel elements add.

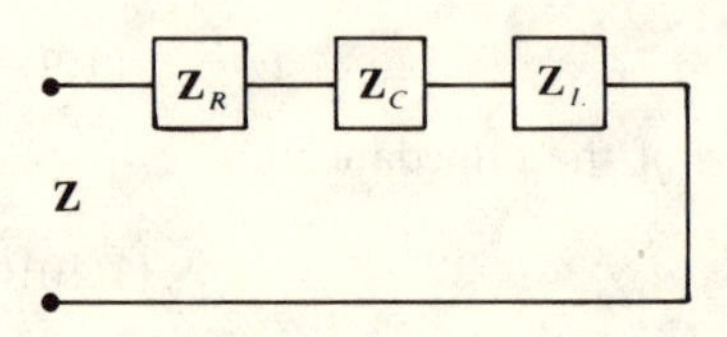

Figure 1.7 Impedances of series elements add.

The expressions for component admittances which enter into (1.12) are obtained from the component impedance expressions (1.1):

$$\mathbf{Y}_L = \frac{1}{\mathbf{Z}_L} = \frac{1}{j\omega L} = -jB_L$$

$$\mathbf{Y}_C = \frac{1}{\mathbf{Z}_C} = j\omega C = jB_C$$

$$\mathbf{Y}_R = \frac{1}{\mathbf{Z}_R} = \frac{1}{R} = G \tag{1.14}$$

It follows that the imput admittance of the network, or equivalently, the admittance "seen" by the generator, may be written

$$\mathbf{Y} = Ye^{j\phi} \tag{1.15}$$

where

$$Y = |G + j(B_C - B_L)| = \sqrt{\frac{1}{R^2} + \left(\omega C - \frac{1}{\omega L}\right)^2} \tag{1.16}$$

and

$$\tan\phi = \frac{B_C - B_L}{G} \tag{1.17}$$

(See Fig. 1.8.) The imaginary part B of $\mathbf{Y}$ is called the *susceptance*. For the parallel circuit being considered, B has the value

$$B = \omega C - \frac{1}{\omega L} \tag{1.18}$$

The real part of Y is called the *conductance*:

$$G = \frac{1}{R} \tag{1.19}$$

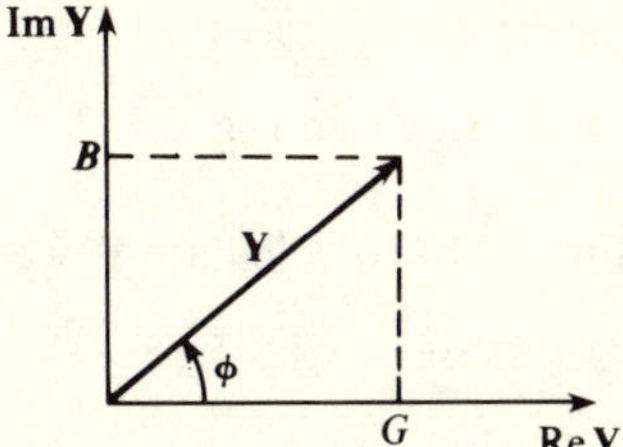

Figure 1.8 Components of the admittance **Y**.

Substituting (1.15) into (1.9), we obtain

$$\mathbf{I}(t) = \mathbf{V}_S Y e^{j\phi} = V_S Y e^{j(\omega t + \phi)} \tag{1.20}$$

Taking the real part of both sides of this equation gives

$$I(t) = V_S Y \cos(\omega t + \phi) \tag{1.21}$$

The current leads the voltages by the phase ϕ given by (1.17). On the complex **V**, **I** plane this displacement appears as shown in Fig. 1.9.

Equation (1.21) may be rewritten

$$I(t) = I_S \cos(\omega t + \phi) \tag{1.22}$$

where the amplitude of current I_S is given by

$$I_S = V_S Y$$

$$I_S = V_S \sqrt{G^2 + (B_C - B_L)^2}$$

$$I_S = V_S \sqrt{\frac{1}{R^2} + \left(\omega C - \frac{1}{\omega L}\right)^2} \tag{1.23}$$

At resonance, $B_C = B_L$ and current amplitude is *minimum*:

$$I_{S,\,\text{res}} = V_S/R \tag{1.24}$$

From (1.17) we see that at this frequency, the current is in phase with the source voltage.

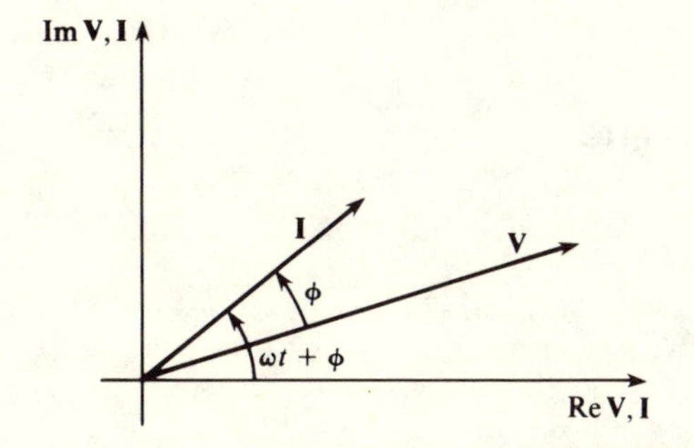

Figure 1.9 The current leads the voltage by ϕ radians in the parallel *LRC* circuit.

Example 1.2
What are the resonant frequencies of the network shown in Fig. 1.10?

Ans. First redraw the network as shown in Fig. 1.11. To calculate $\mathbf{Y}_2$ we set

$$\mathbf{Y}_2 = \frac{1}{\mathbf{Z}_2} = \frac{1}{j\left(\omega L - \dfrac{1}{\omega C_2}\right)}$$

whereas $\mathbf{Y}_1 = j\omega C_1$. Now we can draw the equivalent network shown in Fig. 1.12. Setting

$$\mathbf{Z} = \frac{1}{\mathbf{Y}} = \frac{1}{\mathbf{Y}_1 + \mathbf{Y}_2} = \frac{1}{j\omega C_1 - \dfrac{j}{\omega L - \dfrac{1}{\omega C_2}}}$$

gives the equivalent network shown in Fig. 1.13, corresponding to total

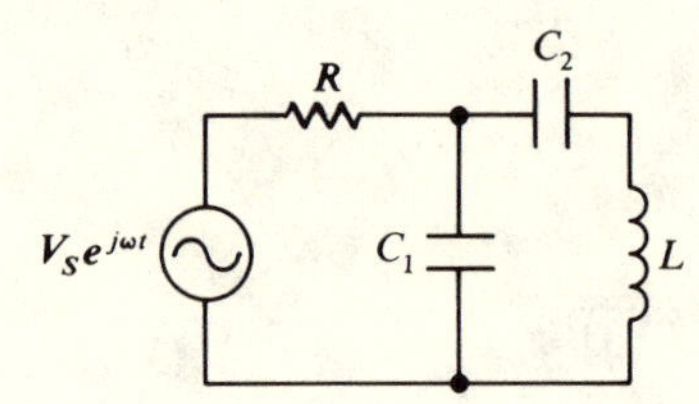

Figure 1.10

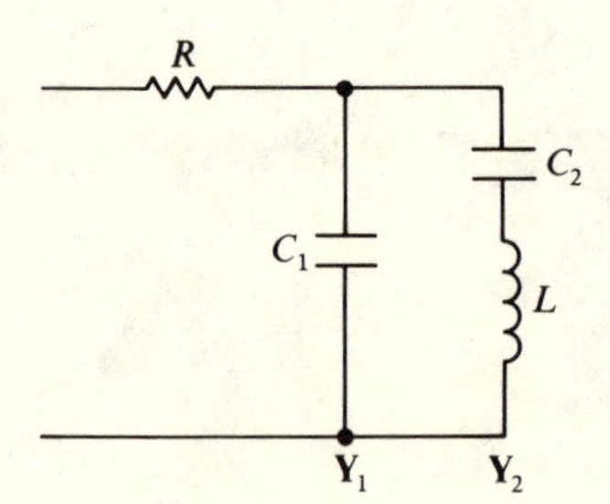

Figure 1.11

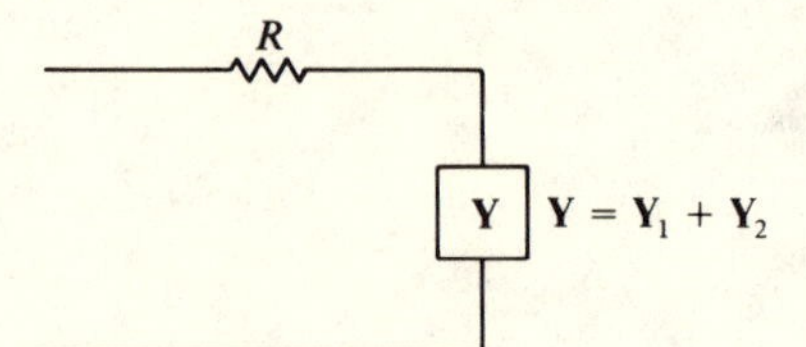

Figure 1.12

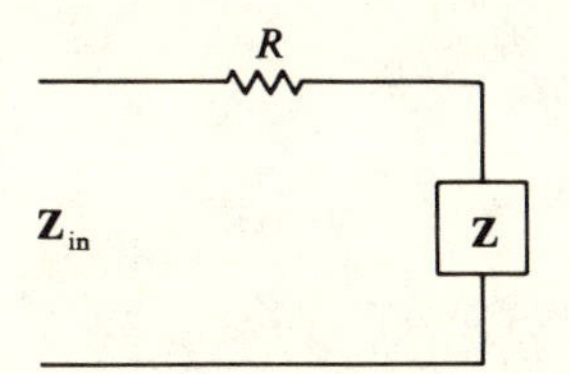

Figure 1.13

input impedance

$$\mathbf{Z}_{in} = R + jX$$

$$X = \frac{1}{\dfrac{1}{\omega L - \dfrac{1}{\omega C_2}} - \omega C_1} = \frac{\omega^2 L C_2 - 1}{\omega\left[C_2 - C_1\left(\omega^2 L C_2 - 1\right)\right]}$$

Series resonance occurs when $X = 0$. For the network shown this occurs at the frequency

$$\omega^2 = \frac{1}{LC_2}$$

Parallel resonance occurs when $X = \infty$. For the network shown, we obtain

$$\omega^2 = \frac{1 + \dfrac{C_2}{C_1}}{LC_2}$$

1.3 POWER FLOW

For any of the circuits considered previously there is a net power flow from the generator to the network. Consider a network with input impedance **Z**. (See Fig. 1.14.) The instantaneous power delivered to the network is

$$P = IV \tag{1.25}$$

where V is the source voltage

$$V = V_S \cos \omega t$$

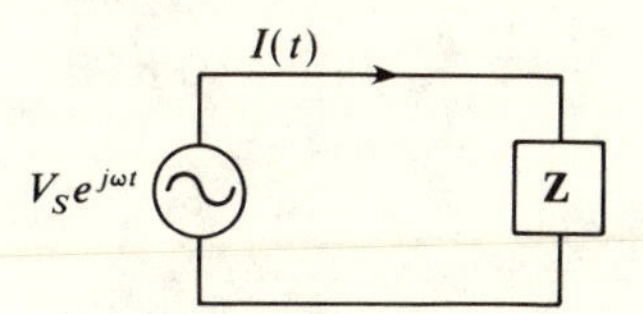

Figure 1.14

and I is the current through the generator

$$I = I_S \cos(\omega t - \phi)$$

Substituting these expressions into (1.25) gives

$$P(t) = I_S V_S \cos \omega t \cos(\omega t - \phi)$$

Expanding the second cos term allows this expression to be rewritten

$$\begin{aligned} P(t) &= I_S V_S (\cos^2 \omega t \cos \phi + \cos \omega t \sin \omega t \sin \phi) \\ &= I_S V_S (\cos^2 \omega t \cos \phi + \tfrac{1}{2} \sin 2\omega t \sin \phi) \end{aligned} \tag{1.26}$$

were we have used the relation

$$\cos \omega t \sin \omega t = \tfrac{1}{2} \sin 2\omega t$$

To obtain average power flow into the network we form the integral

$$\langle P \rangle = \frac{1}{T} \int_0^T P(t)\, dt$$

where $T = 1/f = 2\pi/\omega$ is the period of oscillation. Integrating (1.26) we obtain (writing $\langle\ \rangle$ for time average)

$$\begin{aligned} \langle P \rangle &= I_S V_S \cos \phi \langle \cos^2 \omega t \rangle \\ &= \frac{I_S V_S}{2} \cos \phi \qquad \left(\cos \phi = \frac{R}{Z} \right) \end{aligned}$$

or equivalently

$$\begin{aligned} \langle P \rangle &= I_{\text{rms}} V_{\text{rms}} \cos \phi = I_{\text{rms}}^2 Z \cos \phi \\ &= \frac{V_{\text{rms}}^2}{Z} \cos \phi \end{aligned} \tag{1.27}$$

At resonance Z is purely resistive, $\phi = 0$, and $\langle P \rangle$ is maximum. For a purely reactive network $\phi = \pm\pi/2$ and $\langle P \rangle = 0$. There is no average energy transfer.

Complex Power-Flow Formula

An alternate expression for the average power flow (1.27) is given by

$$\langle P \rangle = \tfrac{1}{2} \operatorname{Re} \mathbf{I}\mathbf{V}^* \tag{1.28}$$

where $\mathbf{V}^*$ represents the complex conjugate of $\mathbf{V}$. To demonstrate the equivalence of this expression and (1.27) we first note that

$$\mathbf{I}\mathbf{V}^* = \frac{\mathbf{V}\mathbf{V}^*}{\mathbf{Z}} = \frac{V_S^2}{\mathbf{Z}}$$

Multiplying numerator and denominator by $\mathbf{Z}^*$ gives

$$\mathbf{IV}^* = \frac{V_S^2 \mathbf{Z}^*}{Z^2}$$

so that

$$\tfrac{1}{2}\operatorname{Re} I\mathbf{V}^* = \frac{1}{2}\frac{V_S^2}{Z^2}R = \frac{V_{\text{rms}}^2}{Z}\cos\phi$$

which is the same as (1.27).

Example 1.3

a. What is the resonant frequency of the network shown in Fig. 1.15?
b. What is the average power flow to the network at resonance for values $V_S = 100$ volts, $L = 1.5\ \mu$H, $C = 2.5$ pF and $R = 1$ megohm?

Ans.

a. We must calculate the input impedance of the network. The resonant frequency is obtained by setting $\operatorname{Im} Z = 0$. The admittance of the RC branch is

$$\mathbf{Y} = j\omega C + \frac{1}{R}$$

corresponding to the impedance

$$\mathbf{Z} = \frac{R}{1 + j\omega CR}$$

It follows that the input impedance has the value

$$\mathbf{Z}_{\text{in}} = j\omega L + \frac{R}{1 + j\omega CR}$$

$$= j\omega L + \frac{R(1 - j\omega CR)}{1 + (\omega CR)^2}$$

$$= \frac{j\omega L\left[1 + (\omega RC)^2\right] + R - j\omega CR^2}{1 + (\omega CR)^2}$$

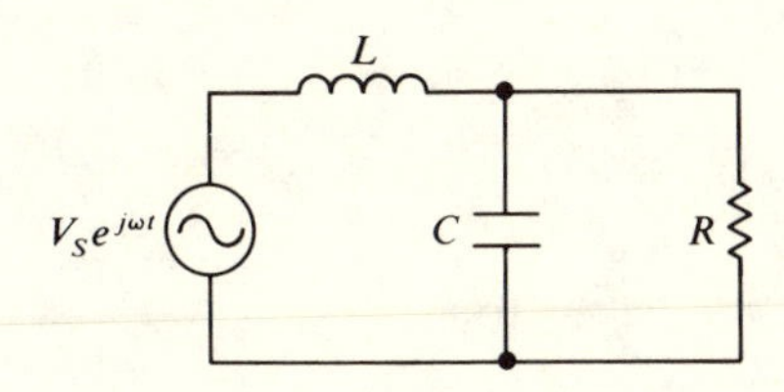

Figure 1.15

Resonance occurs when $\operatorname{Im} \mathbf{Z}_{\text{in}} = 0$,

$$\left[1 + (\omega CR)^2\right] \operatorname{Im} \mathbf{Z}_{\text{in}} = \omega L\left[1 + (\omega CR)^2\right] - \omega CR^2 = 0$$

corresponding to the frequency

$$\omega = \frac{1}{RC}\sqrt{\frac{RC}{L/R} - 1}$$

$$RC = 10^6 \times 2.5 \times 10^{-12} = 2.5 \times 10^{-6} \text{ s}$$

$$\frac{L}{R} = \frac{1.5 \times 10^{-6}}{10^6} = 1.5 \times 10^{-12} \text{ s}$$

$$\omega = \frac{10^6}{2.5}\sqrt{\frac{2.5 \times 10^{-6}}{1.5 \times 10^{-12}} - 1} = 0.52 \times 10^9$$

$$= 5.2 \times 10^8 \text{ rad/s}$$

$$f = \frac{\omega}{2\pi} = 83 \text{ MHz}$$

b. At this frequency the impedance of the network is

$$Z_{\text{in}} = \operatorname{Re} \mathbf{Z}_{\text{in}} = \frac{R}{1 + (\omega RC)^2}$$

$$= \frac{10^6}{1 + (5.2 \times 10^8 \times 2.5 \times 10^{-6})^2} = 0.59 \text{ ohms}$$

so that

$$\langle P \rangle = V_{\text{rms}}^2 / Z$$

$$= \frac{(100)^2}{2} \times \frac{1}{0.59} = 8475 \text{ watts}$$

1.4 REACTIVE AND RESISTIVE ENERGY

Consider the simple *LRC* circuit shown in Fig. 1.16. Equating the sum of the voltage drops across the three elements to zero, by Kirchhoff's law we obtain the equation

$$L\frac{dI}{dt} + \frac{q}{C} + RI = 0 \tag{1.29}$$

Multiplying through by I gives

$$\frac{d}{dt}\left(\tfrac{1}{2}LI^2\right) + \frac{d}{dt}\left(\frac{1}{2}\frac{q^2}{C}\right) + RI^2 = 0$$

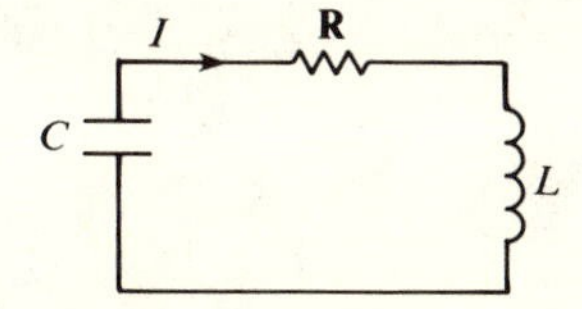

Figure 1.16 The series *LRC* circuit.

or equivalently

$$\frac{d}{dt}(W_L + W_C) + P_J = 0 \tag{1.30}$$

Here we have equated

$$W_L = \tfrac{1}{2}LI^2, \qquad W_C = \frac{1}{2}\frac{q^2}{C}$$

which respectively represent inductive and capacitive energies. The sum $W_L + W_C$ represents the *reactive* or *stored* energy of the circuit, whereas P_J represents the rate of which energy is lost from the circuit to Joule heat. If there is no resistance in the circuit, then $W_C + W_L$ is constant in time. Under these conditions the circuit oscillates at the frequency

$$\omega_0 = \frac{1}{\sqrt{LC}}$$

If the capacitor is charged to the potential V_0 at $t = 0$, then at subsequent time

$$W_C = W_0 \cos^2 \omega_0 t$$
$$W_L = W_0 \sin^2 \omega_0 t \tag{1.31}$$

where

$$W_0 = \tfrac{1}{2}CV_0^2$$

There is a periodic transfer of reactive energy between the capacitor and the inductor. When capacitive energy is maximum, inductive energy is minimum and vice versa.

1.5 DAMPED OSCILLATIONS

Let us return to the *LRC* circuit considered above, whose equation of motion is given by (1.29). Differentiating this equation with respect to time gives the current equation

$$L\frac{d^2I}{dt^2} + R\frac{dI}{dt} + \frac{I}{C} = 0 \tag{1.32}$$

The solutions to this equation are obtained by substituting the trial solution

$$I = Ae^{-\nu t}$$

into (1.32). We obtain

$$L\nu^2 - R\nu + \frac{1}{C} = 0$$

which has the two solutions

$$\nu_{\pm} = \frac{R \pm \sqrt{R^2 - 4L/C}}{2L}$$

It follows that the general solution to (1.32) is

$$I(t) = Ae^{-\nu_+ t} + Be^{-\nu_- t} \tag{1.33}$$

If $R^2 - 4L/C < 0$ then $\nu_{\pm}$ are complex conjugates of each other and $I(t)$ suffers damped oscillations. Under these conditions we may write

$$\nu_+ = \alpha + j\bar{\omega}, \qquad \nu_- = \alpha - j\bar{\omega} \tag{1.34}$$

where

$$\alpha \equiv R/2L$$

$$\omega_0^2 = 1/LC$$

$$\bar{\omega}^2 = \omega_0^2 - \alpha^2 \tag{1.35}$$

The solution (1.33) may be written in the three equivalent forms

$$I(t) = Ae^{-(\alpha + j\bar{\omega})t} + Be^{-(\alpha - j\bar{\omega})t} \tag{1.36a}$$

$$I(t) = e^{-\alpha t}[(B + A)\cos\bar{\omega}t + j(B - A)\sin\bar{\omega}t] \tag{1.36b}$$

$$I(t) = I_0 e^{-\alpha t}\cos(\bar{\omega}t + \phi) \tag{1.36c}$$

all of which represent damped oscillations.

1.6 THE Q OF A CIRCUIT

If the decay constant α in the solution (1.36) is much less than the natural frequency ω_0, then as may be seen from (1.35)

$$\bar{\omega} \simeq \omega_0, \qquad \alpha \ll \omega_0$$

The current oscillates at nearly constant frequency and amplitude. Let us obtain in this limit an expression for the *ratio of energy stored in the circuit to energy lost per cycle of oscillation*. This ratio is called the Q of the circuit. From (1.31) we see that the energy in the circuit may also be calculated when it is entirely inductive:

$$W = \tfrac{1}{2}LI_{\max}^2$$

From (1.36c) we may equate $I_{max} = I_0 e^{-\alpha t}$, so that

$$W = \tfrac{1}{2} L I_0^2 e^{-2\alpha t} \tag{1.37}$$

The rate at which this stored energy decreases, $\langle P_{loss} \rangle$, is obtained by differentiating the last equation:

$$\langle P_{loss} \rangle = -\frac{dW}{dt} = 2\alpha W, \qquad \alpha = \frac{\langle P_{loss} \rangle}{2W} \tag{1.38}$$

For slowly varying stored energy,

$$\text{energy lost per cycle} \simeq \langle P_{loss} \rangle \frac{2\pi}{\omega_0}$$

Substituting this value into the definition

$$Q \equiv \frac{2\pi \text{ (energy stored in circuit)}}{\text{energy lost per cycle}}$$

gives

$$Q = \frac{\omega_0 W}{\langle P_{loss} \rangle} \tag{1.39}$$

With (1.38) we may write

$$Q = \frac{\omega_0}{2\alpha}$$

If $Q \gg 1$ then the stored energy in the circuit remains approximately constant over many cycles of oscillation. With the preceding equation and (1.37) we may write

$$W = W_0 e^{-\omega_0 t/Q}$$

For $Q = 1000$, after 100 oscillations ($t = 100 \times 2\pi/\omega_0$), W decreases from W_0 to $0.53W_0$.

Example 1.4

Show that the Q of the series LRC circuit (see Fig. 1.16) is given by

$$Q = \frac{\omega_0 L}{R} = \frac{1}{\omega_0 CR}$$

Ans.

The total energy stored in the circuit is given by either $W = \frac{1}{2}LI^2$ or $\frac{1}{2}CV^2$ and $\langle P_{loss} \rangle = \frac{1}{2}I^2 R$. Since the value of I is given, the Q is found to be

$$Q = \frac{\omega_0 W}{\langle P_{loss} \rangle} = \frac{\omega_0 \frac{1}{2} L I^2}{\frac{1}{2} I^2 R} = \frac{\omega_0 L}{R}$$

The circuit resonates at ω_0 so that $\omega_0 L = 1/\omega_0 C$ and

$$Q = \frac{1}{\omega_0 CR}$$

1.7 RESONANCE

For the series *LRC* circuit discussed in Sec. 1.1, the average power transfer from the generator to the circuit is given by (1.27):

$$\langle P \rangle = \frac{V_{\mathrm{rms}}^2}{Z} \cos\phi$$

Equivalently we may write

$$\frac{\langle P \rangle}{V_{\mathrm{rms}}^2} = \frac{R}{R^2 + X^2} \tag{1.40}$$

A sketch of $\langle P \rangle$ vs circuit reactance X is shown in Fig. 1.17 and is seen to be maximum when

$$X = \omega L - \frac{1}{\omega C} = 0$$

which gives the resonant frequency

$$\omega_0 = \sqrt{\frac{1}{LC}}$$

At this frequency there is maximum transfer of energy to the circuit. Furthermore, since $X = 0$ at this frequency, the circuit is purely resistive. Since

$$I_{\mathrm{rms}} = \frac{V_{\mathrm{rms}}}{Z} = \frac{V_{\mathrm{rms}}}{Z} = \frac{V_{\mathrm{rms}}}{\sqrt{R^2 + X^2}} \tag{1.41}$$

we see that, as noted previously, the rms current is maximum at resonance

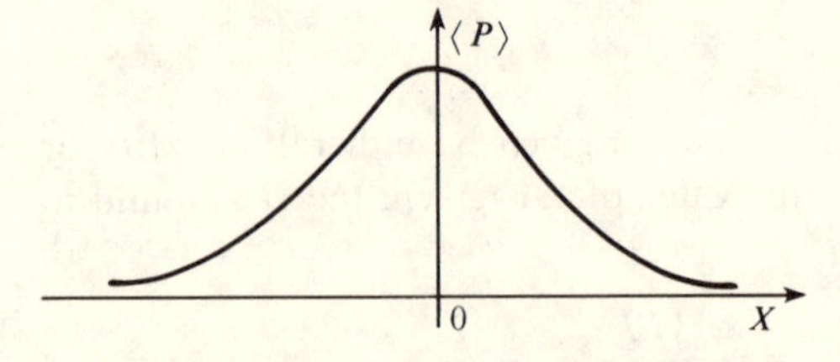

Figure 1.17 Average power to an *LRC* circuit as a function of reactance.

and has the value

$$I_{\text{rms, res}} = \frac{V_S}{\sqrt{2}\,R} \tag{1.42}$$

The instantaneous current through the generator is in phase with the generator voltage and has the value

$$I(t)_{\text{res}} = \frac{V_S}{R}\cos\omega t \tag{1.43}$$

A sketch of some of these parameters vs ω are shown in Table 1.1.

For the parallel LRC circuit discussed in Sec. 1.2, the instantaneous power flow to the circuit is given, referring to (1.21), by

$$P(t) = IV = YV_S^2 \cos\omega t \cos(\omega t + \phi)$$

which gives the time average

$$\langle P \rangle = V_{\text{rms}}^2 Y \cos\phi \tag{1.44}$$

But $\cos\phi = G/Y$ [see (1.17), (1.18)]. It follows that

$$\langle P \rangle = V_{\text{rms}}^2 G = \frac{V_{\text{rms}}^2}{R}$$

Thus we find that for the parallel LRC network, the power flow to the

Table 1.1 Frequency dependence of parameters for the series *LRC* circuit

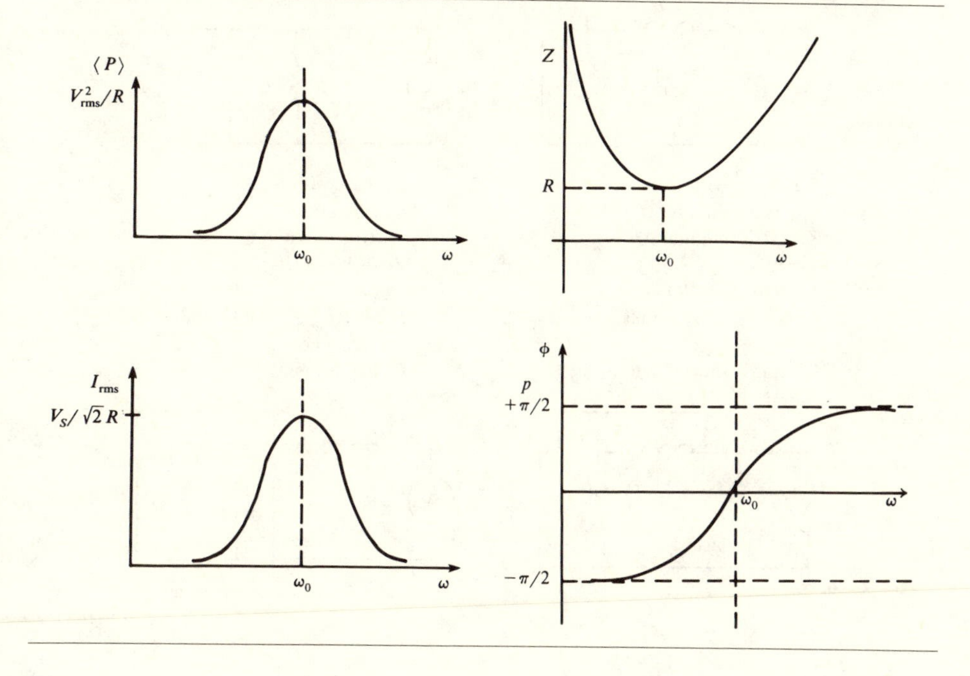

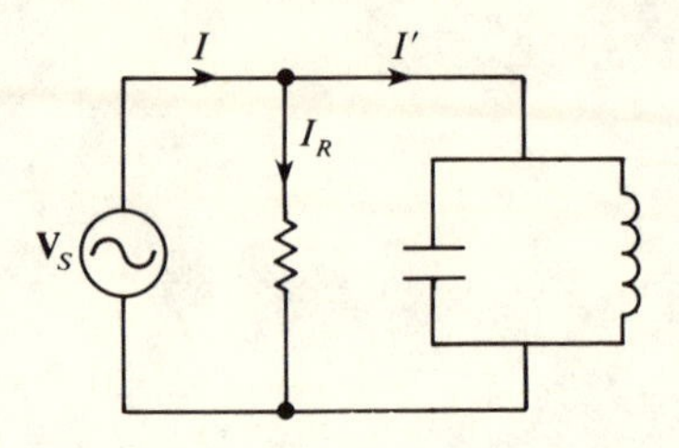

Figure 1.18 Division of current in the parallel *LRC* network.

circuit is *independent of frequency*. However, a peculiar effect does occur at the frequency $1/\sqrt{LC}$. Namely, in the circuit at hand let us call the current to the *LC* branch I' and the admittance of this branch Y'. (See Fig. 1.18.) Then

$$\mathbf{I}' = \mathbf{Y}'\mathbf{V}_S$$

At resonance

$$\mathbf{Y}' = j\omega C + \frac{1}{j\omega L} = 0$$

Table 1.2 Resonance properties

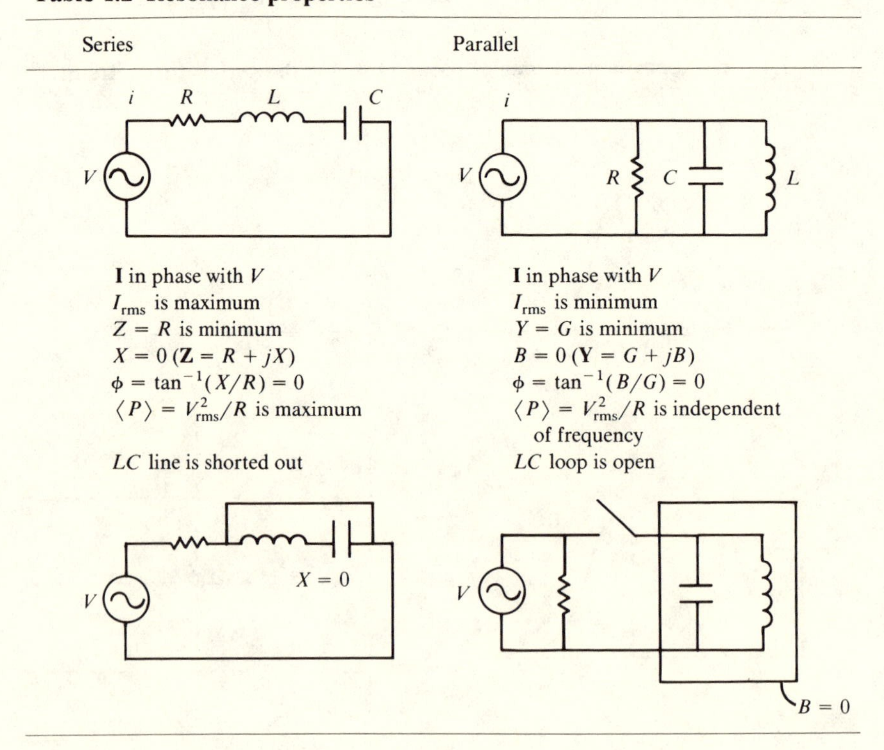

Series	Parallel
I in phase with V	**I** in phase with V
I_{rms} is maximum	I_{rms} is minimum
$Z = R$ is minimum	$Y = G$ is minimum
$X = 0$ ($\mathbf{Z} = R + jX$)	$B = 0$ ($\mathbf{Y} = G + jB$)
$\phi = \tan^{-1}(X/R) = 0$	$\phi = \tan^{-1}(B/G) = 0$
$\langle P \rangle = V_{\mathrm{rms}}^2/R$ is maximum	$\langle P \rangle = V_{\mathrm{rms}}^2/R$ is independent of frequency
LC line is shorted out	*LC* loop is open

and there is no current to the LC branch. The current through the generator at resonance is

$$I(t)_{\text{res}} = V_S Y = \frac{V_S \cos \omega t}{R}$$

with corresponding rms value

$$I_{\text{rms, res}} = \frac{V_{\text{rms}}}{R}$$

which is *minimum*.

A review of these resonance properties is given in Table 1.2.

Example 1.5

A transmission line operating at the frequency ω is terminated in a load comprised of an inductance and a capacitance. What combination of these two elements:

a. effects an open circuit load?
b. effects a short circuit load?

Ans.

a. An open circuit is effected by placing the inductance in parallel with the capacitance at values such that $LC = \omega^{-2}$. See Fig. 1.19.
b. A short circuit is effected by placing the inductance in series with the capacitance, again at values such that $LC = \omega^{-2}$. See Fig. 1.20.

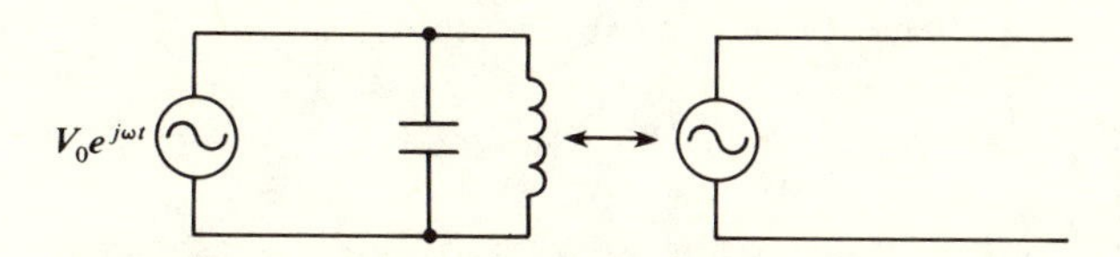

Figure 1.19 Equivalent open circuit; $\omega = 1/\sqrt{LC}$.

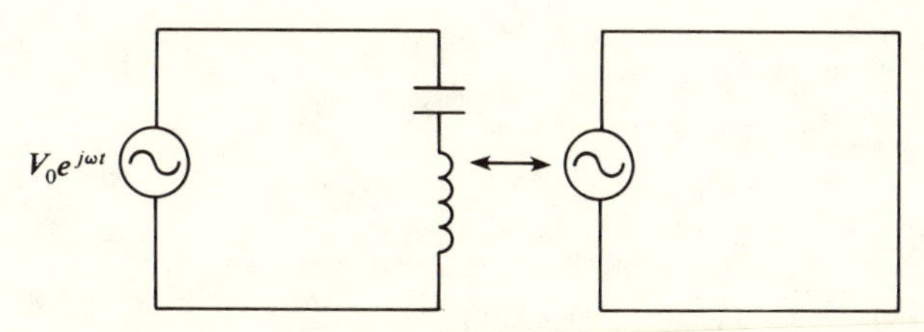

Figure 1.20 Equivalent short circuit; $\omega = 1/\sqrt{LC}$.

Example 1.6

A parallel LC network is driven by a constant current generator. The circuit is shown in Fig. 1.21. Obtain an expression for the total energy W stored in the LC loop, as a function of time, at resonance ($\omega^2 = 1/LC$). The switch S is closed at $t = 0$. Show that for $\omega t \gg 1$, $W \propto (\omega t)^2$.

Ans. We must compute

$$W = W_C + W_L$$

where

$$W_C = \tfrac{1}{2}CV^2$$
$$W_L = \tfrac{1}{2}LI_L^2$$

The voltage across the capacitor is V, and I_L is the current through the inductor. Kirchhoff's current law gives the equation

$$I_S \sin \omega t = I_C + I_L$$

where I_C is the current to the capacitor. Equating voltages across the two elements gives

$$V = L\frac{dI_L}{dt} = \frac{\int I_C\,dt}{C}$$

Differentiating with respect to time, we obtain

$$LC\frac{d^2 I_L}{dt^2} = I_C$$

Substituting this relation into the above current equation gives

$$\frac{d^2}{dt^2}I_L + \omega^2 I_L = \omega^2 I_S \sin \omega t$$

Here we have used the fact that the network is driven at resonance, $\omega^2 = 1/LC$. The solution to this differential equation is comprised of a homogeneous and an inhomogeneous part. For the homogeneous solution

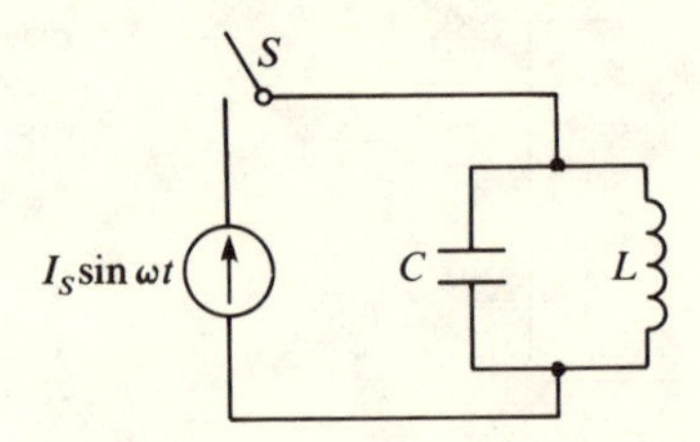

Figure 1.21 Circuit for Example 1.6.

we obtain

$$I_{L,h} = A\cos\omega t + B\sin\omega t$$

To find the inhomogeneous solution we substitute the trial solution $Ct^n\cos\omega t$ and find that only $n = 1$ gives an equality. There results

$$I_{L,i} = -\frac{I_S}{2}\omega t\cos\omega t$$

The total solution is given by the sum $I_{L,h} + I_{L,i}$. The initial conditions that $I_L = V = 0$ at $t = 0$ serve to determine A and B. We obtain finally

$$I_L = \frac{I_S}{2}(\sin\omega t - \omega t\cos\omega t)$$

The voltage across the capacitor is

$$V = L\frac{dI_L}{dt} = \frac{I_S L\omega^2}{2}(t\sin\omega t)$$

These two expressions for I_L and V permit calculation of the stored energy:

$$W_L = W_0(\sin\omega t - \omega t\cos\omega t)^2$$

$$W_C = W_0(\omega t\sin\omega t)^2$$

and

$$W_0 \equiv \frac{LI_S^2}{8}$$

For $\omega t \gg 1$, the total stored energy is seen to grow as

$$W \sim W_0(\omega t)^2$$

PROBLEMS

1.1. The components of a series LRC circuit have the following values:

$$R = 10 \text{ ohms}$$
$$L = 100 \text{ mH}$$
$$C = 409\ \mu\text{F}$$

At time $t = 0$, a peak current of 3 mA flows in the circuit.

a. Show that the current suffers oscillatory decay.
b. What is the frequency of oscillation?
c. What is the peak current after one period of oscillation?
d. What fraction of energy contained in the circuit is lost to heat in one period of oscillation?

1.2.

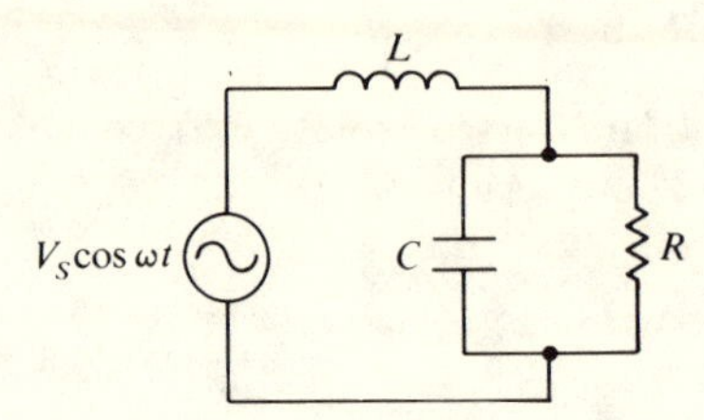

Figure 1.22 Network in Problem 1.2.

a. Show that the complex impedance of the network depicted in Fig. 1.22 is given by

$$\mathbf{Z} = \frac{R^{-1} + j\left(\omega L Y_2^2 - \omega C\right)}{Y_2^2}$$

Here we have written

$$Y_2^2 = R^{-2} + (\omega C)^2$$

for the admittance of the RC network.

b. Resonance occurs for this circuit when $\mathbf{Z}$ is purely resistive. Use this fact to show that the resonant frequency of the circuit is

$$\omega_0^2 = \frac{1}{LC} - \frac{1}{(RC)^2}$$

1.3. Prove the *power-flow theorem*. Namely, show that maximum power is transferred to a load impedance $\mathbf{Z}$ from a voltage source $\mathbf{V}$ with internal resistance $\mathbf{z}$ if $\mathbf{Z} = \mathbf{z}^*$. The circuit is shown in Fig. 1.23.

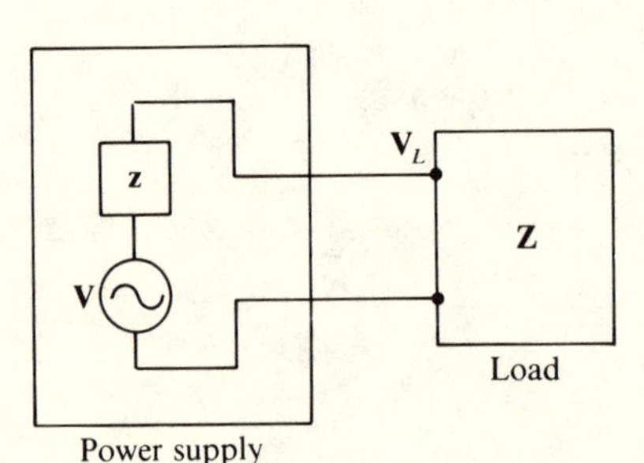

Figure 1.23 Circuit relevant to the power-flow theorem.

CHAPTER
TWO

ELEMENTARY TRANSMISSION-LINE CONCEPTS

2.1 THE IDEAL TRANSMISSION LINE

A transmission line transmits electrical energy from a generator to a load. In general the transmission line is composed of two conductors separated by a dielectric material. We consider an *ideal*, *uniform* transmission line. The line is *ideal* or *lossless* if no electrical power is dissipated along the system. The transmission line is *uniform* if a cross section of the line drawn in a plane normal to the power flow is the same for all points on the line. Cross sections of some typical transmission lines are shown in Fig. 2.1.

At any point along the line, an equal but oppositely directed current flows in the two conductors and a voltage difference exists across the line. The variation of current and voltage along an ideal line may be ascertained by examining a distributed equivalent circuit for the line. This equivalent line is composed of a distributed series inductance L (H/m) and distributed shunt capacitance C (F/m). (See Fig. 2.2.)

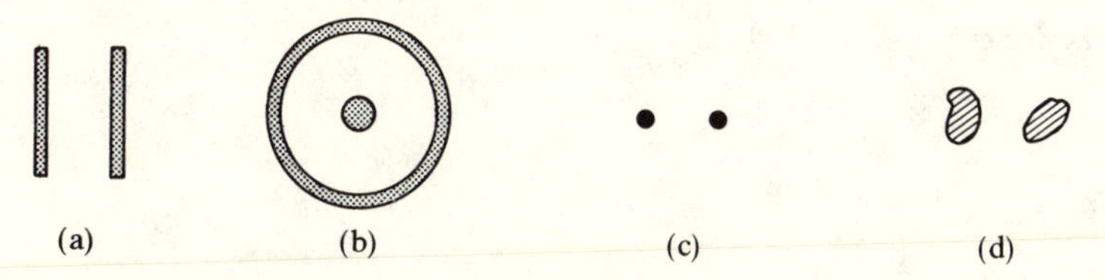

Figure 2.1 Four examples of transmission lines.

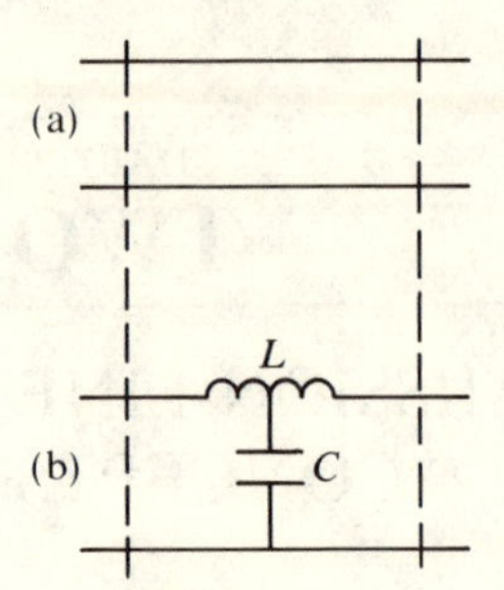

Figure 2.2 (a) The line. (b) Equivalent circuit element.

2.2 WAVE EQUATIONS FOR *V* AND *I*

Let the line be parallel to the z-axis. The voltage change δV in the length dz is the product of the inductance in this section $(L\,dz)$ and the time rate of change of current:

$$\delta V = -L\,dz\,\frac{\partial I}{\partial t}$$

The minus sign indicates that there is a voltage drop due to the series inductance if $\partial I/\partial t$ is positive. The change δV may also be written

$$\delta V = \frac{\partial V}{\partial z}\,dz$$

Equating these two relations gives

$$\frac{\partial V}{\partial z}\,dz = -L\,dz\,\frac{\partial I}{\partial t} \tag{2.1}$$

The change in current, δI, in the length dz is due to the rate of charging the capacitor:

$$\delta I = -C\,dz\,\frac{\partial V}{\partial t}$$

Again we may set

$$\delta I = \frac{\partial I}{\partial z}\,dz$$

to obtain

$$\frac{\partial I}{\partial z}\,dz = -C\,dz\,\frac{\partial V}{\partial t} \tag{2.2}$$

These effects are depicted in Fig. 2.3.

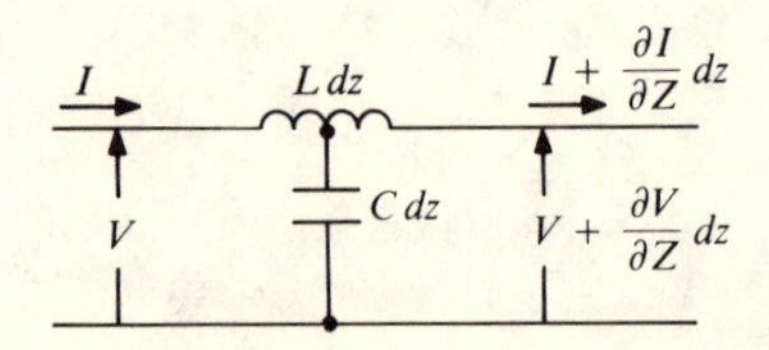

Figure 2.3 Current and voltage changes along a transmission line.

Canceling the common factor dz in (2.1) and (2.2) gives two coupled equations for V and I:

$$\frac{\partial V}{\partial z} = -L\frac{\partial I}{\partial t}$$

$$\frac{\partial I}{\partial z} = -C\frac{\partial V}{\partial t} \tag{2.3}$$

Differentiating the first of these with respect to t and the second with respect to z and multiplying the first by C allows elimination of V. We obtain

$$\frac{\partial^2 I}{\partial z^2} = LC\frac{\partial^2 I}{\partial t^2} \tag{2.4}$$

In like manner, elimination of I gives

$$\frac{\partial^2 V}{\partial z^2} = LC\frac{\partial^2 V}{\partial t^2} \tag{2.5}$$

2.3 TRAVELING WAVES

Equations (2.4) and (2.5) are called wave equations and imply that currents and voltages initiated in some finite domain along the line will propagate away from that domain. Consider the voltage equation (2.5). Let

$$v^2 \equiv \frac{1}{LC} \tag{2.6}$$

Then (2.5) may be rewritten

$$\frac{\partial^2 V}{\partial z^2} - \frac{1}{v^2}\frac{\partial^2 V}{\partial t^2} = 0 \tag{2.7}$$

This equation, which is called the *wave equation*, has the general solution

$$V = V_+(z - vt) + V_-(z + vt) \tag{2.8}$$

Here $V_+(z - vt)$ represents an arbitrary function of $z - vt$ such as, for example,

$$V_+ = \ln(z - vt), \qquad V_+ = \sin(z - vt)$$

$$V_+ = \arctan(z - vt), \qquad V_+ = e^{j(z-vt)}$$

What is the significance of the + subscript in V_+? It designates that the spatial form, V_+ at any instant, propagates in the direction of increasing z. Suppose $V_+(z - vt)$ at the instant $t = 0$ has the form shown in Fig. 2.4a. At $t > 0$, V_+ has the form shown in Fig. 2.4b. The reason for this behavior is as follows. Consider specific values z' and t' for which $z' - vt' = 8$. Suppose $V_+(8) = 3$. Now V_+ is a function only of $z - vt$. So at all $z - vt = 8$, $V_+ = 3$. It follows that the value $V_+ = 3$ is propagated on the

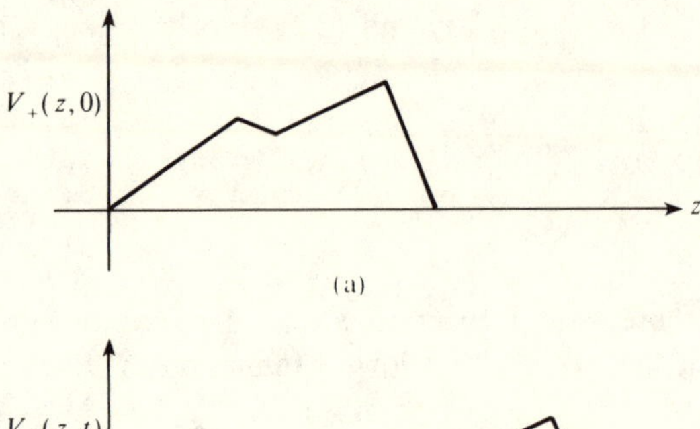

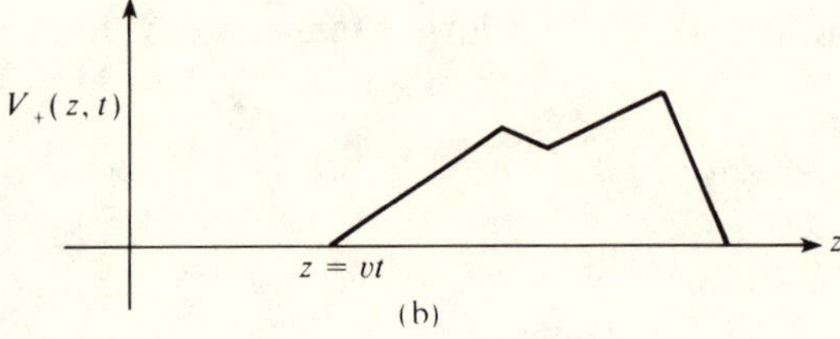

Figure 2.4 The voltage wave V_+: (a) at $t = 0$; (b) at $t > 0$.

curve

$$z = vt + 8$$

With increasing t, the value $V_+ = 3$ moves to values of increasing z with the speed v. The voltage V_+ is called the *forward wave*, whereas the voltage V_- is called the *reflected wave*.

2.4 CHARACTERISTIC IMPEDANCE

The current on the line is given by the general solution to (2.4),

$$I = I_+(z - vt) + I_-(z + vt) \tag{2.9}$$

The forward wave I_+ and reflected wave I may be related to the forward and reflected voltages, V_+ and V_-, through the first relation in (2.3):

$$-L\frac{\partial I}{\partial t} = \frac{\partial V}{\partial z} = \frac{\partial V_+}{\partial z} + \frac{\partial V_-}{\partial z} \tag{2.10}$$

Due to the specific functional form of V_+ and V_- we may write

$$\frac{\partial V_+}{\partial z} = -\frac{1}{v}\frac{\partial V_+}{\partial t}, \qquad \frac{\partial V_-}{\partial z} = +\frac{1}{v}\frac{\partial V_-}{\partial t}$$

Substituting into (2.10) gives

$$-L\frac{\partial I}{\partial t} = -\frac{1}{v}\left(\frac{\partial V_+}{\partial t} - \frac{\partial V_-}{\partial t}\right)$$

which may be integrated to yield

$$I = \frac{1}{vL}(V_+ - V_-) \equiv \frac{1}{Z_0}(V_+ - V_-) \tag{2.11}$$

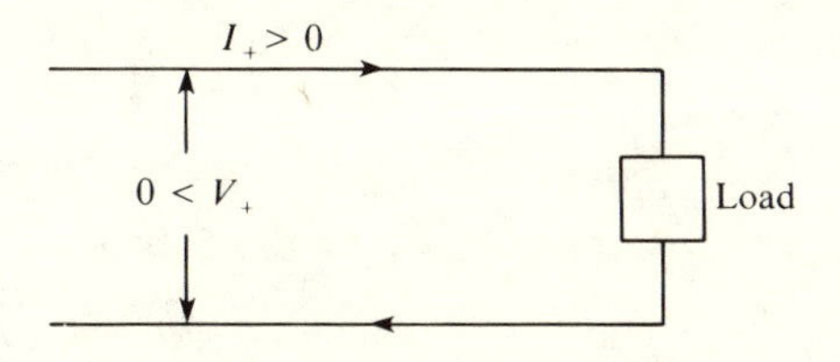

Figure 2.5 Current-wave convention: $I_+ = +V_+/Z_0$.

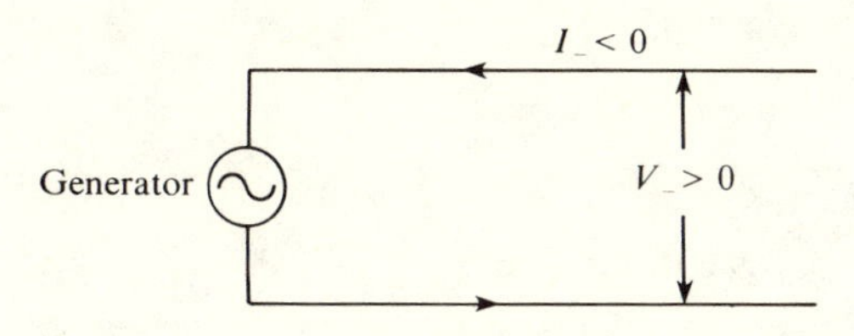

Figure 2.6 Current-wave convention: $I_- = -V_-/Z_0$.

(neglecting a constant of integration). The parameter

$$Z_0 = vL = \sqrt{\frac{L}{C}}$$

is called the *characteristic impedance* of the line. Since the functional forms of V_+ and I_+ are identical, as are those of V_- and I_-, we may further write, comparing (2.10) and (2.11),

$$Z_0 = \sqrt{\frac{L}{C}} = \frac{V_+}{I_+} = -\frac{V_-}{I_-} \tag{2.12}$$

If V_+ is known at some z and t, then given Z_0, one also knows I_+ at the same position and time.

The significance of the positive and negative signs in the last equation is as follows. Suppose V_+ is positive at some point on the line. Then our convention is that the "top" lead is at a positive potential with respect to the reference potential of the "bottom" lead. This means that the voltage drop V_+ will drive the *current flow* in the I_+ wave toward the load in the top lead and away from the load on the bottom lead. (See Fig. 2.5.)

Suppose V_- is also positive. In which direction is current driven due to V_-? Now we must examine the current wave that moves toward the generator. (See Fig. 2.6.)

Both descriptions are seen to be consistent with (2.12). A positive current indicates that the current in the top lead is from generator to source —or, with respect to our conventions for drawing figures, from left to right.

Example 2.1

A 300-Ω ideal line running at steady state carries the forward current

$$I_+ = I_0 \cos(\beta z - \omega t)$$

where β is a constant wavenumber, ω is a constant angular frequency, and the amplitude $I_0 = 0.9$ A.

a. What is the form of the forward voltage wave?
b. If $\beta = 2\pi \text{ m}^{-1}$ and $f = 1$ GHz, what is the velocity of the forward wave?
c. What are the values of L and C?

Ans.

a.
$$\begin{aligned} V_+ &= Z_0 I_+ = Z_0 I_0 \cos(\beta z - \omega t) \\ &= 300 \times 0.9 \cos(\beta z - \omega t) \\ &= 270 \cos(\beta z - \omega t) \text{ volts} \end{aligned}$$

b. We note that I_+ may be rewritten

$$I_+ = I_0 \cos \beta \left(z - \frac{\omega}{\beta} t \right)$$

Which allows the identification

$$\frac{\omega}{\beta} = v = \frac{1}{\sqrt{LC}} = 10^9 \text{ m/sec}$$

c. We are also given

$$Z_0 = \sqrt{\frac{L}{C}} = 300\ \Omega$$

Combining these values gives

$$L = 0.30\ \mu\text{H/m}$$
$$C = 3.3\ \text{pF/m}$$

2.5 WAVE SPEED AND INTRINSIC IMPEDANCE

An important property of an ideal homogeneous transmission line with leads of arbitrary cross section is as follows. Consider that the leads are encased in a medium of permittivity ϵ and permeability μ. Then the speed of wave propagation is[1]

$$v = \frac{1}{\sqrt{LC}} = \frac{1}{\sqrt{\mu\epsilon}} \tag{2.13}$$

which is the same as the speed of light in an homogeneous medium with

[1] J. D. Jackson, "*Classical Electrodynamics*," 2nd ed. (Wiley, New York, 1975).

parameters ϵ, μ. If the leads are separated by air, then v has the value

$$v = \frac{1}{\sqrt{\mu_0 \epsilon_0}} \equiv c = 3 \times 10^8 \text{ m/sec}$$

The characteristic impedance of the transmission line may be written

$$Z_0 = \eta F \tag{2.14}$$

where F is some dimensionless function and η is the so-called *intrinsic impedance*. It has the value

$$\eta = \sqrt{\frac{\mu}{\epsilon}} \tag{2.15}$$

If the leads are separated by air then

$$\eta_0 = \sqrt{\frac{\mu_0}{\epsilon_0}} = 120\pi = 377\ \Omega \tag{2.16}$$

which is called the *impedance of free space*.

Example 2.2

a. For a transmission line with identical circular leads of radius a separated by the distance s and encased in a medium with parameters μ, ϵ, and assuming $s \gg a$, show that

$$Z_0 = \frac{\eta}{\pi} \ln\left(\frac{s}{a}\right)$$

b. If $\epsilon = 2\epsilon_0$, $\mu = \mu_0$, and $s = 35a$, what is Z_0?

c. Show that $v = 1/\sqrt{LC} = 1/\sqrt{\mu\epsilon}$ for this system, and obtain a value for v.

Ans.

a. Consider a section of line 1 meter long. The cross section of the line appears as shown in Fig. 2.7. Assume that equal and opposite amounts of charge $\pm Q$ are on the leads of this unit section. Similarly, equal and

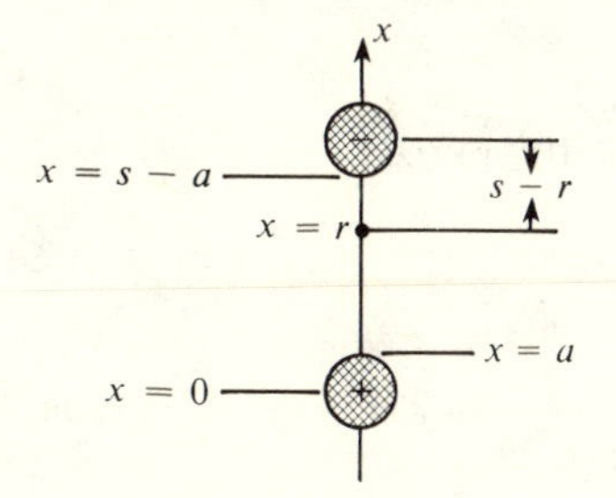

Figure 2.7 Cross section of a twin-lead transmission line.

opposite currents of magnitude I flow in the leads. Ampere's law[2] gives the magnetic fields from the two leads, at $z = r$:

$$2\pi r B_+ = \mu I, \qquad 2\pi(s - r)B_- = \mu I$$

These magnetic fields are in the same direction at the point $z = r$, and we obtain

$$B = B_+ + B_- = \frac{\mu I}{2\pi}\left[\frac{1}{r} + \frac{1}{s - r}\right]$$

The magnetic flux through the unit-length rectangle between leads is

$$\Phi = \int_a^{s-a} B\,dr = \frac{\mu I}{2\pi}\left[\ln r - \ln(s - r)\right]_a^{s-a}$$

$$= \frac{\mu I}{\pi}\ln\left(\frac{s - a}{a}\right) = LI$$

This gives the inductance per unit length

$$L = \frac{\mu}{\pi}\ln\left(\frac{s - a}{a}\right) \simeq \frac{\mu}{\pi}\ln\frac{s}{a}$$

Gauss's law gives the electric field from the two leads:[3]

$$2\pi r E_+ = \frac{Q}{\epsilon}, \qquad 2\pi(s - r)E_- = \frac{Q}{\epsilon}$$

These electric fields are in the same direction at $z = r$, and we obtain

$$E = E_+ + E_- = \frac{Q}{2\pi\epsilon}\left[\frac{1}{r} + \frac{1}{s - r}\right]$$

The potential difference between the leads is

$$V = \int_a^{s-a} E\,dr = \frac{Q}{\epsilon\pi}\ln\left(\frac{s - a}{a}\right) = \frac{Q}{C}$$

This gives the capacitance per unit length[3]

$$C = \frac{\pi\epsilon}{\ln\left(\dfrac{s - a}{a}\right)} \simeq \frac{\pi\epsilon}{\ln\left(\dfrac{s}{a}\right)}$$

Substituting these results into the expression $Z_0 = \sqrt{L/C}$, we obtain the expression given in the statement of the problem.

b. $Z_0 = 302\ \Omega$

c.

$$v = \frac{1}{\sqrt{LC}} = \frac{1}{\sqrt{\mu\epsilon}} = 2.12 \times 10^8 \text{ m/sec}$$

---

[2]Ampere's and Gauss's laws are discussed in Sec. 6.2.

[3]Note that these relations are valid in the domain $s \gg a$. Only in this limit may one assume that fields from each lead are rotationally symmetric.

2.6 REFLECTION COEFFICIENT

The standard transmission-line configuration involving a generator and load is represented by the diagram shown in Fig. 2.8. The load may represent a second section of line with different characteristics than the first section. The entire second section has been replaced in the diagram by an equivalent load.

The net voltage across the load is V_L. If V_+ and V_- are the forward and reflected voltage values at the load, then the voltage across the line at the load is $V_+ + V_-$, so that

$$V_L = V_+ + V_- \tag{2.17}$$

Likewise the current flow into the load, I_L, is equal to the current in the line at the load, $I_+ + I_-$, and we have

$$I_L = I_+ + I_- \tag{2.18}$$

The *load impedance* Z_L is defined as the ratio

$$Z_L = \frac{V_L}{I_L} \tag{2.19}$$

so that

$$I_L = \frac{V_L}{Z_L}$$

Substituting this expression into (2.18), we have

$$I_+ + I_- = \frac{V_L}{Z_L}$$

which, with (2.12) gives

$$\frac{V_+}{Z_0} - \frac{V_-}{Z_0} = \frac{V_L}{Z_L} \tag{2.20}$$

The reflection coefficient at the load is defined by

$$\rho = \frac{V_-}{V_+} \tag{2.21}$$

The reflected voltage is equal to ρ times the forward or incident voltage. The reflection coefficient represents the fraction of voltage in the forward wave

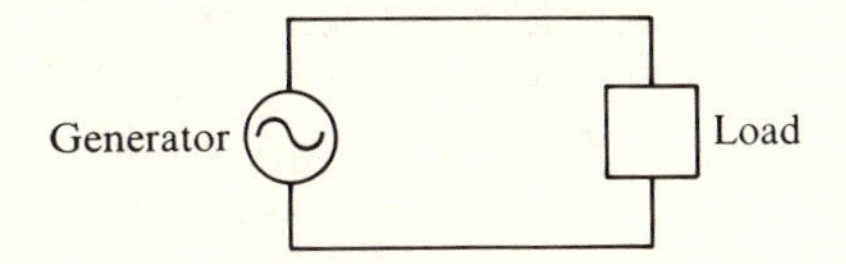

Figure 2.8 Transmission-line configuration.

that is present in the reflected wave. An equivalent expression for ρ is given by the current ratio

$$\rho = -\frac{I_-}{I_+} \tag{2.22}$$

Let us obtain an equation for ρ in terms of characteristic and load impedances. Equating the two respective expressions for V_L as given by (2.17) and (2.20) and setting $V_- = \rho V_+$, we find

$$\rho = \frac{Z_L - Z_0}{Z_L + Z_0} \tag{2.23}$$

A very significant fact in transmission-line theory is evident in this expression. Namely, if $Z_L = Z_0$ then $\rho = 0$ and there is no reflected voltage and, by (2.22), no reflected current. Thus there is no reflected power. All power flows into the load. The line is *matched* to the load.

The condition $\rho = 0$ is satisfied at any point on an ideal homogeneous line of infinite length. (See Fig. 2.9.) At A there is only a forward wave, since there is no discontinuity in the line or terminating load after A to cause a reflected wave. Therefore $V_- = 0$ at A, so that

$$\rho = \frac{V_-}{V_+} = 0$$

It follows that a line matched to a given load Z_L is not distinguishable from an infinite homogeneous line with characteristic impedance $Z_0 = Z_L$.

The transmission coefficient τ is given by

$$\tau = \frac{V_L}{V_+} = \frac{2Z_L}{Z_L + Z_0} \tag{2.24}$$

It represents the fraction of voltage in the forward wave that appears across the load. If the line is terminated in a matched load ($Z_L = Z_0$), then $\tau = 1$. In this case there is no reflected wave, so that $V_- = 0$ and $\rho = 0$. Combining (2.23) and (2.24) gives the general relation

$$\tau - \rho = 1 \tag{2.25}$$

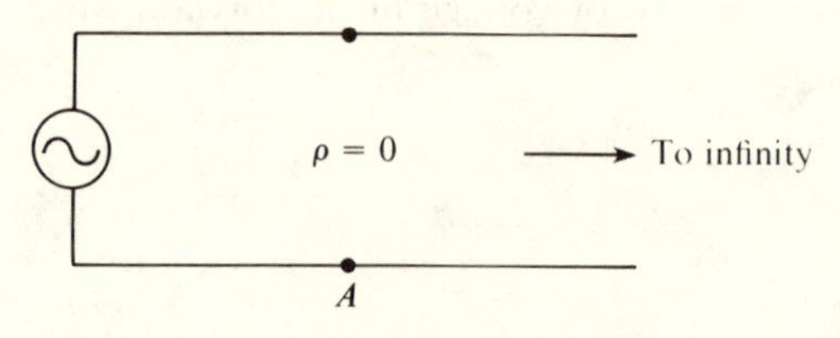

Figure 2.9 On a homogeneous line, $\rho = 0$.

Example 2.3
What are the values of the reflection coefficient for:

a. A short circuit?
b. An open circuit?

Ans.
a. Since the voltage drop across a short circuit is zero,

$$V_L = 0 = V_+ + V_-$$

so that $V_- = -V_+$ and

$$\rho_{sc} = -1$$

(see Fig. 2.10).

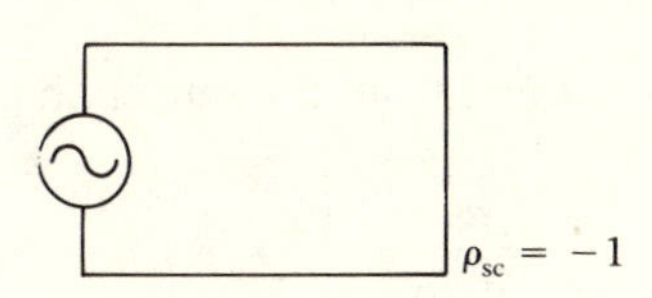

Figure 2.10 The short-circuit line.

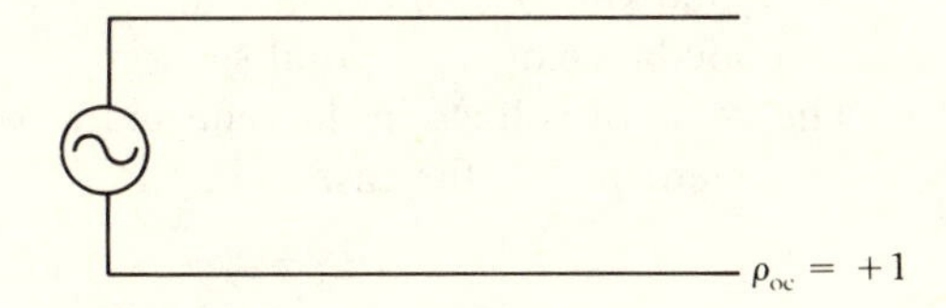

Figure 2.11 The open-circuit line.

b. Since there is no current through an open circuit,

$$I_L = 0 = I_+ + I_- = \frac{1}{Z_0}(V_+ - V_-)$$

so that $V_+ = V_-$ and

$$\rho_{oc} = +1$$

(see Fig. 2.11).

Example 2.4
A twin-lead transmission line of infinite length is encased in dielectric with permittivity $2\epsilon_0$ for $z < 0$ and permittivity $3\epsilon_0$ for $z > 0$. (See Fig. 2.12.) A voltage wave propagates toward the junction from the region $z < 0$ carrying voltage V_+.

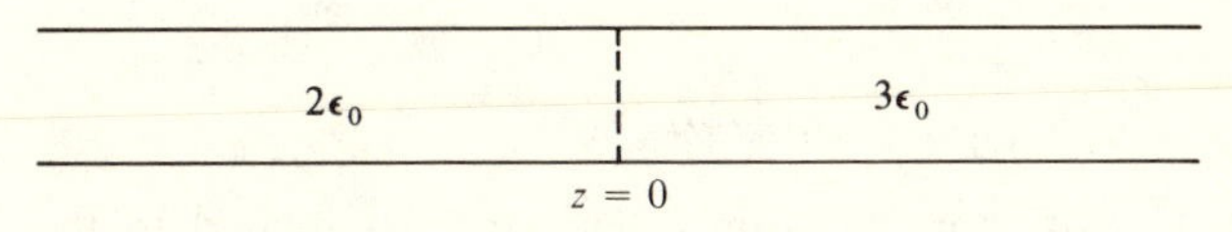

Figure 2.12 Configuration for Example 2.4.

a. What is the value of voltage V_T carried in the transmitted wave in terms of V_+?
b. What is the value of the voltage V_- in the reflected wave in terms of V_+?

Ans.

a. The value of V_T is obtained from the transmission coefficient τ. Using the results of Example 2.2, we find

$$\tau = \frac{2}{1 + \dfrac{Z_0}{Z_L}} = \frac{2}{1 + \sqrt{\frac{3}{2}}}$$

so that

$$V_T = \tau V_+ = 0.90 V_+$$

Note that the load impedance at the junction is given by the characteristic impedance of the second section.

b. The value of voltage in the reflected wave is obtained from the reflection coefficient ρ. For the case at hand

$$\rho = \frac{1 - Z_0/Z_L}{1 + Z_0/Z_L} = \frac{1 - \sqrt{\frac{3}{2}}}{1 + \sqrt{\frac{3}{2}}} = -0.10$$

and

$$V_- = -0.10 V_+$$

corresponding to the equality

$$V_T = V_+ + V_-$$

Note also that

$$I_- = -\frac{V_-}{Z_0} = +0.10 \frac{V_+}{Z_0}$$

2.7 POWER FLOW

In complex representation the forward voltage wave

$$V_+ = V_0 \cos(\beta z - \omega t)$$

assumes the form

$$\mathbf{V}_+ = V_0 e^{i(\beta z - \omega t)}$$

so that $V_+ = \mathrm{Re}\, \mathbf{V}_+$. In this representation other variables related to the transmission line, such as the impedance and reflection coefficient, are complex as well.

Referring to our results in Section 1.3, we find the expression for average power flow (1.28)

$$\langle P \rangle = \tfrac{1}{2}\mathrm{Re}\,\mathbf{IV}^* \tag{1.28}$$

It follows that the power flow in the forward wave is

$$\langle P_+ \rangle = \tfrac{1}{2}\mathrm{Re}\,\mathbf{V}_+\mathbf{I}_+^* \tag{2.26}$$

whereas power flow carried by the reflected wave is

$$\begin{aligned}\langle P_- \rangle &= -\tfrac{1}{2}\mathrm{Re}\,\mathbf{V}_-\mathbf{I}_-^* \\ &= +\tfrac{1}{2}\mathrm{Re}\,|\rho|^2\mathbf{V}_+\mathbf{I}_+^* \end{aligned}\tag{2.27}$$

and we may equate

$$\langle P_- \rangle = |\rho|^2\langle P_+ \rangle \tag{2.28}$$

or equivalently

$$\langle P_+ \rangle + \langle P_- \rangle = \langle P_+ \rangle(1 + |\rho|^2) \tag{2.29}$$

For the average power delivered to the load we find

$$\begin{aligned}\langle P_L \rangle &= \tfrac{1}{2}\mathrm{Re}\,V_L I_L^* \\ &= \tfrac{1}{2}\mathrm{Re}\,(\mathbf{V}_+ + \mathbf{V}_-)(\mathbf{I}_+^* + \mathbf{I}_-^*) \\ &= \tfrac{1}{2}\mathrm{Re}\,(\mathbf{V}_+ + \rho\mathbf{V}_+)(\mathbf{I}_+^* - \rho^*\mathbf{I}_+^*) \\ &= \tfrac{1}{2}\mathrm{Re}\,\mathbf{V}_+\mathbf{I}_+^*(1 + \rho)(1 - \rho^*)\end{aligned}$$

This product may be expanded to obtain

$$\langle P_L \rangle = \tfrac{1}{2}\mathrm{Re}\,\mathbf{V}_+\mathbf{I}_+^*(1 - |\rho|^2) + \tfrac{1}{2}\mathrm{Re}\,\mathbf{V}_+\mathbf{I}_+^*(\rho - \rho^*) \tag{2.30}$$

Introducing the characteristic impedance (2.12) permits (2.30) to be rewritten

$$\langle P_L \rangle = \tfrac{1}{2}\mathrm{Re}\,|\mathbf{I}_+|^2\mathbf{Z}_0(1 - |\rho|^2) + \tfrac{1}{2}\mathrm{Re}\,|\mathbf{I}_+|^2\mathbf{Z}_0(\rho - \rho^*) \tag{2.31}$$

For a lossless transmission line, $\mathbf{Z}_0$ is real, so that the second term in (2.31), which includes the purely imaginary difference $\rho - \rho^*$, vanishes, and we obtain

$$\langle P_L \rangle = \langle P_+ \rangle(1 - |\rho|^2) \tag{2.32}$$

Comparison with (2.28) leads to the equation

$$\langle P_+ \rangle = \langle P_L \rangle + \langle P_- \rangle \tag{2.33}$$

The power expended in the load and that carried by the reflected wave are provided by the power carried in the forward wave. Note also that with Z_0 given by (2.12), we may rewrite (2.26) as

$$\langle P_+ \rangle = \frac{|V_+|^2}{2Z_0} \tag{2.34}$$

This permits (2.32) to be rewritten in the alternative forms:

$$\langle P_L \rangle = \frac{|V_+|^2}{2Z_0}(1 - |\rho|^2) \tag{2.35a}$$

$$\langle P_L \rangle = \frac{1}{2Z_0}(|V_+|^2 - |V_-|^2) \tag{2.35b}$$

$$\langle P_L \rangle = \frac{1}{2Z_0} V_{\max} V_{\min} \tag{2.35c}$$

Example 2.5

A twin-lead transmission line operating in steady state is terminated in a load comprising an inductance L in parallel with a capacitance C, as shown in Fig. 2.13. The complex forward voltage wave has the form

$$\mathbf{V}_+ = V_0 e^{j(\beta z - \omega t)}$$

where $V_0 = 100$ ohms, and $\omega = 20$ rad/sec.

a. What is the average power in the forward wave?
b. What is the power expended in the load?

Ans.

a.
$$\langle P_+ \rangle = \frac{1}{2}\frac{V_0^2}{Z_0} = \frac{1}{2}\frac{(100)^2}{50} = 100 \text{ watts}$$

b. For the case at hand, the load admittance has the value

$$\begin{aligned} Y_L &= j\omega C + \frac{1}{j\omega L} = j\omega\left(C - \frac{1}{\omega^2 L}\right) \\ &= j20 \times 10^3\left(10^{-6} - \frac{1}{4 \times 10^8 \times 5 \times 10^{-6}}\right) \\ &= -j9.98 \end{aligned}$$

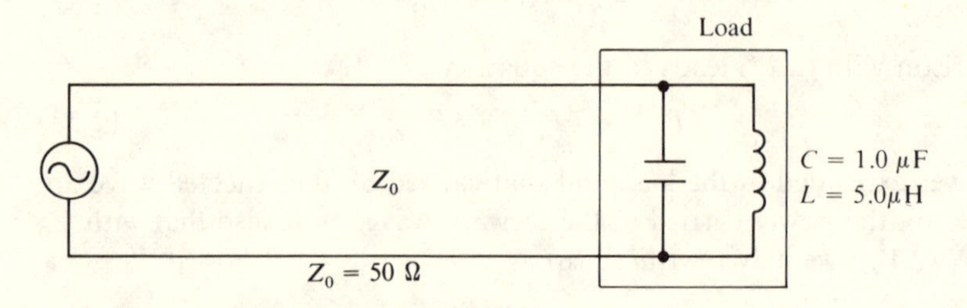

Figure 2.13 Configuration for Example 2.5.

corresponding to the load impedance

$$Z_L = \frac{1}{Y_L} = j0.10$$

which we see is pure imaginary. It follows that the reflection coefficient

$$\boldsymbol{\rho} = \frac{\mathbf{Z}_L - Z_0}{\mathbf{Z}_L + Z_0} = \frac{j0.10 - 50}{j0.10 + 50}$$

has the form

$$\boldsymbol{\rho} = -\frac{\mathbf{z}}{\mathbf{z}^*}$$

where $\mathbf{z}$ is a complex number. Writing $\mathbf{z}$ in polar form $\mathbf{z} = re^{j\theta}$, we find $\mathbf{z}^* = re^{-j\theta}$ and $\boldsymbol{\rho} = e^{j2\theta}$, or equivalently

$$\boldsymbol{\rho} = -(\cos 2\theta + j\sin 2\theta)$$

The reflection coefficient has magnitude

$$|\boldsymbol{\rho}|^2 = \cos^2 2\theta + \sin^2 2\theta = 1$$

so that there is no power expended in the load:

$$\langle P_L \rangle = \langle P_+ \rangle (1 - |\boldsymbol{\rho}|^2) = 0$$

This is always the case for a purely reactive load.

Example 2.6

A complex voltage wave $V_+(z, t)$ propagates on a transmission line of characteristic impedance Z_{01}, toward a junction which couples the line to a second transmission line with characteristic impedance Z_{02}. (See Fig. 2.14.) Use the conservation equation (2.33) to derive the relation

$$1 = |\rho|^2 + |\tau|^2 \frac{Z_{01}}{Z_{02}}$$

Ans. Identifying the transmitted power as P_L, we rewrite (2.33) as

$$\langle P_+ \rangle = \langle P_- \rangle + \langle P_T \rangle$$

With (1.28) we obtain, for an ideal line,

$$\langle P \rangle = \tfrac{1}{2}\mathrm{Re}\, VI^* = \tfrac{1}{2}\mathrm{Re}\, |V|^2/Z_0$$
$$= \tfrac{1}{2}|V|^2/Z_0$$

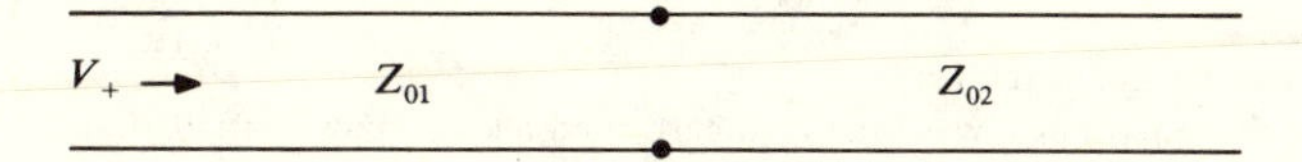

Figure 2.14 Configuration for Example 2.6.

Thus, the above power equality becomes

$$\frac{|V_+|^2}{Z_{01}} = \frac{|V_-|^2}{Z_{01}} + \frac{|V_T|^2}{Z_{02}}$$

Dividing by $|V_+|^2/Z_{01}$ gives the desired equation

$$1 = |\rho|^2 + \frac{Z_{01}}{Z_{02}}|\tau|^2$$

where for complex voltages we have written (2.24) in the form

$$|\tau| = \frac{V_T}{V_+}$$

PROBLEMS

2.1. A cylindrical coaxial transmission line contains a central cable of radius a and an outer cylindrical cable of inner radius b. A dielectric medium with permittivity ϵ separates the two leads. (See Fig. 2.15.)

a. Show that the characteristic impedance of this line is given by

$$Z_0 = \sqrt{\frac{\mu_0}{\epsilon}} \ln\frac{b}{a}$$

b. What is the intrinsic impedance of this line?

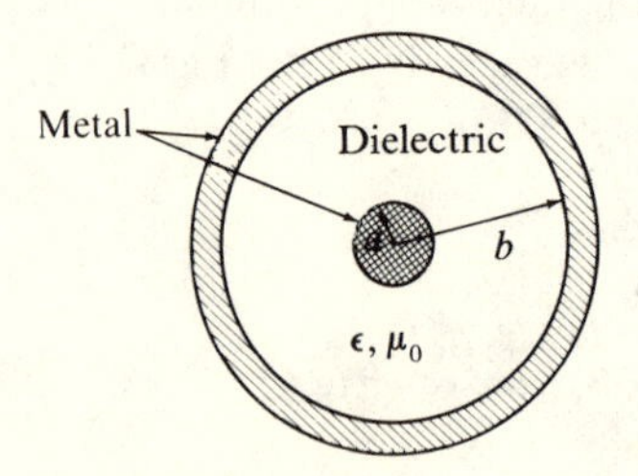

Figure 2.15 Coaxial transmission line.

2.2. A 75-ohm line is terminated in a load which has the equivalent network shown in Fig. 2.16, where

$$2\pi\omega = 62.8 \text{ kHz}$$
$$L_1 = L_2 = 20 \text{ mH}$$
$$C = 15\ \mu\text{F}$$

a. What is the reflection coefficient at the load?
b. The voltage wave which propagates toward the load carries an average power of 30 watts. What is the average power carried by the reflected wave?

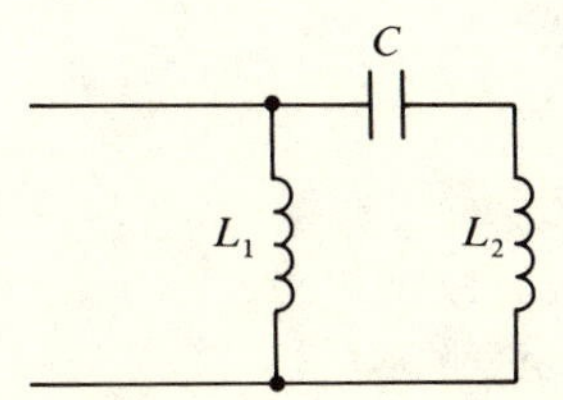

Figure 2.16

2.3. A standing voltage wave of the form

$$V(z,t) = 100\cos(\beta z - \omega t) + 50\cos(\beta z + \omega t)$$

is carried by a 75-ohm transmission line which is terminated in a load.

a. What is the expression for the forward current wave?
b. What is the reflection coefficient at the load?
c. What is the average power expended in the load?

CHAPTER

THREE

DC PULSES

3.1 THE FORWARD VOLTAGE WAVE

The problem we consider at this point concerns a transmission line l meters long which is terminated in a purely resistive load and which has a battery of voltage V_S at its source. (See Fig. 3.1.) The line has characteristic impedance Z_0, whereas the battery has internal resistance R_S. When the switch is closed, a forward voltage wave is launched toward the load. At this instant the starting current "sees" only the characteristic impedance and battery resistance. After the switch is closed, it takes a finite time for the forward voltage wave to reach the load and interact with it. Prior to the time of this interaction the presence of the load has no effect on the signal.

An equivalent circuit describing the initial situation is shown in Fig. 3.2. The starting current has the value

$$I_+ = \frac{V_S}{R_S + Z_0} \tag{3.1}$$

The amplitude of the forward voltage wave is the voltage across Z_0.

$$V_+ = I_+ Z_0 = \left(\frac{Z_0}{R_S + Z_0} \right) V_S \tag{3.2}$$

The current I_+ and voltage V_+ propagate toward the load with the speed given by (2.6):

$$v = \frac{1}{\sqrt{LC}}$$

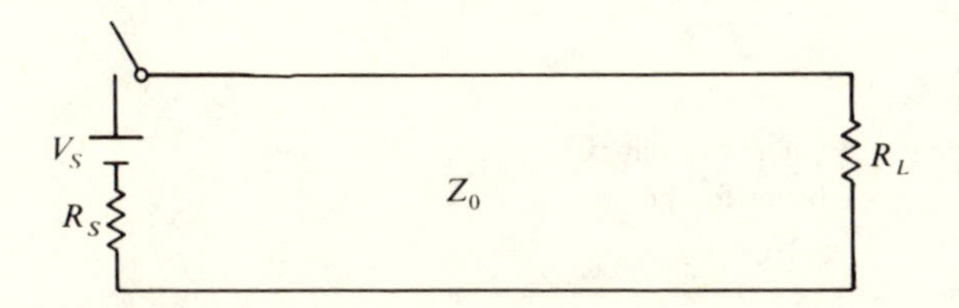

Figure 3.1 The DC line with a resistive load.

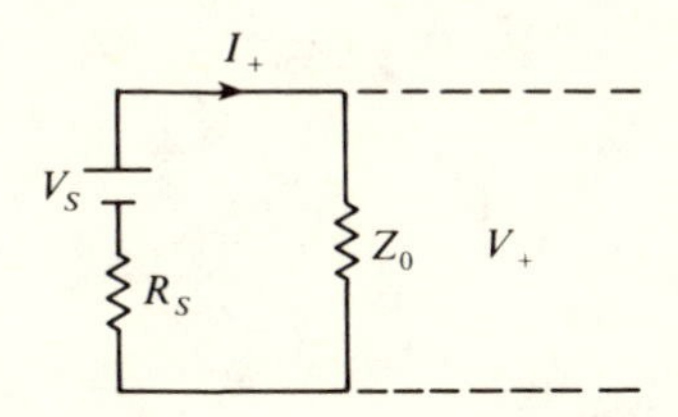

Figure 3.2 The equivalent starting circuit.

At time $vt < l$, the voltage and current on the line appear as shown in Fig. 3.3.

If one measures the voltage at $z = l/3$ during the time $vt < l$, then a graph is obtained as shown in Fig. 3.4.

An alternate description of the voltage on the line is obtained by plotting the displacement of the leading edge of the wave (or equivalently vt) on the vertical axis and displacement z on the horizontal axis. (See Fig. 3.5.) Once the wave passes a point on the transmission line, the voltage at the point is left at the value V_+. So at the point $l/2$, say, for $vt < l/2$, the

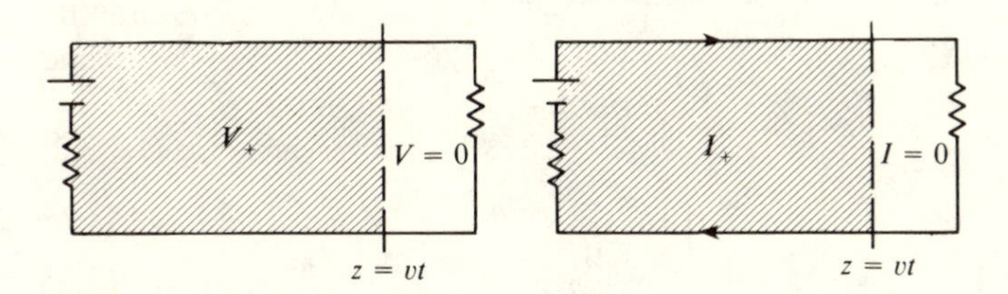

Figure 3.3 Voltage and current propagation.

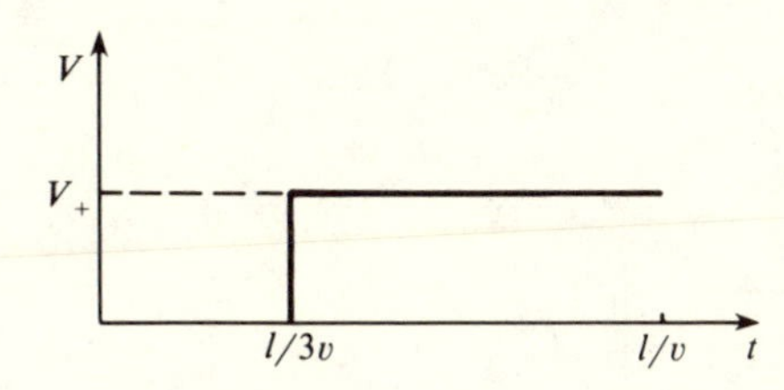

Figure 3.4 Voltage at $z = l/3$ in the interval $vt < l$.

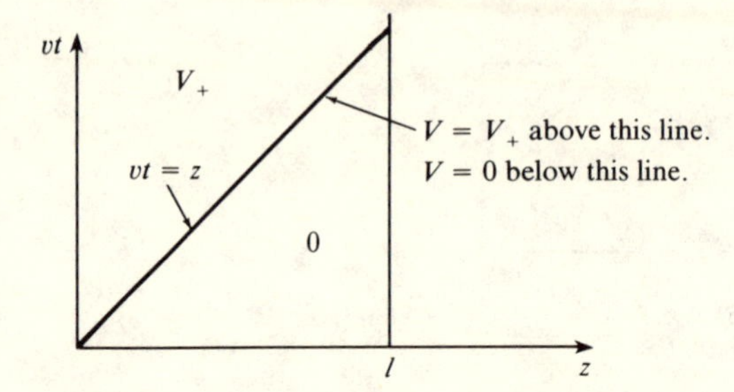

Figure 3.5

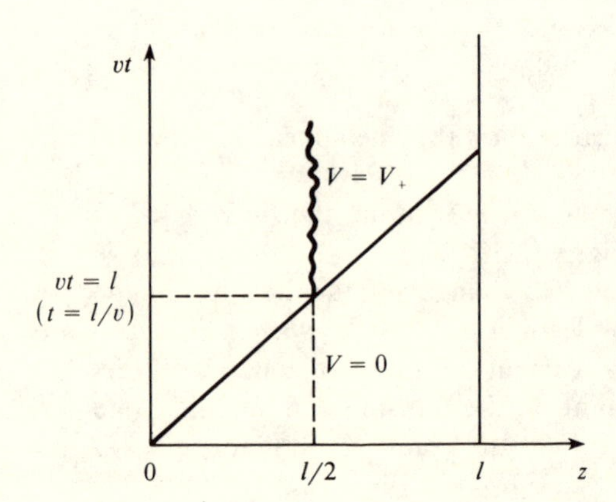

Figure 3.6 Voltage ahead of and behind the voltage wave.

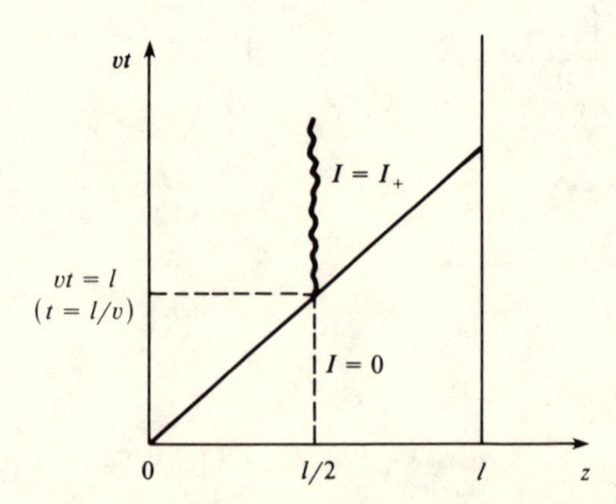

Figure 3.7 Current ahead of and behind the current wave.

voltage $V = 0$. For $vt \geq l/2$, $V = V_+$. This situation is depicted in Fig. 3.6. An identical situation applies to the current wave $I_+ = V_0/Z_0$ (see Fig. 3.7).

In the event the line is matched to a load resistance Z_0, there is no reflected current or voltage. The asymptotic values

$$V_{\text{asm}} = V_S$$
$$I_{\text{asm}} = V_S/Z_0$$

are attained in the time l/v, where l is the length of the line. The bounce diagram for, say, the current has only one leg, as shown in Fig. 3.8. Since

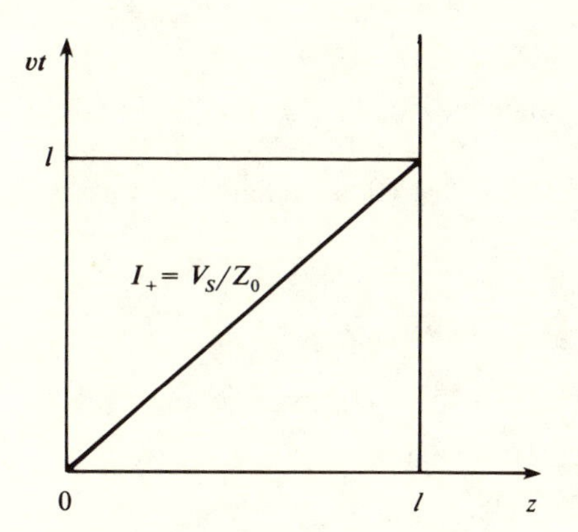

Figure 3.8 Bounce diagram has only one leg for a perfectly matched load.

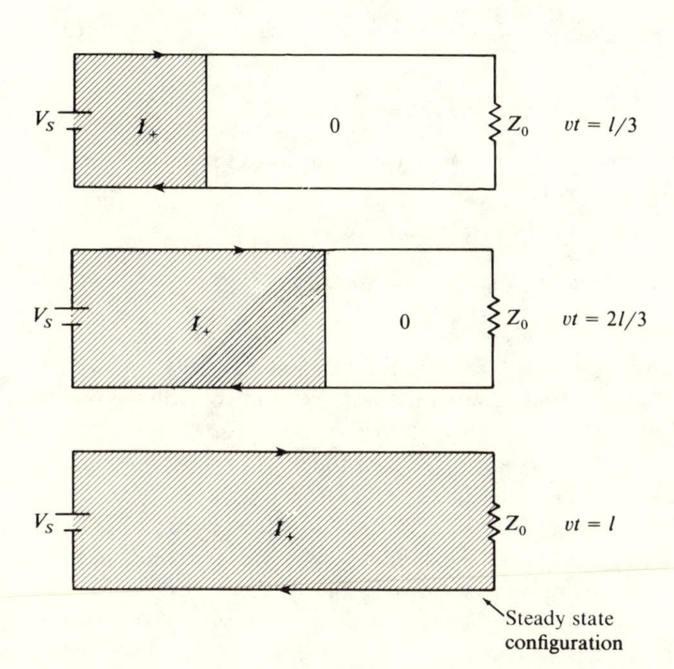

Figure 3.9 Current wave on a matched DC line. Characteristic impedance of line in Z_0.

there is no reflected wave, when the forward current wave encounters the load, the load current is $I_L = I_+$. Now we must recall that I_+ has *both* forward and return current components. In passing through the load, the forward component of I_+ joins the return component of I_+, thereby establishing the steady state.

It is revealing to follow I_+ to the load through a time sequence of diagrams as shown in Fig. 3.9. In the steady state I_+ flows through the current loop comprised of transmission line, source, and load.

3.2 THE REFLECTED WAVE

Let us return to the case of an unmatched load, $R_L \neq Z_0$. When the forward wave (2.2) reaches the load at $t = l/v$, a reflected wave is launched which propagates toward the battery. This reflected wave has the amplitude (writing R for R_L)

$$V_- = \rho V_+ = \left(\frac{R - Z_0}{R + Z_0}\right)\left(\frac{Z_0 V_S}{R_S + Z_0}\right) \tag{3.3}$$

For the case at hand, both R and Z_0 are positive numbers, so that

$$|R - Z_0| \leq |R + Z_0|$$

and

$$|\rho| \leq 1 \tag{3.4}$$

Thus

$$|V_-| \leq |V_+|$$

As the V_- wave propagates toward the battery it adds to the voltage already on the line, namely V_+, to yield

$$V = V_+ + \rho V_+ \tag{3.5}$$

in back of the leading edge of V_-, and the voltage

$$V = V_+$$

ahead of the leading edge of V_-. At the time $vt = 3l/2$, the V_- wave has traveled half the distance back toward the battery, and the voltage on the line at this time, for $R > Z_0$, appears as shown in Fig. 3.10.

If the voltage at $z = l/2$ is measured as a function of time, one obtains the graph shown in Fig. 3.11.

This variation of voltage is more directly obtained in the vt-vs-z diagram of the leading edge of the voltage wave. In the time interval $0 \leq vt \leq 2l$, this diagram appears as shown in Fig. 3.12.

Consider the point $z = l/2$. When the forward wave V_+ passes this point at $t = l/2v$, the voltage changes from $V = 0$ to $V = V_+$. The voltage

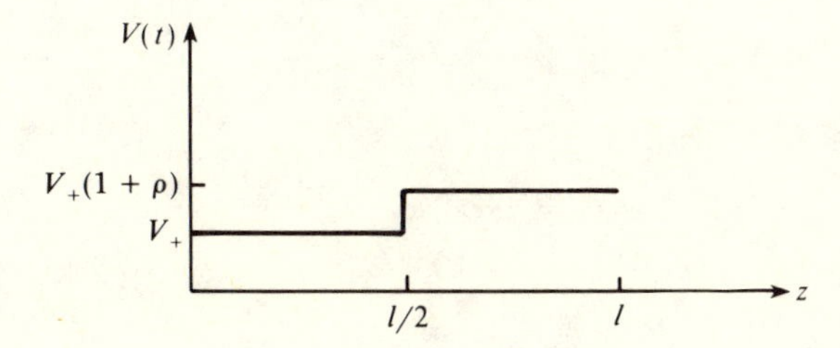

Figure 3.10 Voltage at $vt = 3l/2$.

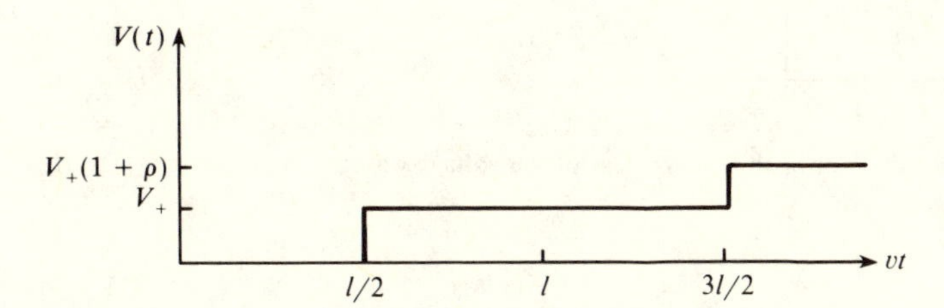

Figure 3.11 Voltage at $z = l/2$.

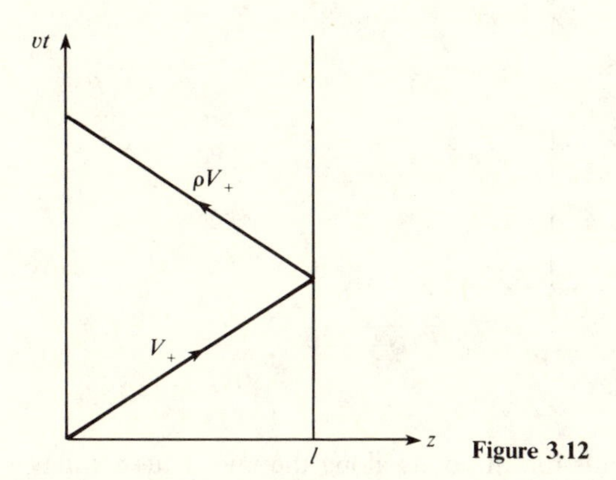

Figure 3.12

maintains this value until $t = 3l/2v$, at which time the reflected wave adds to V_+ to give the value $V = V_+ + \rho V_+$. To obtain this information from the figure we draw a vertical line at $Z = l/2$. This line represents different times at this single point $z = l/2$. (See Fig. 3.13.) As one proceeds upward on the line $z = l/2$ (to greater times), intersection is first made with the leading edge of the forward wave V_+ at $t = l/2v$. At this instant the voltage at $z = l/2$ changes from zero to V_+. It maintains this value until the vertical line pierces the leading edge of the reflected wave ρV_+ at $t = 3l/2v$. At this time the voltage at $z = l/2$ changes from V_+ to $V_+ + \rho V_+$.

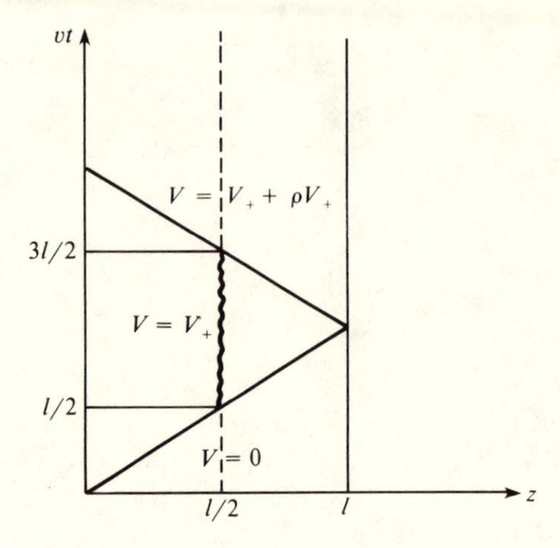

Figure 3.13 Bounce diagram with first two legs of the voltage wave.

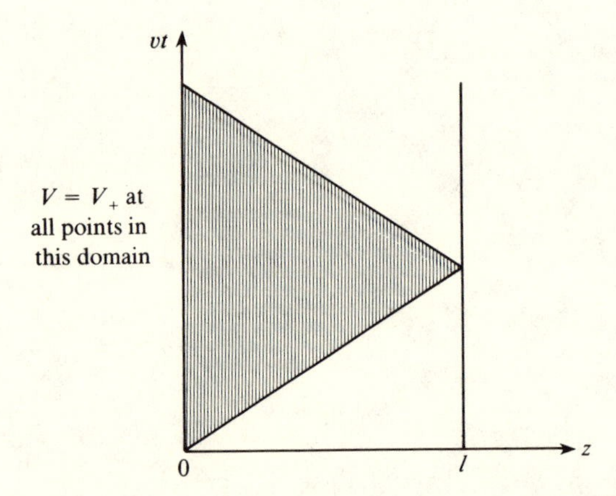

Figure 3.14

A similar construction holds for all points along the line. Thus we may conclude that at all space-time points in the first wedge of our *bounce diagram*, $V = V_+$. (See Fig. 3.14.) In like manner, the current on the line at all values of z and t in the first wedge of the bounce diagram is I_+.

3.3 SECOND- AND HIGHER-ORDER REFLECTIONS

When the first reflected wave V_- reaches the battery, it is re-reflected. The new reflected wave V'_+ has the value

$$V'_+ = \rho_S V_- \tag{3.6}$$

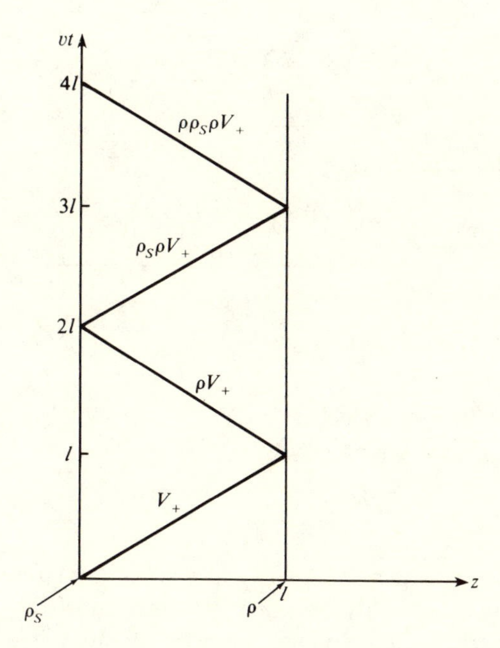

Figure 3.15 Two complete reflections of the voltage wave.

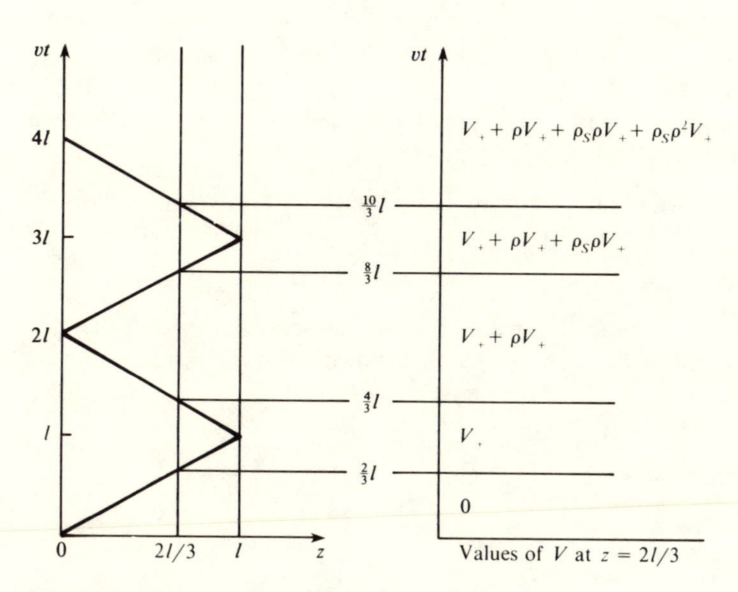

Figure 3.16 Voltages at $z = 2l/3$.

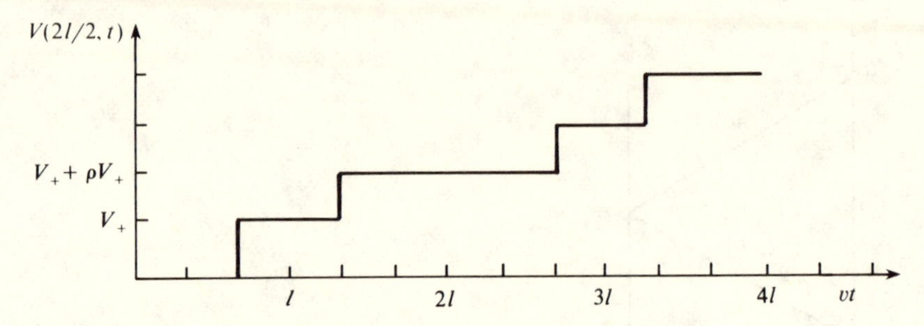

Figure 3.17 Voltage values at $z = 2l/3$.

where ρ_S is the reflection coefficient at the battery:

$$\rho_S = \frac{R_S - Z_0}{R_S + Z_0} \tag{3.7}$$

This reflection coefficient comes into play at $t = 2l/v$. The value of the voltage on the line for later times is readily obtained with the aid of our bounce diagram of the leading edge of the voltage wave. Let us draw this diagram up to the second reflection at the source. (See Fig. 3.15.)

What is the voltage at the point $z = 2l/3$ as a function of time? To obtain the answer, draw a vertical line at this point and add up the voltages as this line pierces different reflected wave segments, as shown in Fig. 3.16. From these data we may easily sketch the voltage at the point $z = 2l/3$ vs time vt (with ρ and ρ_S positive): see Fig. 3.17.

Example 3.1

Suppose that a lossless battery acts as source to an ideal transmission line l meters long terminated in a short circuit.

a. Obtain the current on the line at $z = l/2$ in the time interval $0 \le vt < 4l$.
b. Obtain the voltage on the line at $z = l/2$ in this same time interval.

Ans.

a. The line appears as shown in Fig. 3.18. Constructing the bounce diagram for the current, we obtain the diagram shown in Fig. 3.19. Here we have

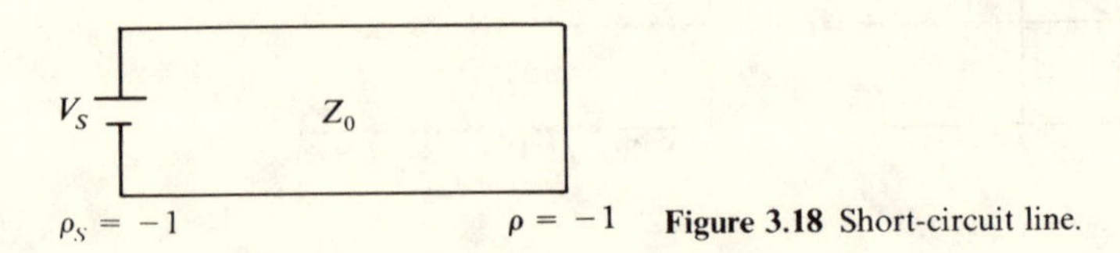

Figure 3.18 Short-circuit line.

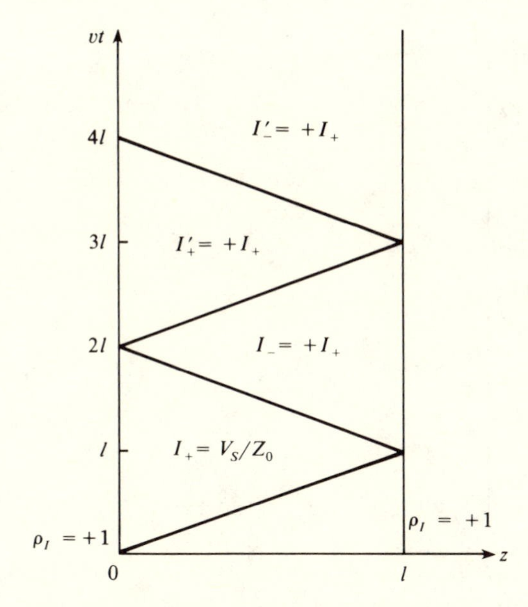

Figure 3.19 Current bounces on a short-circuit line.

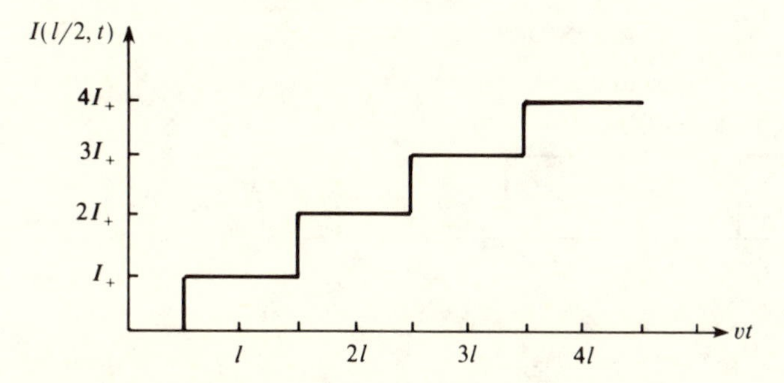

Figure 3.20 Growth of current on a short-circuit line.

introduced the current reflection coefficient,

$$\rho_I = -\rho.$$

$$I_- = \rho_I I_+$$

[compare with Eq. (2.22)]. The above bounce diagram gives the desired graph (see Fig. 3.20). This graph illustrates the growth to infinity of the current on a shorted line.

b. The bounce diagram for the voltage appears as shown in Fig. 3.21. From this diagram we obtain the desired graph of V vs time at $z = l/2$. (See

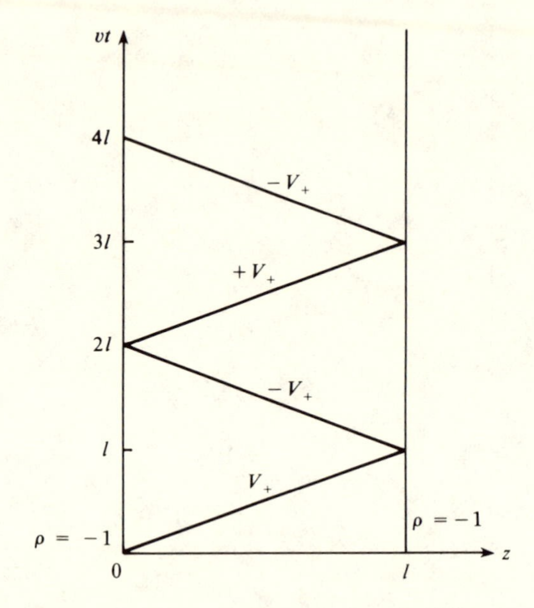

Figure 3.21 Voltage bounces on a short-circuit line.

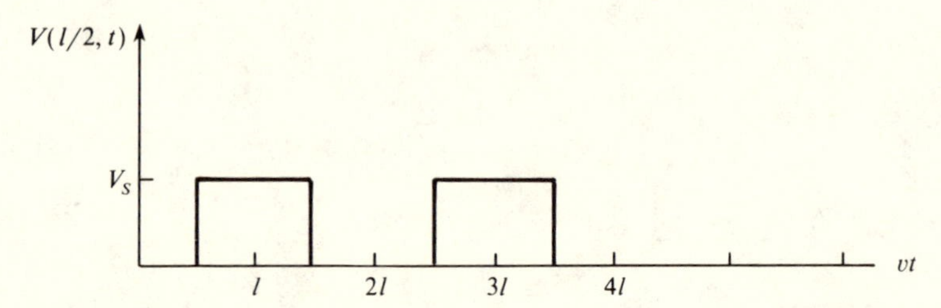

Figure 3.22

Fig. 3.22.) The voltage at $z = l/2$ oscillates and has the average value

$$\langle V \rangle = V_S/2$$

Example 3.2

Repeat the preceding example for the case that the line is terminated in an open circuit.

Ans.

a. The circuit appears as shown in Fig. 3.23. First we construct the bounce diagram for the current, as in Fig. 3.24. This gives the desired graph at $z = l/2$ as shown in Fig. 3.25. The current at $z = l/2$ has the average

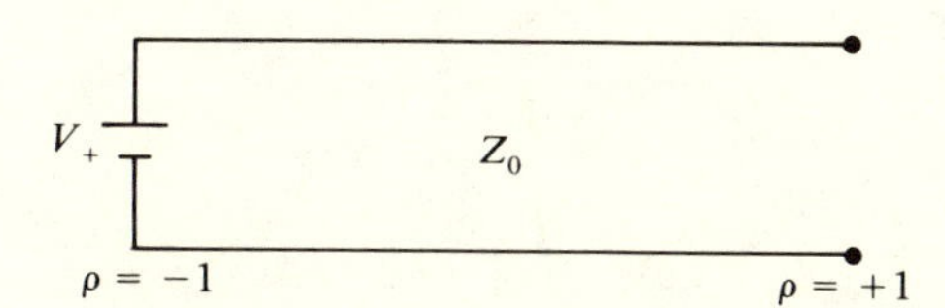

Figure 3.23

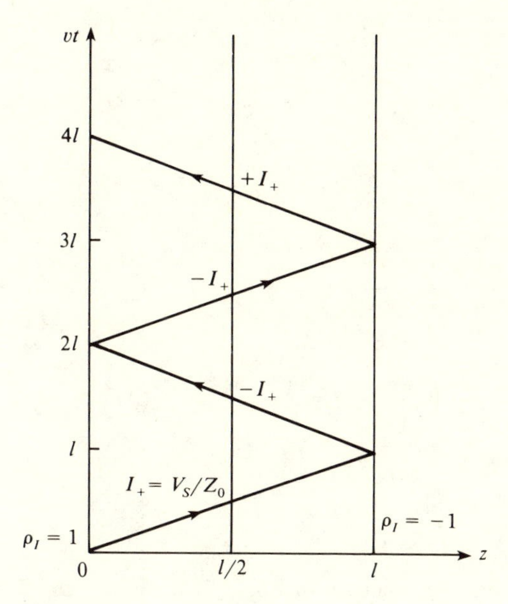

Figure 3.24

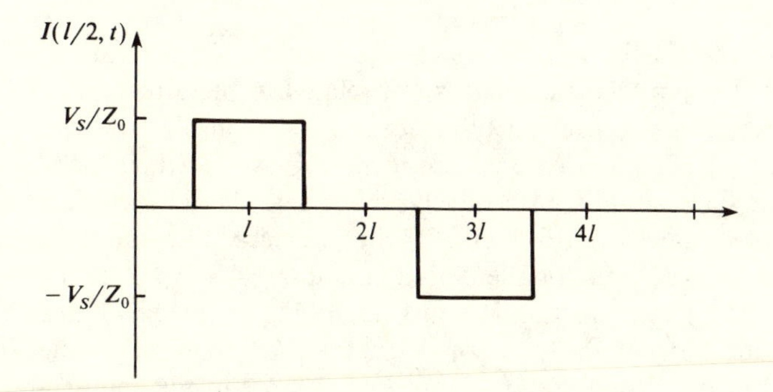

Figure 3.25 The current curve for Example 3.2.

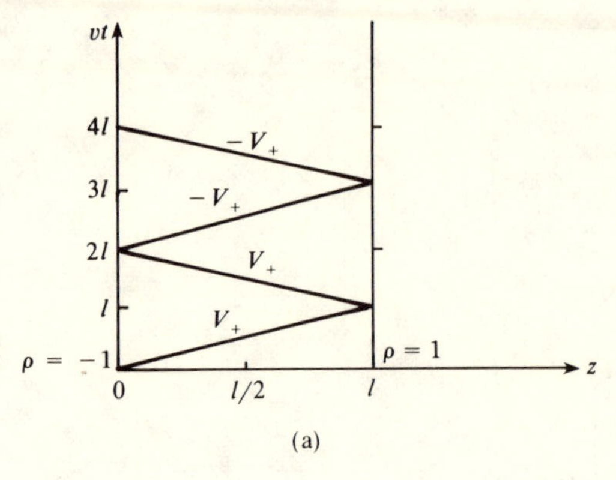

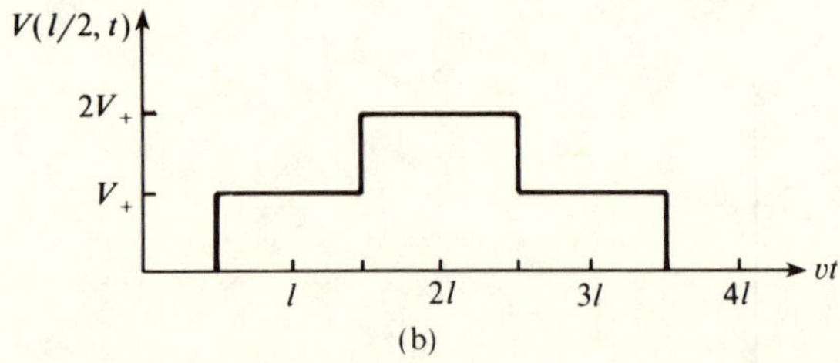

Figure 3.26 (a) Bounce diagram and (b) voltage at $z = l/2$ for Example 3.2.

value

$$\langle I \rangle = 0$$

b. The bounce diagram for the voltage is shown in Fig. 3.26a. The graph of V vs vt at $z = l/2$ has the form shown in Fig. 3.26b. The voltage at $z = l/2$ has the open-circuit average

$$\langle V \rangle = V_S$$

Utilizing the results of the last two examples, let us offer a clear-cut demonstration that the sign of the current is not related to the direction of the current wave, as discussed previously in Section 2.2. Consider the two cases depicted in Fig. 3.27. The two lines are identical save for their loads which are open and short circuits, respectively. After S is closed, a current wave $I_+ = V_S/Z_0$ propagates toward the load. Upon reflection, for the open circuit we obtain $I_- = -I_+$ and for the closed circuit $I_- = +I_+$. Both I_- waves propagate in the same direction (toward the battery), but their signs are opposite. The meaning of the signs of I_- is that $I_- < 0$ flows in the counterclockwise direction (to cancel I_+) and $I_- > 0$ flows in the counterclockwise direction (to enhance I_+).

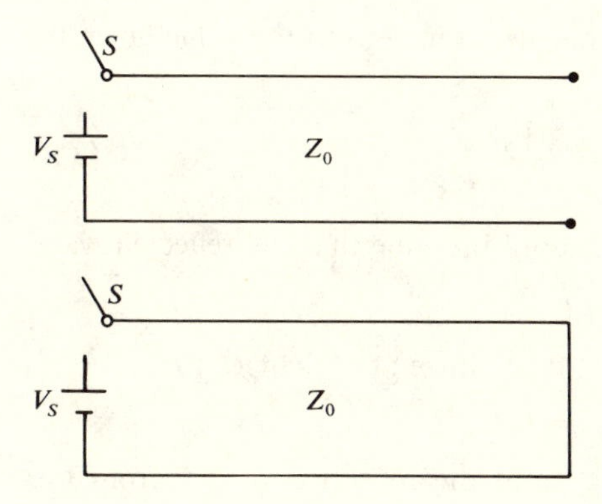

Figure 3.27

3.4 ASYMPTOTIC VALUES

In this section we wish to demonstrate the following important fact. For a transmission line with a resistive load R_L, and a battery with internal resistance R_S, the voltage on the line approaches the uniform constant value

$$V = \frac{V_S R_L}{R_S + R_L}$$

as time approaches infinity.

To obtain this result we must sum the infinite number of contributions of the reflected voltage waves. The transmission line we are considering has the constituents shown in Fig. 3.28. The switch is closed at $t = 0$. At any point z on the line the voltage is initially zero. When the forward voltage

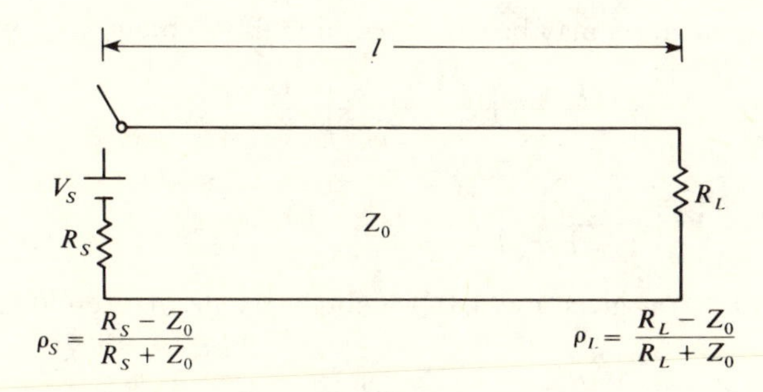

Figure 3.28

wave passes this point, the voltage changes from zero to the value given by (2.2):

$$V = V_+ = \frac{V_S Z_0}{Z_0 + R_S} \tag{3.8}$$

This is the value of V at the point z until the time that the reflected wave

$$V_- = \rho_L V_+$$

passes the point. At this same instant the voltage at z changes to

$$V = V_+ + \rho_L V_+$$

This value of V is maintained at z until the reflection of V_- from the source reaches z. When this occurs the voltage at z changes to

$$V = V_+ + \rho_L V_+ + \rho_S(\rho_L V_+)$$

Now there is a forward wave $\rho_S \rho_L V_+$ traveling toward the load. Its reflection has the amplitude $\rho_L(\rho_S \rho_L V_+)$. When this reflected wave passes the point z, the voltage at z changes to

$$V = V_+ + \rho_L V_+ + \rho_S(\rho_L V_+) + \rho_L(\rho_S \rho_L V_+)$$

Continuing, we find the sixth value of voltage at the point z is the sum

$$V = V_+ + \rho_L V_+ + (\rho_S \rho_L) V_+ + \rho_L(\rho_S \rho_L) V_+ + (\rho_L \rho_S)^2 V_+ + \rho_L(\rho_L \rho_S)^2 V_+$$

We may conclude that the voltage at the point z for large time ($t \gg v/l$) has the form

$$V = V_+\left\{\left[1 + (\rho_L \rho_S) + (\rho_L \rho_S)^2 + \cdots + (\rho_L \rho_S)^n + \cdots\right] + \rho_L\left[1 + (\rho_L \rho_S) + (\rho_L \rho_S)^2 + \cdots + (\rho_L \rho_S)^n + \cdots\right]\right\} \tag{3.9}$$

Now we recall the summation

$$1 + x + x^2 + \cdots = \frac{1}{1 - x}, \qquad |x| < 1$$

Since $|\rho_L \rho_S| < 1$, this summation may be employed in (3.9) to obtain

$$V = V_+\left(\frac{1}{1 - \rho_L \rho_S} + \frac{\rho_L}{1 - \rho_L \rho_S}\right)$$

$$= V_+\left(\frac{1 + \rho_L}{1 - \rho_L \rho_S}\right)$$

Substituting the value of V_+ as given by (3.8), we obtain the desired result,

$$V \sim \frac{V_S R_L}{R_S + R_L}, \qquad t \gg l/v$$

The current has the asymptotic value

$$I \sim \frac{V_S}{R_S + R_L}, \qquad t \gg l/v$$

These are the values of V and I, due to the forcing DC voltage V_S turned on at $t = 0$, long after the transients due to the distributed reactive L, C components of the system have decayed away.

Example 3.3

A battery with zero internal impedance has an open-circuit voltage of 100 volts. At time $t = 0$ it is switched into a 50-ohm-impedance air-dielectric ($\epsilon = \epsilon_0$) cable via a 150-ohm resistor. The cable is 300 m long and is terminated in a load of 33.333 ohms.

a. Draw a bounce diagram to show the pulse propagation through the 300-meter section of cable. Extend your calculations to cover the first 4 μsec after the switch closes. Show voltage values on each leg of the bounce diagram.
b. Draw graphs of the voltage V_L across the load and the current I_L through the load as a function of time.
c. What are the asymptotic values of V_L and I_L in the limit $t \to \infty$?

Ans.

a. The circuit, together with reflection-coefficient values, is shown (with switch in open position) in Fig. 3.29. The voltage wave propagates with speed $v = 1/\sqrt{\mu\epsilon} = 3 \times 10^8$ m/sec, so that the time of flight from battery to load is $\tau = 300/(3 \times 10^8) = 1$ μsec. The amplitude of the first voltage pulse is obtained from the equivalent circuit shown in Fig. 3.30:

$$V_+ = \frac{50}{200} \times 100 = 25 \text{ volts}$$

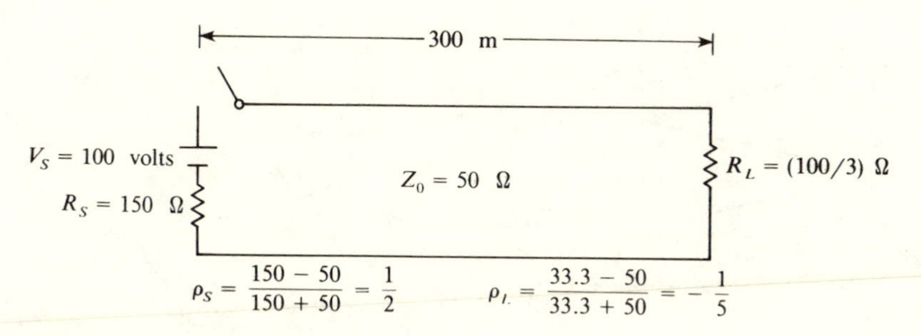

Figure 3.29 Configuration for Example 3.3.

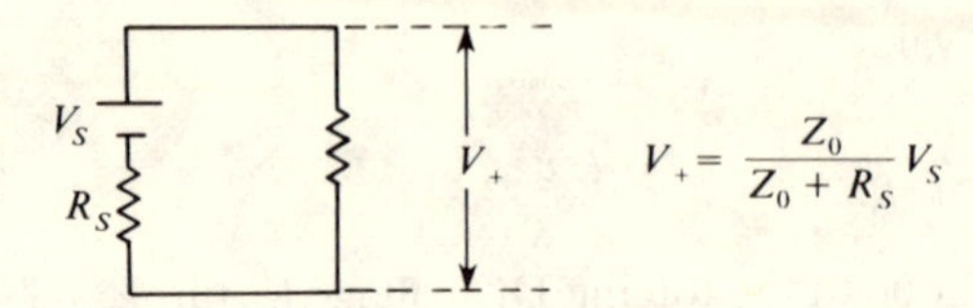

Figure 3.30

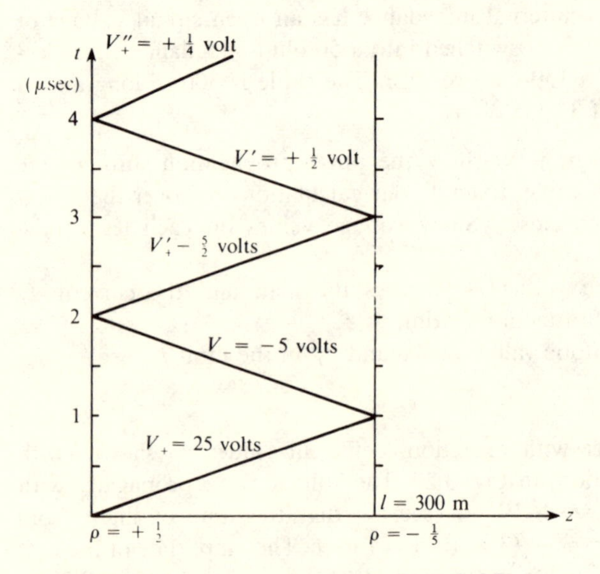

Figure 3.31 Voltage bounce diagram for Example 3.3.

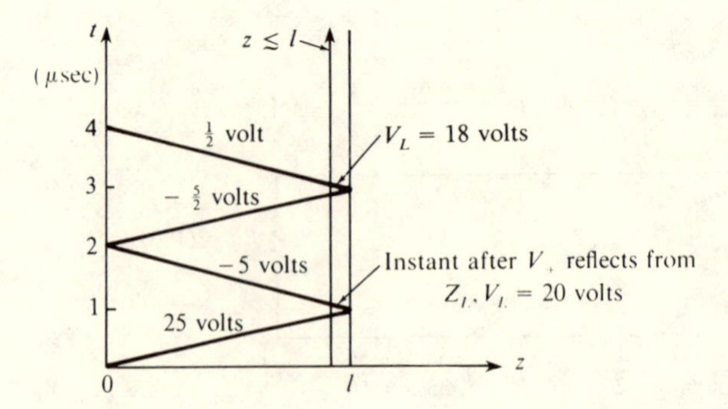

Figure 3.32

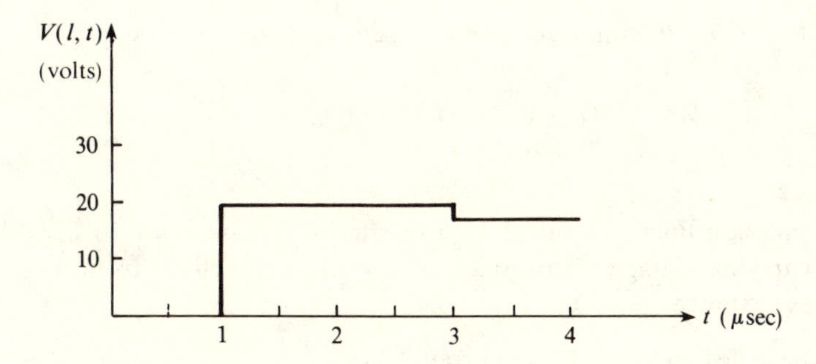

Figure 3.33 Voltage at the load for Example 3.3.

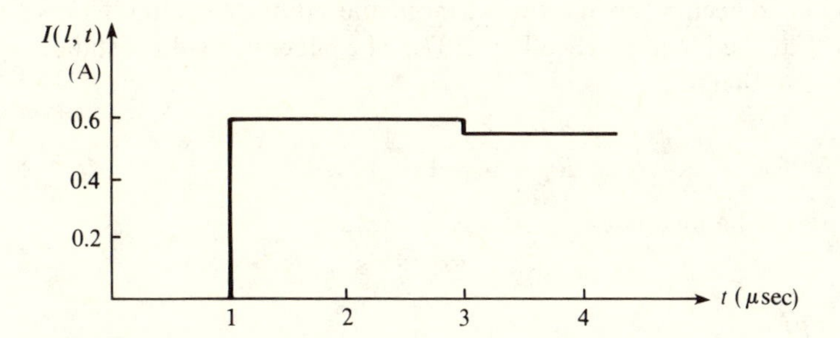

Figure 3.34 Current through the load for Example 3.3.

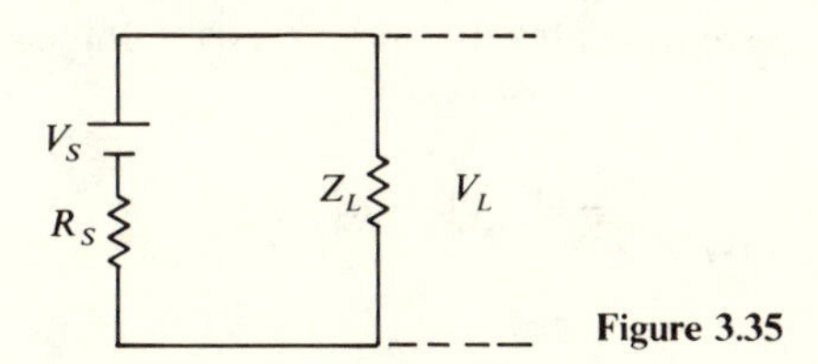

Figure 3.35

With these values at hand we are able to construct the bounce diagram shown in Fig. 3.31.

b. To obtain the voltage through the load, we draw a vertical line close to the load and add up values as the voltage wave crosses this line. (See Fig. 3.32.) In this manner we obtain the load voltage graph shown in Fig. 3.33. The current through the load is given by Ohm's law,

$$I_L = V_L/Z_L$$

which has the graph of Fig. 3.34.

c. In the limit $t \to \infty$ all transients decay to zero and the circuit is purely resistive (see Fig. 3.35):

$$V_L^{\text{asm}} = 18.2\ \text{V}, \qquad I_L^{\text{asm}} = 0.55\ \text{A}$$

Example 3.4

Two transmission lines are joined by a resistance R_L as shown in Fig. 3.36. A wave carrying voltage V_+ propagates toward the junctions from the Z_{01} side of the resistance.

a. What are the reflected voltage and current?
b. What are the transmitted voltage and current?

Ans.

a. To obtain the reflection coefficient we note the equivalent circuit presented to V_+ at the junction. (See Fig. 3.37.) The effective load resistance is $R_L + Z_{02}$, so that

$$\rho = \frac{R_L + Z_{02} - Z_{01}}{R_L + Z_{02} + Z_{01}}$$

and $V_- = \rho V_+$. The incident current has the value

$$I_+ = V_+/Z_{01}$$

and $I_- = -\rho I_+$.

b. To obtain the transmitted voltage we first note the voltage across the load:

$$V_L = V_+ + V_- = V_+(1 + \rho)$$

The equivalent load circuit drawn above indicates that it acts as a voltage divider, so that

$$V_T = \frac{Z_{02}}{R_L + Z_{02}} V_L = \frac{Z_{02}}{R_L + Z_{02}} V_+(1 + \rho)$$

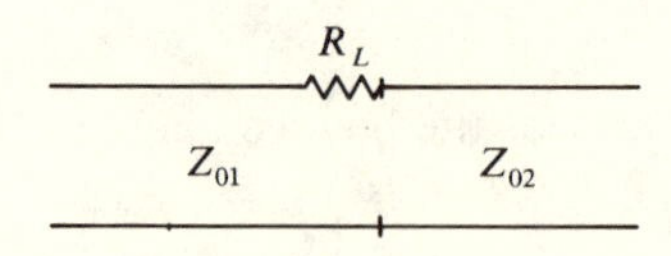

Figure 3.36 Configuration for Example 3.4.

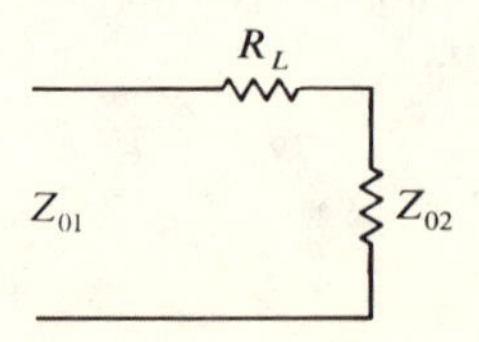

Figure 3.37

Example 3.5

Two transmission lines are joined by a resistance R_L as shown in Fig. 3.38. A wave carrying voltage V_+ propagates toward the junction from the Z_{01} side of the resistance.

a. What are the reflected voltage and current?
b. What are the transmitted voltage and current?

Ans.

a. The load impedance presented to V_+ has the equivalent circuit shown in Fig. 3.39, and the value

$$Z_L = \frac{Z_{02} R_L}{Z_{02} + R_L}$$

The reflection coefficient is then given by

$$\rho = \frac{Z_L - Z_{01}}{Z_L + Z_{01}} = \frac{R_L(Z_{02} - Z_{01}) - Z_{01}Z_{02}}{R_L(Z_{02} + Z_{01}) + Z_{01}Z_{02}}$$

(Note that if R_L is replaced by a short circuit, $\rho = -1$.) The reflected voltage is $V_- = \rho V_+$. With

$$I_+ = V_+/Z_{01},$$

we obtain $I_- = -\rho I_+$.

b. To obtain the transmitted current we note the equivalent circuit shown in Fig. 3.40. It follows that

$$I_T = I_L \frac{R_L}{R_L + Z_{02}} = I_+(1 - \rho)\frac{R_L}{R_L + Z_{02}}$$

The transmitted voltage is the voltage across the load,

$$V_T = V_+ + V_- = V_+(1 + \rho)$$

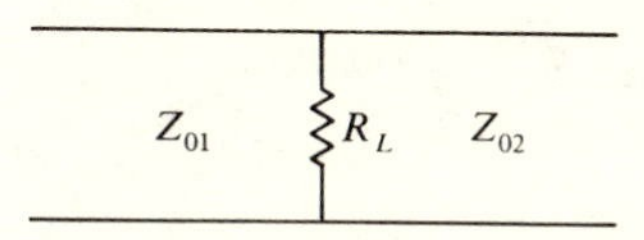

Figure 3.38 Configuration for Example 3.5.

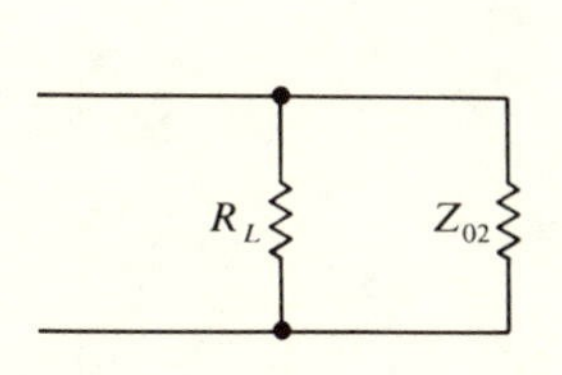

Figure 3.39

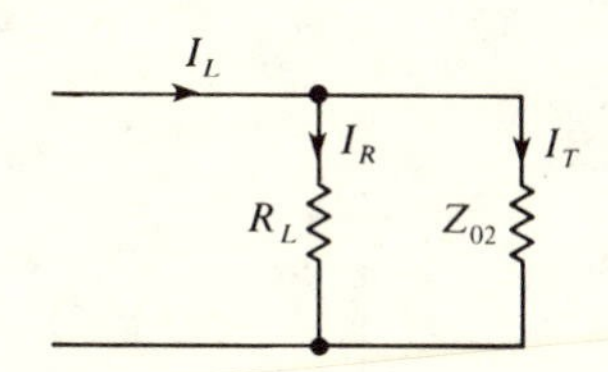

Figure 3.40

3.5 SWITCH AT THE LOAD

Consider the transmission line configuration shown in Fig. 3.41. We wish to obtain current and voltage values after the switch S is closed. The starting voltage (V_+) and current (I_+) values obey the following equations:

$$V_L = V_S + V_+ = I_L R_L$$

$$I_+ = \frac{V_+}{Z_0} = -I_L$$

where I_L represents the current flowing through R_L. Solving for V_+ and I_+, we obtain

$$V_+ = -\frac{V_S Z_0}{Z_0 + R_L}, \qquad I_+ = -\frac{V_S}{R_L + Z_0}$$

With these starting values at hand we may construct a bounce diagram. Let us consider the specific case that $R_L = Z_0$. Then $\rho = 0$ at the load, whereas $\rho = -1$ at the battery, which is assumed to have no internal resistance. The bounce diagram for voltage is as shown in Fig. 3.42. After the time $2l/v$ the voltage on the line is V_S. The current bounce diagram appears as shown in Fig. 3.43. After the time $2l/v$ the current on the line is the DC value

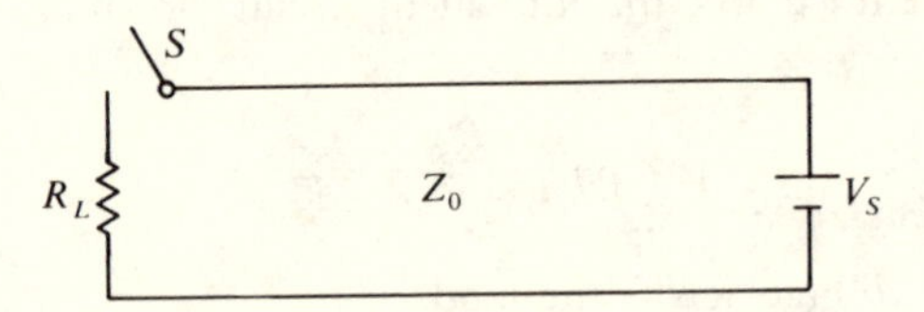

Figure 3.41 Switch at the load.

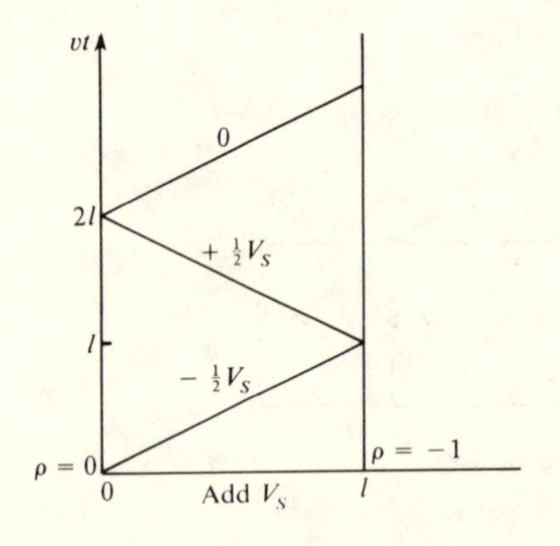

Figure 3.42 Voltage bounce diagram.

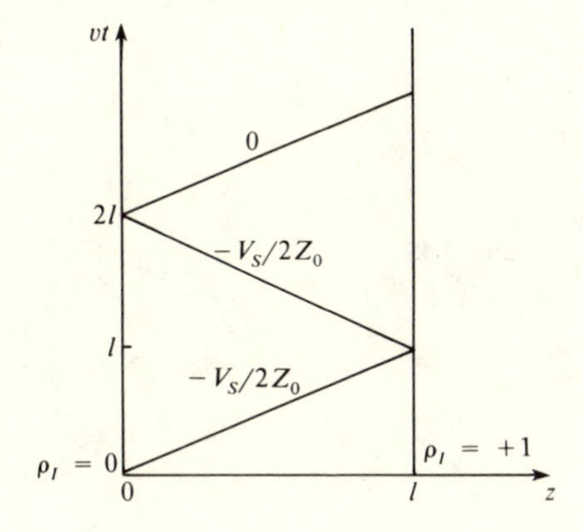

Figure 3.43 Current bounce diagram.

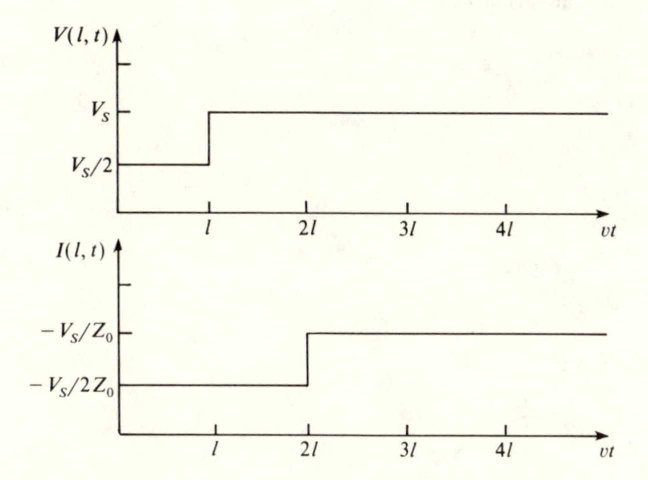

Figure 3.44 Voltage and current curves for a switch at the load.

$I = V_S/R_L = V_S/Z_0$. Sketches of the voltage and current at the load as a function of time are shown in Fig. 3.44.

3.6 SWITCH AT THE MIDPLANE

The next fundamental configuration we wish to consider is shown in Fig. 3.45. The switch S is at the midpoint, $l/2$ meters from the battery. After the switch is closed, current flow is as shown in Fig. 3.46.

That I_1 and I_2 flow in the same direction follows from the law of conservation of current. Namely, if I_2 flows away from S, then I_1 must flow toward S, since there is no source of current at S. A similar argument based

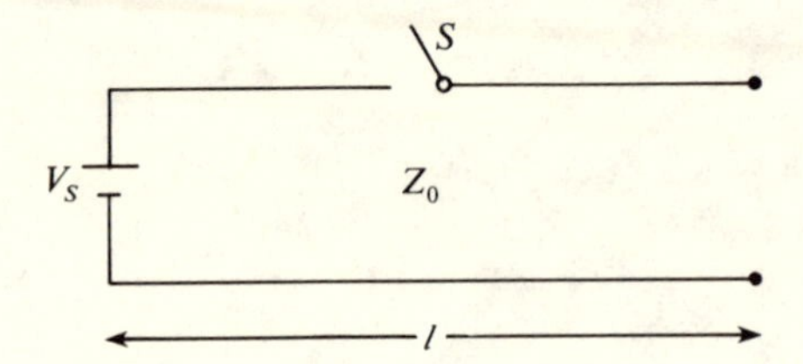

Figure 3.45 Switch at midplane.

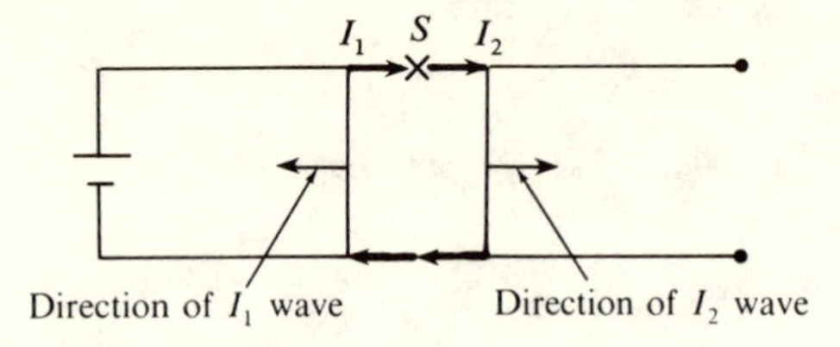

Figure 3.46 Starting current waves.

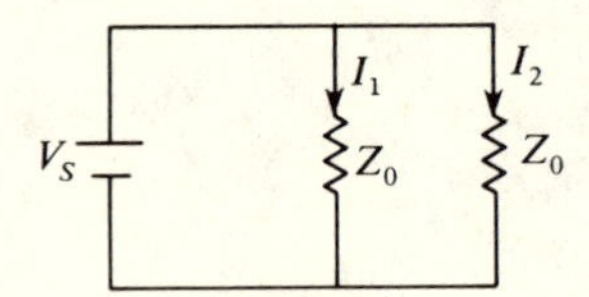

Figure 3.47

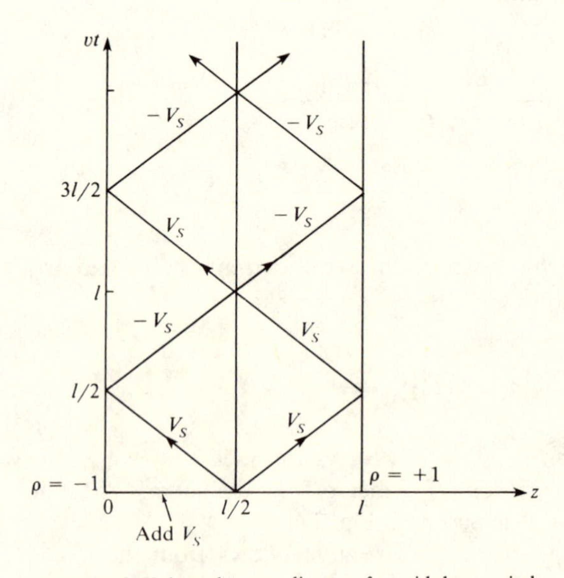

Figure 3.48 Voltage bounce diagram for midplane switch.

on conservation indicates that $I_1 = I_2$. This equality may also be obtained from an equivalent-circuit construction as shown in Fig. 3.47. That is, I_1 and I_2 both "see" impedance Z_0. It follows that $I_1 = I_2$.

The starting voltages are given by

$$V_1 = V_2 = Z_0 I_1 = Z_0 I_2 = V_S$$

We are now prepared to draw the bounce diagram for this configuration: see Fig. 3.48. The voltage across the open circuit vs time has the graph given in Fig. 3.49.

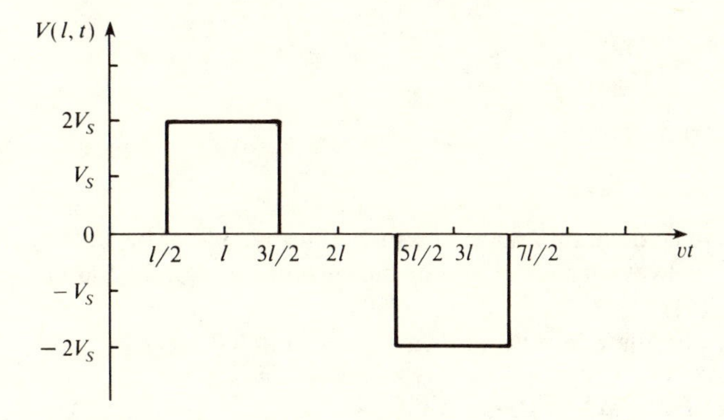

Figure 3.49 Voltage at the open circuit.

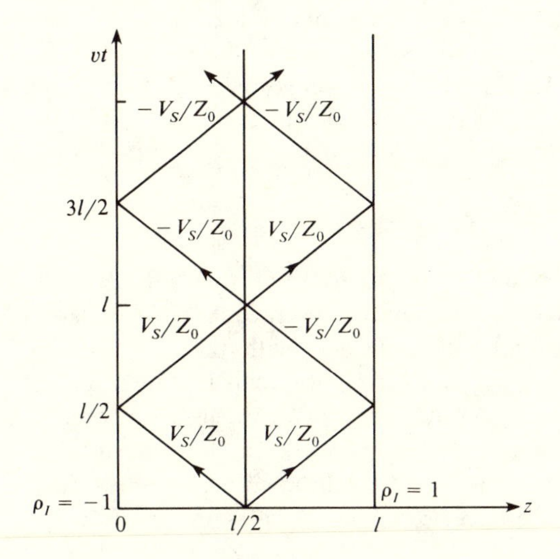

Figure 3.50 Current bounce diagram for midplane switch.

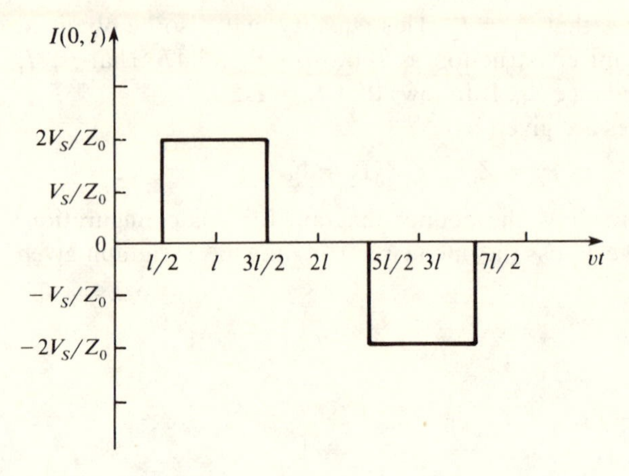

Figure 3.51 Current through the battery.

The current bounce diagram appears as shown in Fig. 3.50. The current at the open circuit is always zero. A graph of the current through the battery is depicted in Fig. 3.51.

Both V_L and I_B oscillate in a square-wave pattern indefinitely.

3.7 THE BLUMLEIN CONFIGURATION

The so-called Blumlein transmission line[1] is used to deliver a high voltage pulse across a load. The circuit for this line appears in Fig. 3.52. The top line is charged to the potential V_0, whereas the middle load has resistance $2Z_0$. After the switch is closed, the net voltage across the short $V_0 + V_+$ must vanish, so that $V_+ = -V_0$. The V_+ wave proceeds toward the load, leaving zero voltage behind it. The reflection coefficient at the load is

$$\rho = \frac{(2Z_0 + Z_0) - Z_0}{2Z_0 + Z_0 + Z_0} = \frac{1}{2}$$

The reflected voltage is $V_- = \rho V_+ = -\frac{1}{2}V_0$. To obtain the voltage of the transmitted wave, first we note that the net voltage incident on the load is $V_L = V_+ + V_- = -\frac{3}{2}V_0$. The load resistance acts with the characteristic impedance Z_0 as a voltage divider. The equivalent circuit is shown in Fig. 3.53. The transmitted voltage has the value $-\frac{1}{2}V_0$. We are now prepared to draw a bounce diagram for this system (Fig. 3.54). In the time interval $l/2v < t < 3l/2v$ the voltage at A is $-\frac{1}{2}V_0$ and the voltage at B is $V_0/2$. In this manner a potential difference V_0 appears across the load for an interval of duration l/v.

[1] Due to A. D. Blumlein.

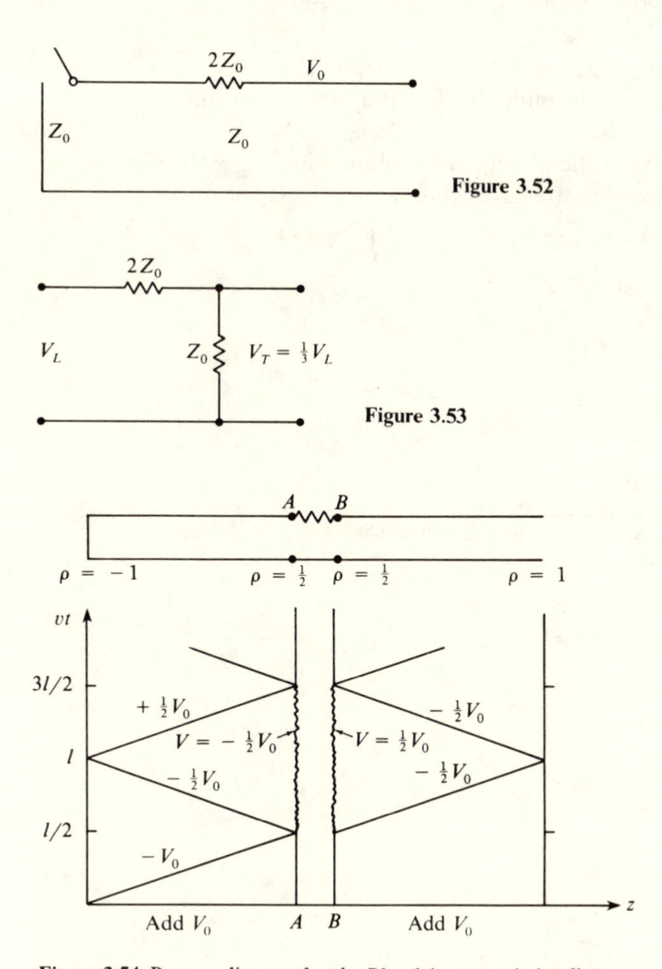

Figure 3.52

Figure 3.53

Figure 3.54 Bounce diagram for the Blumlein transmission line.

Example 3.6

Consider the transmission-line configuration shown in Fig. 3.55. Switch S_1 is closed with S_2 open, thereby raising the potential of the upper line to V_S volts above the bottom line. The S_1 is then opened and S_2 is closed.

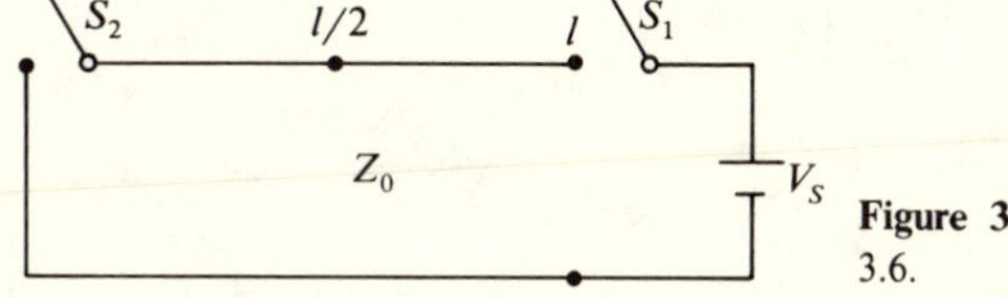

Figure 3.55 Transmission line for Example 3.6.

a. What is the potential at l as a function of time?
b. What is the current at the midpoint $l/2$ as a function of time?

Ans. We begin with the configuration shown in Fig. 3.56. The voltage and current starting values after S_2 is closed obey the equations

$$V_L = 0 = V_S + V_+, \qquad V_+ = -V_S$$

$$I_+ = \frac{V_+}{Z_0} = -\frac{V_S}{Z_0}$$

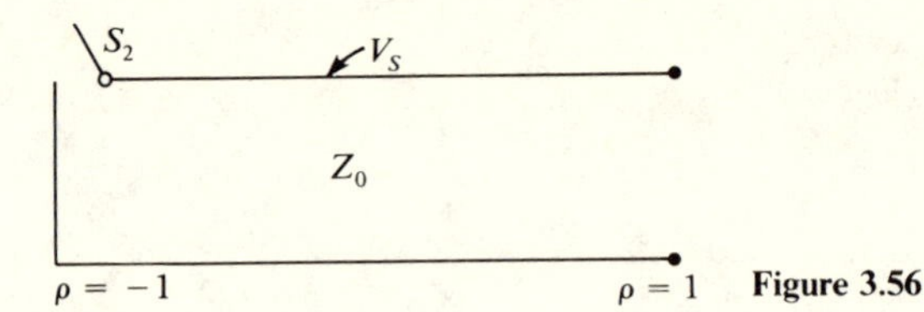

Figure 3.56

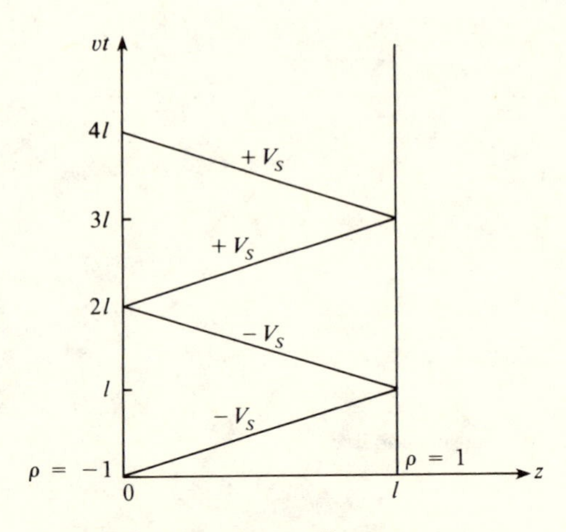

Figure 3.57 Voltage bounce diagram.

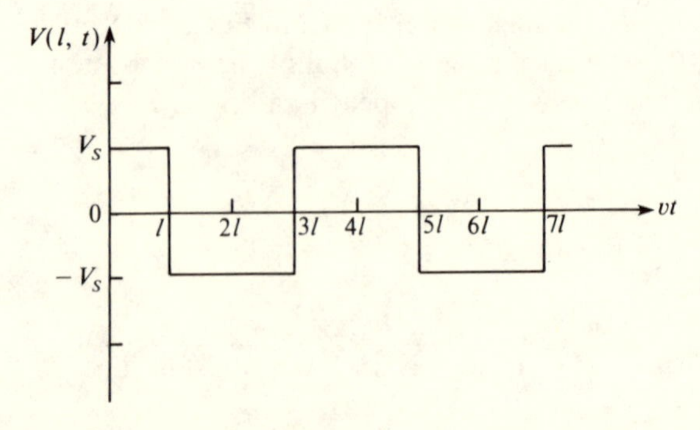

Figure 3.58

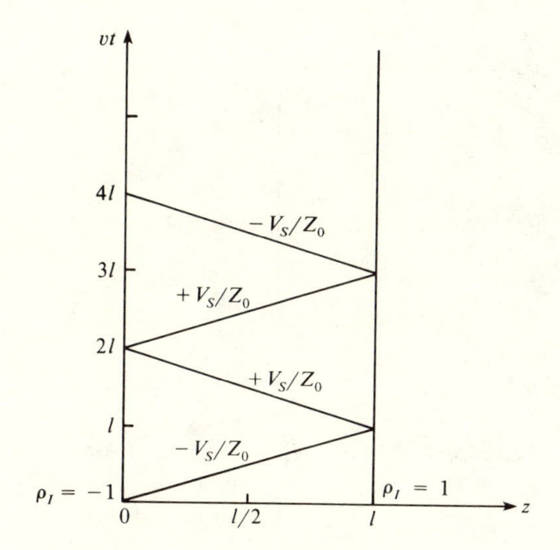

Figure 3.59 Current bounce diagram.

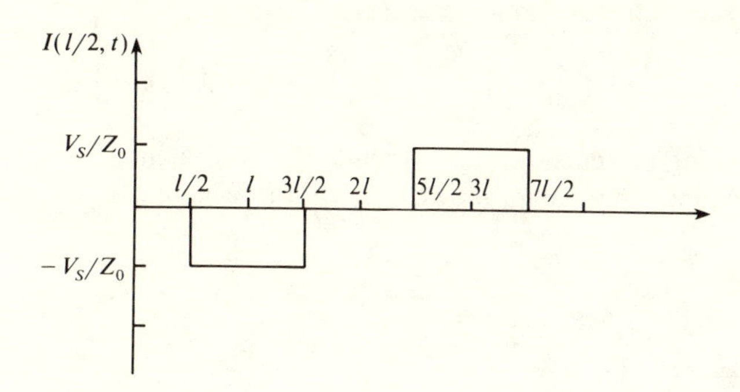

Figure 3.60

This permits construction of the voltage bounce diagram shown in Fig. 3.57. The potential at l has the graph sketched in Fig. 3.58. The bounce diagram for current is shown in Fig. 3.59. The current at the point $z = l/2$ has the graph shown in Fig. 3.60.

Example 3.7

Consider the transmission-line configuration shown in Fig. 3.61. The switch S is closed at $t = 0$.

a. What are the asymptotic values of the currents $I_{1,\text{asm}}$ and $I_{2,\text{asm}}$ all along the line?

b. What is the current through the battery, $I_S(t)$, as a function of time?

c. What is the current through the short, $I_2(l, t)$, as a function of time?

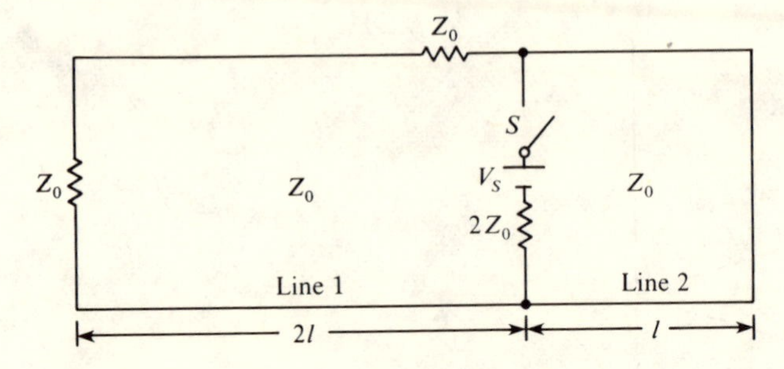

Figure 3.61 Configuration for Example 3.7.

d. What is the current through the load of line 1, $I_1(-2l, t)$, as a function of time?

Ans.

a.

$$I_{2,\text{asm}} = \frac{V_S}{2Z_0} \equiv \frac{I_0}{2}$$

$$I_{1,\text{asm}} = 0$$

b. To obtain the starting currents I_{1+}, I_{2+} we note the equivalent circuits appropriate to the junction domain, shown in Fig. 3.62. We obtain

$$I_S = \frac{3}{8}\frac{V_S}{Z_0} \equiv \tfrac{3}{8}I_0$$

$$I_{1+} = -\frac{Z_0}{3Z_0} I_S = -\tfrac{1}{8}I_0$$

$$I_{2+} = \frac{2Z_0}{3Z_0} I_S = \tfrac{2}{8}I_0$$

The pertinent reflection coefficients are listed in Fig. 3.63. The reflection coefficient presented to the wave that reflects from the short circuit when

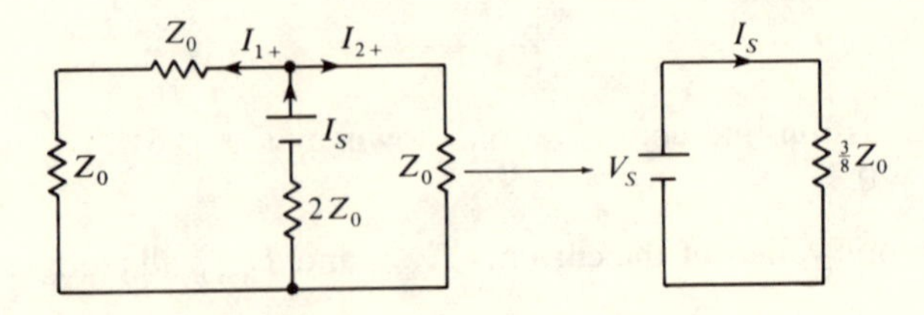

Figure 3.62 Equivalent circuits for Example 3.7.

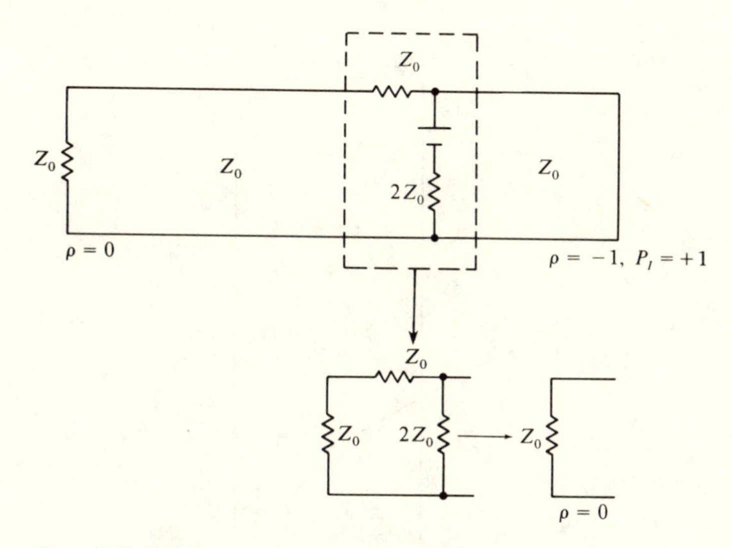

Figure 3.63 Reflection coefficients in Example 3.7.

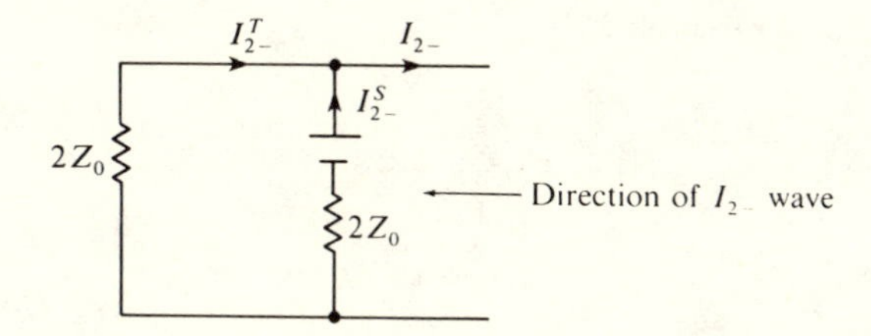

Figure 3.64

it encounters the junction is $\rho = 0$. So I_{2-} does not reflect from the junction. To discover the fate of I_{2-} we note that the junction circuit "seen" by I_{2-} is a current divider. (See Fig. 3.64.) There results

$$I^S_{2-} = \tfrac{1}{2}I_{2-}$$

$$I^T_{2-} = \tfrac{1}{2}I_{2-}$$

(Note that if there were a reflected wave I'_{2+} at the junction, we would have to examine how $I_{2-} + I'_{2+}$ divides.) We are now prepared to draw a current bounce diagram (Fig. 3.65). The transmitted current $\frac{1}{8}I_0$ annihilates the current $-\frac{1}{8}I_0$ already on line 1, leaving behind zero current. At $vt \geq 4l$, there is no current all along line 1. The steady-state current $I = \frac{1}{2}I_0$ flows all along line 2 at $vt \geq 2l$. From our previous calculation we know that the current through the battery starts at the value $\frac{3}{8}I_0$. When

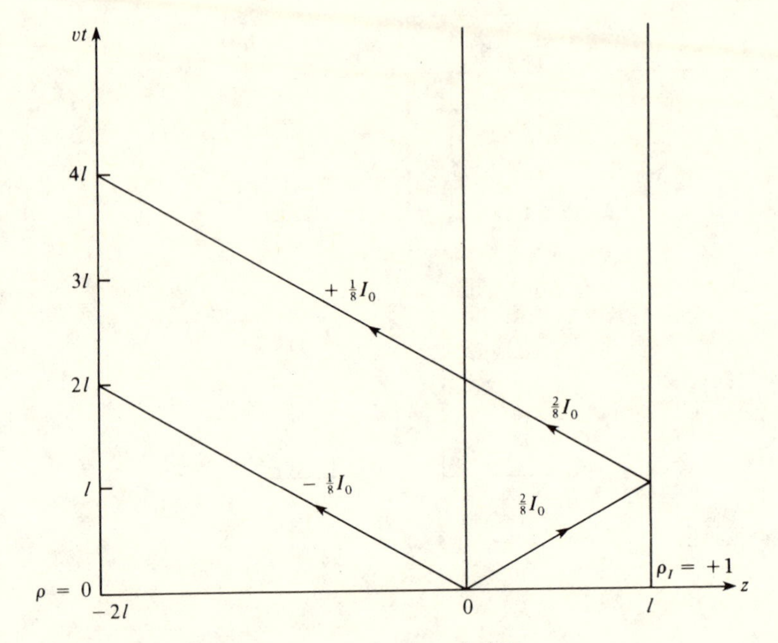

Figure 3.65 Current bounce diagram for Example 3.7.

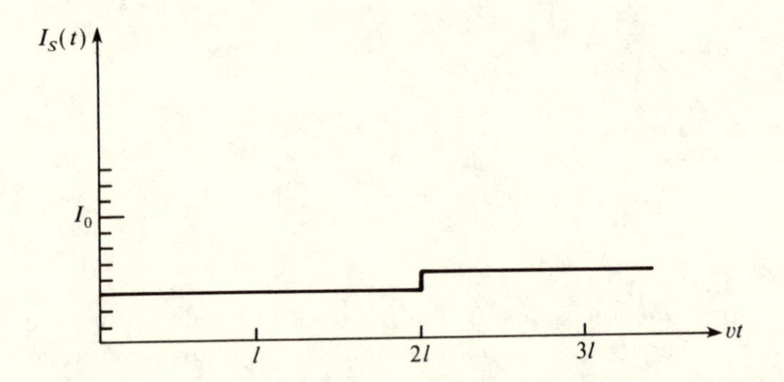

Figure 3.66

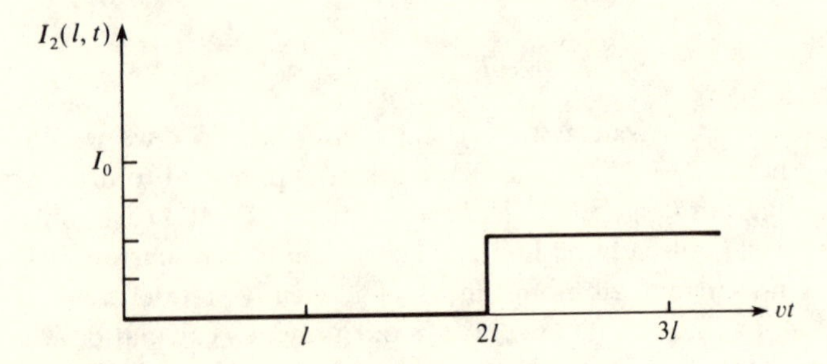

Figure 3.67 Current through the short in Example 3.7.

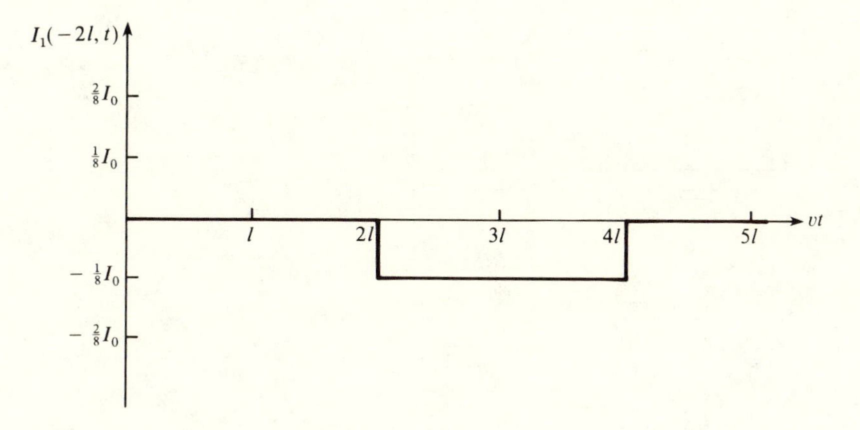

Figure 3.68 Current through the load Z_0 in Example 3.7.

I_{2-} reaches the junction, I_S increases to

$$\tfrac{3}{8}I_0 + \tfrac{1}{8}I_0 = \tfrac{1}{2}I_0$$

which is the asymptotic, steady-state value. (See Fig. 3.66.)

c. The current through the short is obtained directly from the bounce diagram. (See Fig. 3.67.)

d. The current through the load Z_0 of line 1 is likewise directly obtained from the bounce diagram shown in Fig. 3.68.

PROBLEMS

3.1. In the transmission line shown in Fig. 3.69 the switch S is closed for $t < 0$ and opened at $t = 0$. The battery has no internal resistance. Make a plot of the current through the battery as a function of time for $t > 0$.

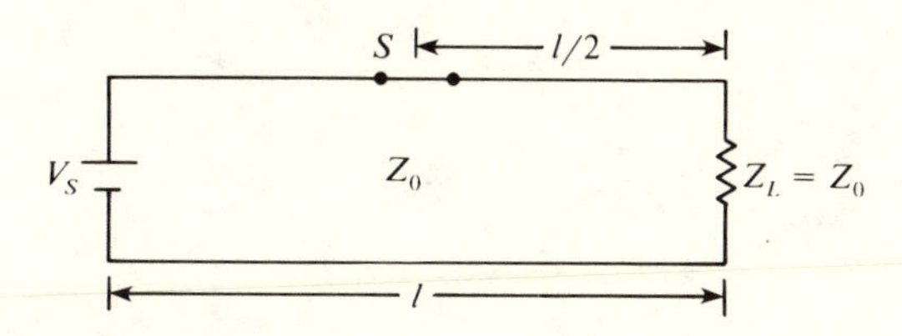

Figure 3.69

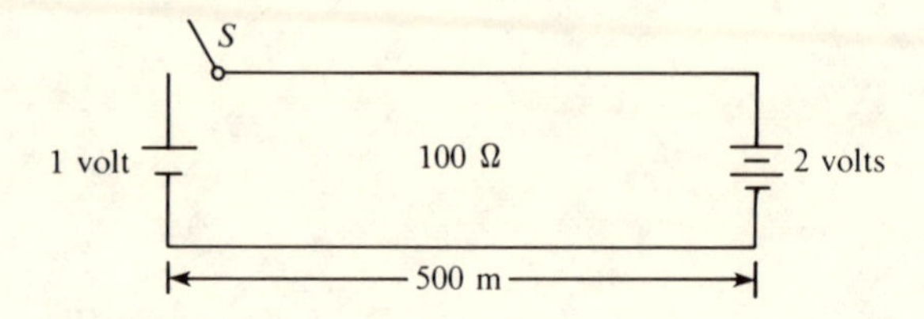

Figure 3.70

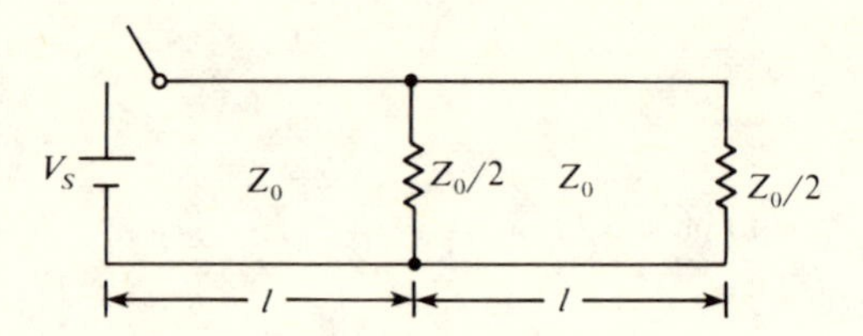

Figure 3.71

3.2. In the transmission line shown in Fig. 3.70, the switch S is closed at $t = 0$. Each battery has 150-Ω internal resistance. Make a plot of the current at the midpoint of the line vs time for $t > 0$.

3.3. Consider the transmission line shown in Fig. 3.71. The battery has internal resistance Z_0. Make a sketch of the voltage across load 2 as a function of time. Take $Z_0 = 200\ \Omega$, $V_S = 100$ volts, and $l = 400$ m.

CHAPTER

FOUR

STANDING-WAVE AC CONFIGURATIONS

4.1 INPUT IMPEDANCE ALONG THE LINE

In practical situations a transmission line serves to transmit electrical energy from an AC generator to a load. A primary problem then is to discover under what conditions maximum power is transferred from the generator to the load. As discussed previously, this condition is realized if the load impedance is matched to the characteristic impedance of the line. But for any but a purely resistive load, the load impedance is complex whereas the characteristic impedance may, for many practical cases, be taken to be purely real. The manner in which this important problem is solved is considered in detail in the following chapter. In the present chapter concepts are developed which are prerequisite to the basic problem of load matching. Our discussion begins with the construction of the impedance at all points along the line.

To this point in the text, we have discussed the characteristic impedance Z_0 (2.12), the intrinsic impedance η (2.15), and the load impedance Z_L (2.19). The impedance at any point along the line is the ratio

$$Z = \frac{V}{I} \tag{4.1}$$

where V and I are the steady state voltage and current at the given point.[1] So to construct Z we must evaluate the ratio V/I at all points on the line.

[1]All variables are complex unless otherwise stated. Boldface type is no longer used to denote complex quantities.

Let the voltage source have angular frequency ω. Our first task is to solve the wave equation (2.7) for the voltage. This solution is composed of a forward voltage wave

$$V_+ e^{j(\omega t - \beta z)}$$

traveling to the load, and a reflected voltage wave

$$V_- e^{j(\omega t + \beta z)}$$

traveling toward the generator. In these expressions, V_+ and V_- are wave amplitudes, and z is displacement along the line. The wavenumber is

$$\beta = 2\pi/\lambda$$

where λ is the wavelength of the forward or reflected voltage waves. In terms of frequency,

$$\omega = \beta v$$

where the wave speed v is given by (2.13):

$$v = \frac{1}{\sqrt{\mu\epsilon}}$$

As demonstrated in obtaining the general solution (2.8), the voltage at any point on the line between the generator and the load is the sum of the forward and reflected waves, so that we may write[2]

$$V(z,t) = e^{j\omega t}\left[V_+ e^{-j\beta z} + V_- e^{j\beta z}\right] \tag{4.2}$$

Similarly for the current

$$I(z,t) = e^{j\omega t}\left[I_+ e^{-j\beta z} + V_- e^{j\beta z}\right] \tag{4.3}$$

Having found, say, the voltage amplitudes V_+, V_-, the current amplitudes are found directly from (2.12):

$$Z_0 = \frac{V_+}{I_+} = -\frac{V_-}{I_-}$$

For example, the current solution (4.3) may be rewritten

$$I(z,t) = \frac{e^{j\omega t}}{Z_0}\left[V_+ e^{-j\beta z} - V_- e^{j\beta z}\right] \tag{4.4}$$

It is conventional in transmission-line analysis to choose the origin of displacement to be at the load, so that $z = 0$ at the load and $z = -l$ at the generator. (See Fig. 4.1.) It also proves convenient to let the time-dependent factor $e^{j\omega t}$ in the expressions above to be tacitly understood. With these

[2] Note that in (2.8) V_+ represents the forward voltage wave, $V_+ = V_+(z,t)$. On the other hand, in (4.2) V_+ is merely the constant amplitude of the forward voltage wave. The (z,t) dependence of this forward wave is included entirely in the phasor $e^{j(\omega t - \beta z)}$.

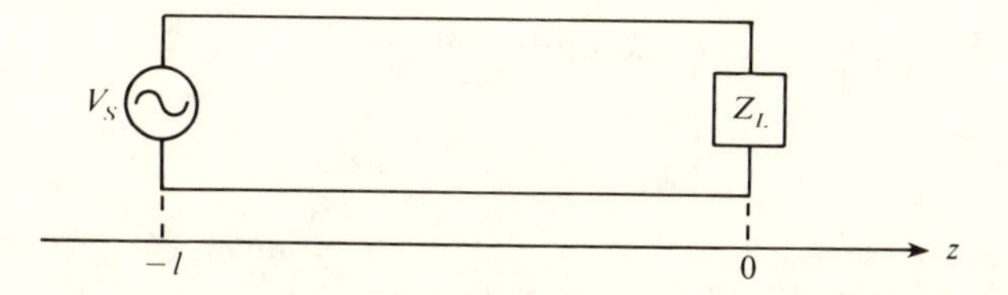

Figure 4.1 Conventional orientation of transmission line.

conventions accepted, the general solutions (4.2)–(4.4) become

$$V(l) = V_+ e^{j\beta l} + V_- e^{-j\beta l} \tag{4.5}$$

$$I(l) = I_+ e^{j\beta l} + I_- e^{-j\beta l} \tag{4.6}$$

$$I(l) = \frac{1}{Z_0}\left(V_+ e^{j\beta l} - V_- e^{-j\beta l}\right) \tag{4.7}$$

With these expressions at hand we turn to the construction of the impedance $Z(l)$ along the line. With (4.5) and (4.7) we obtain

$$Z(l) = Z_0\left(\frac{V_+ e^{j\beta l} + V_- e^{-j\beta l}}{V_+ e^{j\beta l} - V_- e^{-j\beta l}}\right) \tag{4.8}$$

The unknown voltage amplitudes in this expression may be eliminated in terms of the reflection coefficient ρ. Since the forward voltage wave at the load ($l = 0$) is the voltage amplitude V_+ and the reflected voltage at the load is the amplitude V_-, it follows that the reflection coefficient *at the load* is simply

$$\rho_L = \frac{V_-}{V_+}$$

With this relation at hand, we may rewrite (4.5) in terms of known quantities only:

$$Z(l) = Z_0\left(\frac{e^{j\beta l} + \rho_L e^{-j\beta l}}{e^{j\beta l} - \rho_L e^{-j\beta l}}\right) \tag{4.9}$$

This is the impedance l meters from the load. (See Fig. 4.2.)

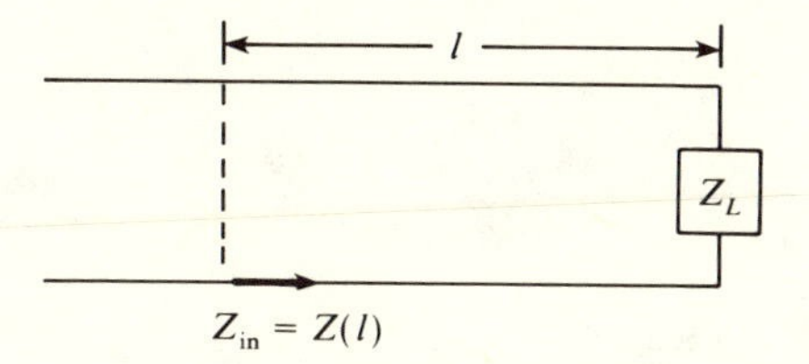

Figure 4.2

At the load, $l = 0$ and (4.9) is seen to return the relation (2.23):

$$Z_L = Z_0\left(\frac{1+\rho_L}{1-\rho_L}\right) \tag{4.10}$$

If ρ_L is rewritten in terms of Z_0 and Z_L, and the exponentials in (4.9) are expanded in terms of $\cos\beta l$ and $\sin\beta l$, one obtains the alternate expression for the impedance,

$$Z(l) = Z_0\left(\frac{Z_L\cos\beta l + jZ_0\sin\beta l}{Z_0\cos\beta l + jZ_L\sin\beta l}\right) \tag{4.11}$$

To find the admittance along the line we merely set $Y_0 = Z_0^{-1}$, $Y_L = Y_L^{-1}$, and $Y(l) = Z(l)^{-1}$ in the preceding equation. These results

$$Y(l) = Y_0\left(\frac{Y_L\cos\beta l + jY_0\sin\beta l}{Y_0\cos\beta l + jY_L\sin\beta l}\right) \tag{4.12}$$

Suppose for the moment that the generator is l meters from the load. (See Fig. 4.3.) What impedance does the generator "see" looking into the line? This is the problem we have just solved in obtaining (4.11). So the expression (4.11) represents the input impedance of a section l meters long terminated in a load Z_l, such as is shown in Fig. 4.4. Equivalently, (4.11) represents the impedance l meters from the load "looking" toward the load, as shown in Fig. 4.5.

An immediate property of $Z(l)$ as given by (4.11) or $Y(l)$ as given by (4.12) is that these are periodic functions of l with period $\lambda/2$. The wavelength $\lambda = 2\pi/\beta$ is the wavelength of either the forward or the reflected voltage or current waves. That $Z(l)$ has period $\lambda/2$ means that

$$Z(l) = Z\left(l + \frac{\lambda}{2}\right) = Z\left(l + \frac{\pi}{\beta}\right). \tag{4.13}$$

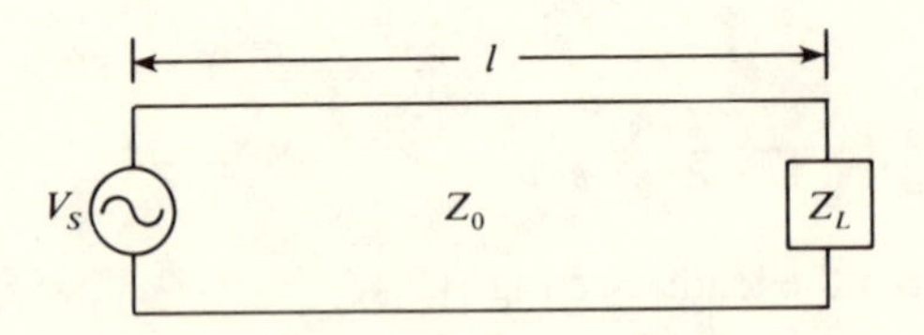

Figure 4.3

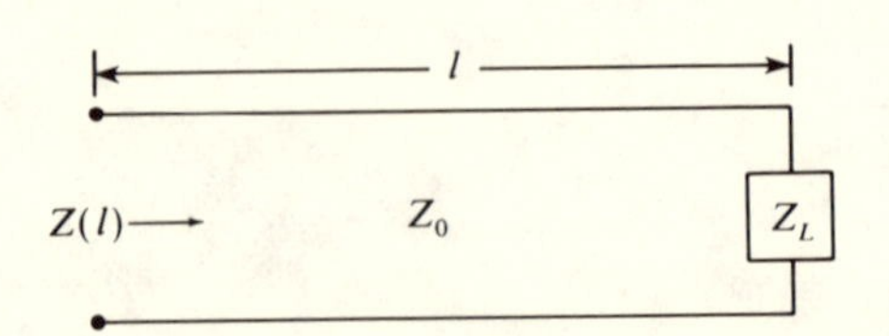

Figure 4.4

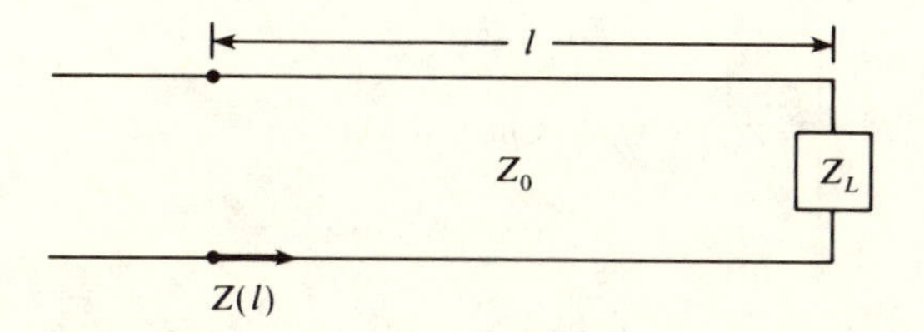

Figure 4.5

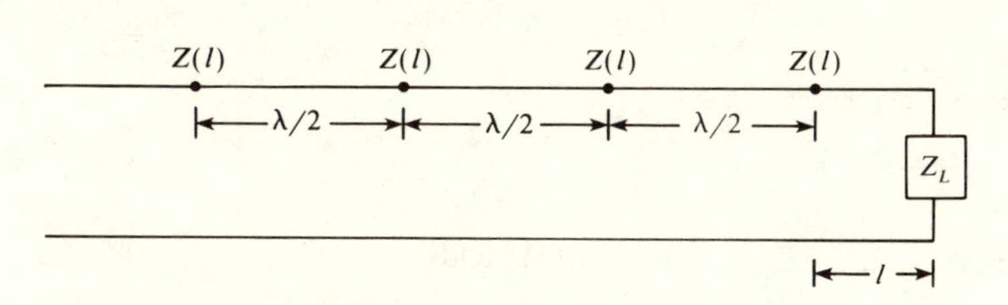

Figure 4.6 Periodicity of a transmission line.

To show this, consider (4.11):

$$Z\left(l + \frac{\lambda}{2}\right) = Z_0 \frac{Z_L \cos(\beta l + \pi) + jZ_0 \sin(\beta l + \pi)}{Z_0 \cos(\beta l + \pi) + jZ_L \sin(\beta l + \pi)}$$

Recall that

$$\cos(\beta l + \pi) = -\cos \beta l$$
$$\sin(\beta l + \pi) = -\sin \beta l$$

Substituting these expressions into the preceding equation returns (4.13). The periodicity of $Z(l)$ is exhibited in Fig. 4.6.

Example 4.1
What is the input impedance on a transmission line l meters from:

a. An open circuit?
b. A short circuit?
c. In what interval of length from open circuit is the input impedance purely capacitive?
d. In what interval of length from a closed circuit is the input impedance purely inductive?

Ans.
a. For an open circuit, $|Z_L| = \infty$ and (4.11) yields

$$Z_{oc}(l) = -jZ_0 \cot \beta l$$

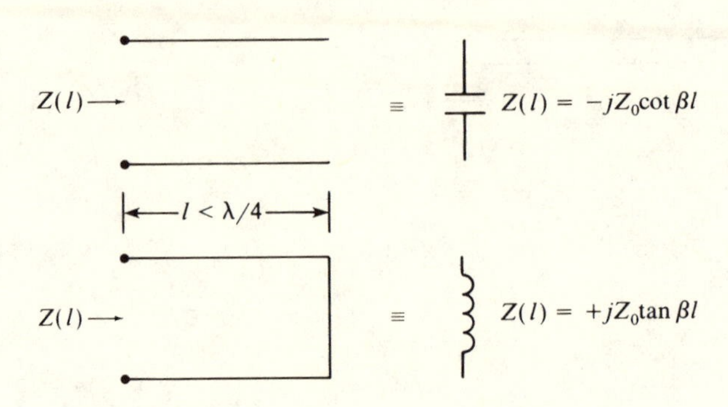

Figure 4.7 Equivalent elements.

b. For a short circuit, $|Z_L| = 0$ and (4.11) yields

$$Z_{cc}(l) = jZ_0 \tan \beta l$$

c. From our answer to part a we see that $\operatorname{Re} Z = 0$ for all l, whereas $\operatorname{Im} Z < 0$ provided $\cot \beta l$ is positive. The first interval in which this occurs is $0 < \beta l < \pi/2$.
d. We must have $\operatorname{Re} Z = 0$ and $\operatorname{Im} Z > 0$. From part b we see that the first interval in which this occurs is $0 < \beta l < \pi/2$. So we have the transmission-line circuit elements shown in Fig. 4.7.

Example 4.2

A 200-Ω twin-lead transmission line with air dielectric is terminated in a load which has impedance

$$Z_L = 100 + j150$$

a. What is the wavelength for the voltage and current waves on the line at 30 MHz?
b. What is the impedance on the line at $l = \lambda/4$?
c. What are the circuit elements of an equivalent network with (i) impedance Z_L, (ii) impedance $Z(\lambda/4)$?

Ans.

a. For air dielectric, $v = c$ and

$$\lambda = \frac{c}{f} = \frac{3 \times 10^8}{3 \times 10^7} = 10 \text{ m}$$

b. At $l = \lambda/4 = 2\pi/4\beta$ and

$$l\beta = \pi/2$$

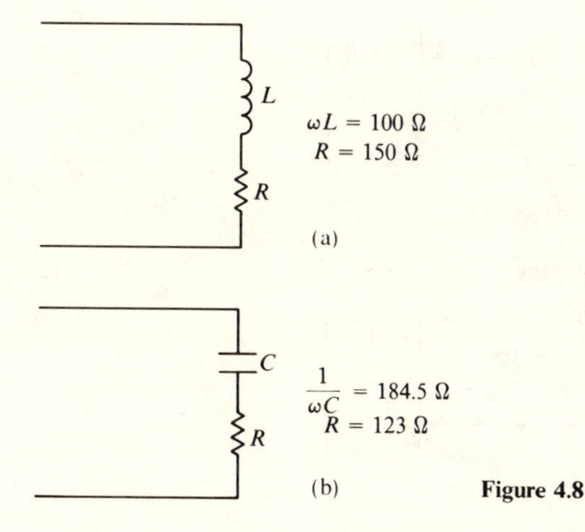

Figure 4.8

At this value (4.11) gives

$$Z\left(\frac{\lambda}{4}\right) = Z_0\left(\frac{jZ_0}{jZ_L}\right) = \frac{Z_0^2}{Z_L}$$

Substituting given values into this expression, we obtain

$$Z\left(\frac{\lambda}{4}\right) = 1233 - j184.5\ \Omega$$

c. (i): The network of the load is shown in Fig. 4.8a. (ii): The network of the impedance $Z(\lambda/4)$ is shown in Fig. 4.8b.

4.2 THE QUARTER-WAVE TRANSFORMER

Suppose a transmission line with characteristic impedance Z_0 is terminated in a matched load, so that $Z_0 = Z_L$. The reflection coefficient $\rho = 0$, and by (2.32) all power stemming from the generator is absorbed by the load. Furthermore, from (4.11) the impedance all along the line is constant and equal to Z_0.

Suppose we wish to join this section of transmission line to a second section with characteristic impedance Z_0', as shown in Fig. 4.9. If the two sections are connected as shown, then $\rho \neq 0$ at the junction and matching to the load is impaired.

A device which remedies this situation, at a given frequency, is called the quarter-wave transformer. It consists of a section of transmission line with characteristic impedance equal to the geometric mean of Z_0 and Z_0'.

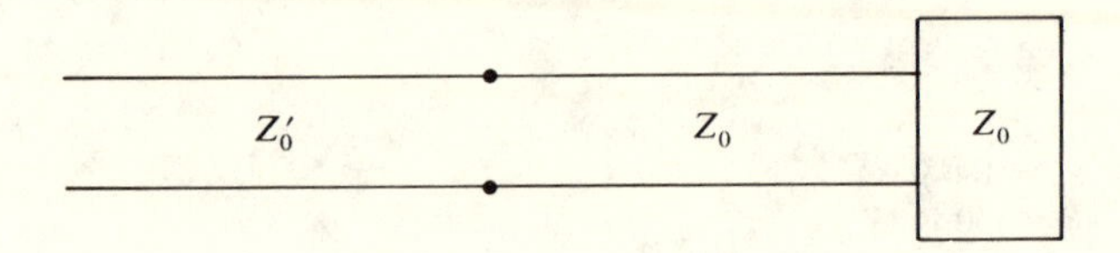

Figure 4.9 Two attached transmission-line sections.

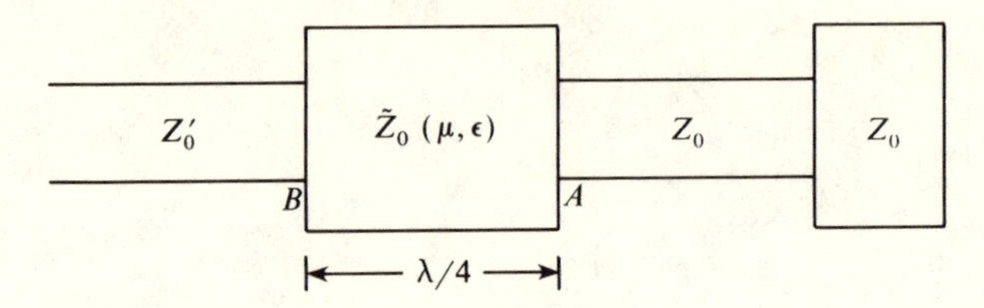

Figure 4.10 The quarter-wave transformer.

That is,

$$\tilde{Z}_0^2 = Z_0 Z_0' \tag{4.14}$$

Furthermore the dielectric material of the quarter-wave section is such that at the given operating frequency, the section is $\lambda/4$ meters long. That is,

$$l = \frac{\lambda}{4} = \frac{\pi}{2\omega\sqrt{\mu\epsilon}} \tag{4.15}$$

The connection is shown in Fig. 4.10.

We must show that the impedance at B is Z_0'. Then the load is matched to Z_0' and there is no reflection of power. The impedance at A is Z_0. With (4.11) we obtain for the impedance at B:

$$Z_B = Z\left(\frac{\lambda}{4}\right) = \frac{\tilde{Z}_0^2}{Z_L} = \frac{\tilde{Z}_0^2}{Z_0}$$

But as stated above in (4.14), $\tilde{Z}_0^2 = Z_0 Z_0'$ and we obtain

$$Z_B = Z_0'$$

so there is no reflection at B and the line is matched to the load.

Example 4.3

A transmission line with 80-Ω characteristic impedance carries voltage oscillating at 50 MHz to a purely resistive load of 300 Ω. What are the length and characteristic impedance of a line with $\epsilon = 1.7\epsilon_0$, $\mu = \mu_0$ which, when connected between the 80-Ω line and the load, will serve to match the load to the 80-Ω line? (See Fig. 4.11.)

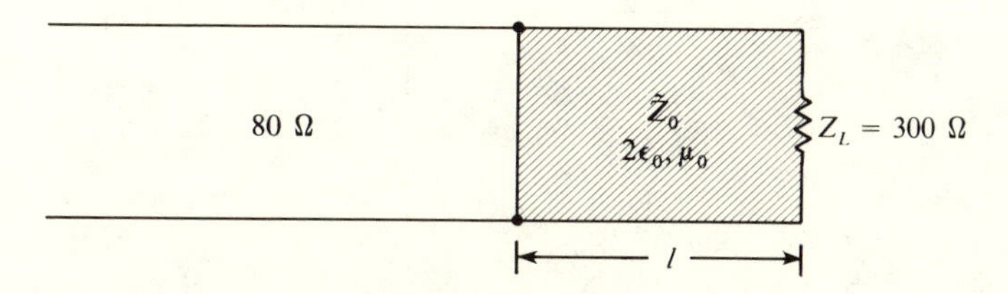

Figure 4.11 Configuration for Example 4.3.

Ans. The impedance $\tilde{Z}_0$ of the quarter-wave transformer is determined from (4.14):

$$\tilde{Z}_0 = \sqrt{Z_0 Z_0'} = \sqrt{80 \times 300} = 155\ \Omega$$

The length of the section is determined by the condition that it is a quarter of a wavelength long:

$$l = \frac{\lambda}{4} = \frac{v}{4f} = \frac{1}{4f\sqrt{\mu\epsilon}} = \frac{c}{4f\sqrt{1.7}} = 1.15\ \text{m}$$

4.3 STANDING-WAVE RATIO

The voltage which exists between the generator and load as given by (4.5) represents a standing wave. Let us find how the amplitude of this standing wave varies with displacement l. With $V_- = \rho_L V_+$ and reinserting the time-dependent factor $e^{j\omega t}$, we obtain

$$V(l, t) = e^{j(\omega t + \beta l)} V_+ \left(1 + \rho_L e^{-j2\beta l}\right) \tag{4.16}$$

Setting

$$\rho_L = |\rho_L| e^{j\theta} \tag{4.17}$$

and taking the modulus of both sides of (4.16) gives

$$|V(l)| = |V_+| \left|1 + |\rho_L| e^{j(\theta - 2\beta l)}\right| \tag{4.18}$$

or equivalently

$$|V(l)| = |V_+| \left[1 + |\rho_L|^2 + 2|\rho_L| \cos(\theta - 2\beta l)\right]^{1/2} \tag{4.19}$$

From this expression we see that the amplitude of the voltage standing wave varies periodically along the line with period given by $2\beta l = 2\pi$, namely, $l = \lambda/2$. (See Fig. 4.12.) As l varies from 0 at the load to large values (toward the generator), the cos term in (4.13) oscillates between $+1$ and

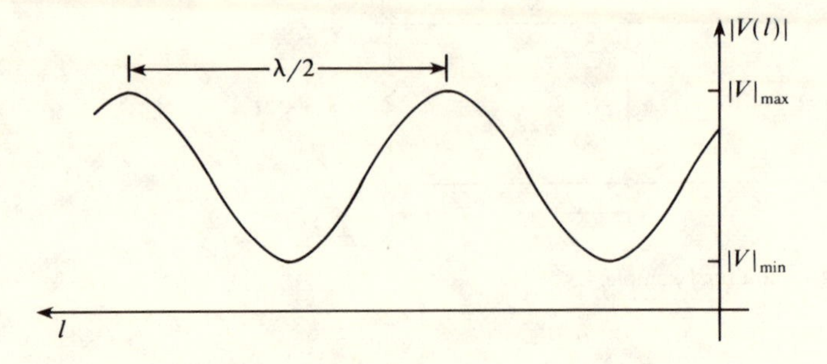

Figure 4.12 Voltage standing-wave pattern.

-1. It follows that

$$|V|_{max} = |V_+|\left[1 + |\rho_L|^2 + 2|\rho_L|\right]^{1/2}$$

or equivalently

$$\begin{aligned} |V|_{max} &= |V_+|(1 + |\rho_L|) \\ &= |V_+| + |V_-| \end{aligned} \tag{4.20}$$

In similar manner we find

$$|V|_{min} = |V_+| - |V_-| \tag{4.21}$$

From the expression (4.7) for the current we obtain

$$I(l,t) = e^{j(\omega t + \beta l)} \frac{V_+}{Z_0}\left(1 - \rho_L e^{-j2\beta l}\right) \tag{4.22}$$

which, with (4.17), gives

$$|I(l)| = \frac{|V_+|}{Z_0}\left[1 + |\rho_L|^2 - 2|\rho_L|\cos(\theta - 2\beta l)\right]^{1/2} \tag{4.23}$$

This expression indicates that the current amplitude is maximum when $\cos(\theta - 2\beta l) = -1$ and minimum when $\cos(\theta - 2\beta l) = +1$. We obtain

$$|I|_{max} = \frac{|V_+| + |V_-|}{Z_0}$$

$$|I|_{min} = \frac{|V_+| - |V_-|}{Z_0} \tag{4.24}$$

At those values of l where $\cos(\theta - 2\beta l) = +1$ we have found that $|V|$ is maximum and $|I|$ is minimum. Furthermore when $\cos(\theta - 2\beta l) = -1$, $|V|$ is minimum and $|I|$ is maximum. The standing-wave voltage and current amplitude have the form shown in Fig. 4.13.

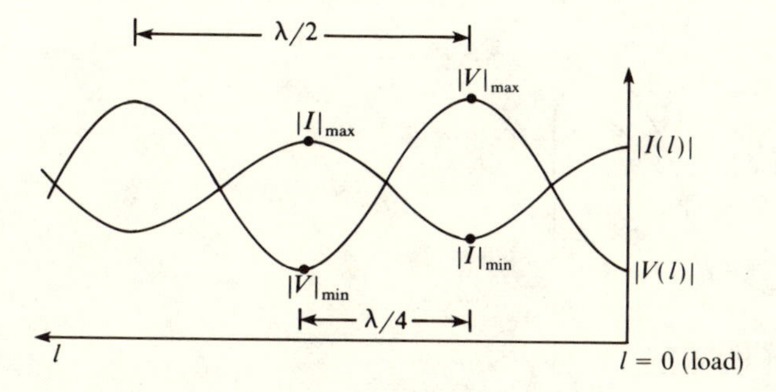

Figure 4.13 Voltage and current standing wave patterns.

Since $|V|$ is maximum at the same values of l where $|I|$ is minimum, it follows that at these same points $|Z|$ is maximum and is given by

$$|Z|_{\max} = \frac{|V|_{\max}}{|I|_{\min}} = Z_0\left[\frac{|V_+| + |V_-|}{|V_+| - |V_-|}\right] = Z_0\left[\frac{1 + |\rho_L|}{1 - |\rho_L|}\right] \tag{4.25}$$

At the points where $|V|$ is minimum and $|I|$ is maximum, $|Z|$ is minimum and has the value

$$|Z|_{\min} = \frac{|V|_{\min}}{|I|_{\max}} = Z_0\left[\frac{|V_+| - |V_-|}{|V_+| - |V_-|}\right] = Z_0\left[\frac{1 - |\rho_L|}{1 + |\rho_L|}\right] \tag{4.26}$$

A very important parameter associated with this standing-wave pattern is the so called *standing-wave ratio* S. It is defined as

$$S = \frac{|V|_{\max}}{|V|_{\min}} > 1 \tag{4.27}$$

or, with (4.20), (4.21),

$$S = \frac{|V_+| + |V_-|}{|V_+| - |V_-|} = \frac{1 + |\rho_L|}{1 - |\rho_L|} \tag{4.28}$$

It follows that the magnitude of the reflection coefficient determines the standing-wave ratio. Likewise

$$|\rho_L| = \frac{S - 1}{S + 1} \tag{4.29}$$

From (4.25) we may write

$$|Z|_{\max} = Z_0\frac{|V_{\max}|}{|V_{\min}|} = Z_0 S \tag{4.30}$$

and from (4.26)

$$|Z|_{\min} = Z_0/S \tag{4.31}$$

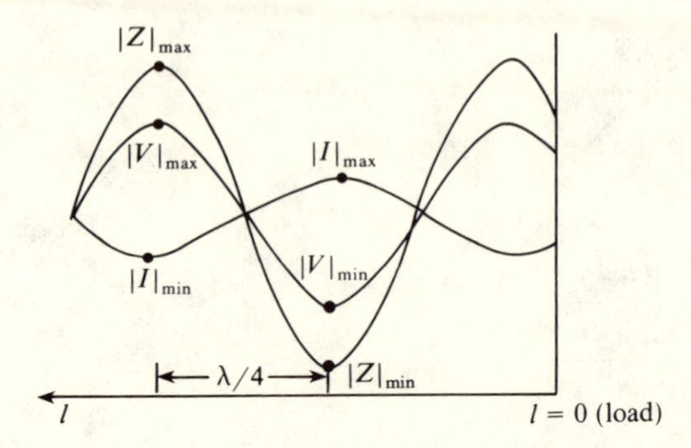

Figure 4.14 Periodic variation of standing-wave patterns away from the load.

Since the value of Z repeats every $\lambda/2$ meters, we may conclude that the location of $|Z|_{min}$ is $\lambda/4$ meters away from $|Z|_{max}$. These positions of impedance are shown together with voltage and current amplitudes in Fig. 4.14.

Consider for example the configuration shown in Fig. 4.15. What is the current amplitude of the generator? The impedance at the short is $|Z|_{min} = 0$. We know from Example 4.1 that there is another point along the line where $|Z| = \infty$. This is the value of $|Z|_{max}$ which occurs $\lambda/4$ meters away from the short. So the equivalent network at the generator is as shown in Fig. 4.16, and $|I| = 0$ at the generator.

It is shown in Example 4.4 to follow that $|Z|_{max}$ and $|Z|_{min}$ are purely resistive. With this property, (4.30) and (4.31) become

$$\begin{aligned} |Z|_{max} &= Z_{max} = Z_0 S > Z_0 \\ |Z|_{min} &= Z_{min} = Z_0/S < Z_0 \end{aligned} \tag{4.32}$$

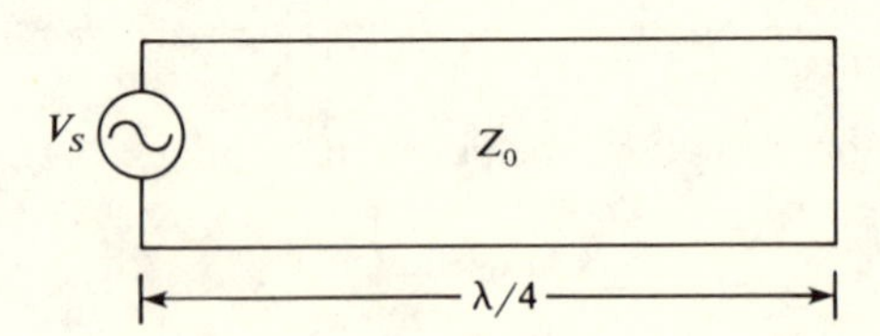

Figure 4.15

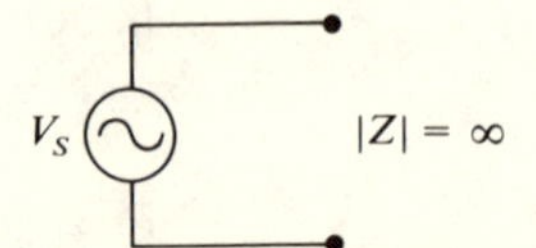

Figure 4.16 The generator circuit.

From these relations we see that

$$Z_{max}Z_{min} = Z_0^2$$

$$Z_{max}/Z_{min} = S^2 \qquad (4.33)$$

Consider the case of a purely resistive load $R_L > Z_0$ which is located $\lambda/4$ meters from a generator. (See Fig. 4.17.) What is the impedance at the generator? Since Z_L is purely resistive, we know that it is either Z_{max} or Z_{min}. We are told that $R_L > Z_0$. From (4.32) we see that this condition implies that R_L is Z_{max}. It follows that the impedance at the generator is $Z_{min} = Z_0^2/Z_{max} = Z_0^2/R_L$. (See Fig. 4.18.)

In the event that $R_L < Z_0$, then R_L becomes Z_{min} and the impedance at the generator is the maximum, Z_0^2/R_L, for the quarter-wavelength line.

From (4.33) we find the following expressions for the standing-wave ratio for the case of purely resistive load:

$$S = R_L/Z_0 \qquad \text{for} \quad R_L > Z_0$$

$$S = Z_0/R_L \qquad \text{for} \quad R_L < Z_0 \qquad (4.34)$$

These properties attendant to the case of the purely resistive load are listed in Table 4.1.

More generally, let the line be terminated in a complex load Z_L. As demonstrated in Example 4.4, at the location on the line where $|Z|$ is either maximum or minimum, Z is purely resistive. Moving toward the generator $\lambda/4$ meters from this point, one again finds Z purely resistive with the extreme values of $|Z|$ reversed. It follows that the results of Table 4.1 still maintain with the resistance R_L replaced by the purely resistive value of Z at the location where $|Z|$ is extreme. These properties are listed in Table 4.2.

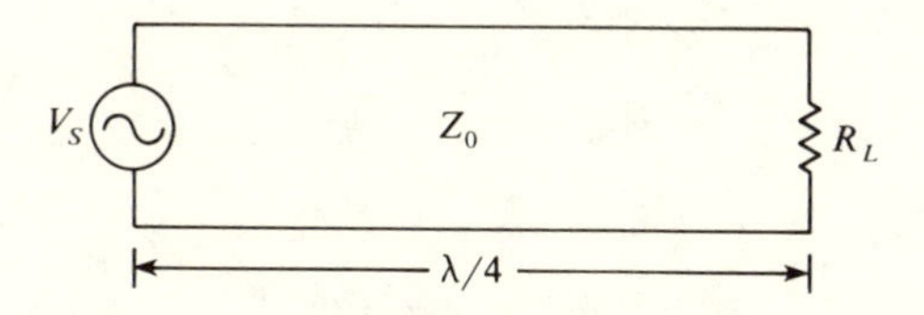

Figure 4.17 A purely resistive load $\lambda/4$ meters away from the generator.

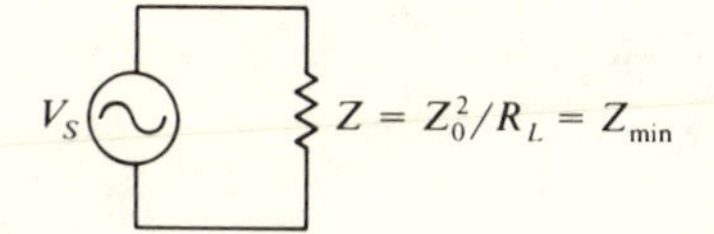

Figure 4.18 The generator circuit.

Table 4.1 Standing-wave properties for the purely resistive load

$l = \lambda/2$	$l = \lambda/4$	$l = 0$
$S = R_L/Z_0$	Z_0	$R_L > Z_0$
$\lvert Z\rvert_{max}$	$\lvert Z\rvert_{min} = \dfrac{Z_0^2}{R_L}$	$\lvert Z\rvert_{max} = R_L$
$\lvert V\rvert_{max}$	$\lvert V\rvert_{min}$	$\lvert V\rvert_{max} = S\lvert V\rvert_{min}$
$\lvert I\rvert_{min}$	$\lvert I\rvert_{max} = \dfrac{\lvert V\rvert_{min}}{\lvert Z\rvert_{min}}$	$\lvert I\rvert_{min} = \dfrac{\lvert V\rvert_{max}}{\lvert Z\rvert_{max}}$

$l = \lambda/2$	$l = \lambda/4$	$l = 0$
$S = Z_0/R_L$	Z_0	$R_L < Z_0$
$\lvert Z\rvert_{min}$	$\lvert Z\rvert_{max} = \dfrac{Z_0^2}{R_L}$	$\lvert Z\rvert_{min} = R_L$
$\lvert V\rvert_{min}$	$\lvert V\rvert_{max}$	$\lvert V\rvert_{min} = \dfrac{\lvert V\rvert_{max}}{S}$
$\lvert I\rvert_{max}$	$\lvert I\rvert_{min} = \dfrac{\lvert V\rvert_{max}}{\lvert Z\rvert_{max}}$	$\lvert I\rvert_{max} = \dfrac{\lvert V\rvert_{min}}{\lvert Z\rvert_{min}}$

Table 4.2 Ideal transmission line with complex load

Case 1. $R > Z_0$, $S = R/Z_0 > 1$

$\lvert Z\rvert_{min}$[a]	$\lvert Z\rvert_{max}$[a]
Toward generator ← ⁞ ←——— $\lambda/4$ ———→ ⁞ → Toward load	
$\lvert Z\rvert_{min} = \dfrac{Z_0^2}{R}$	$\lvert Z\rvert_{max} = R$
$\lvert V\rvert_{min}$	$\lvert V\rvert_{max} = S\lvert V\rvert_{min}$
$\lvert I\rvert_{max} = \dfrac{\lvert V\rvert_{min}}{\lvert Z\rvert_{min}}$	$\lvert I\rvert_{min} = \dfrac{\lvert V\rvert_{max}}{\lvert Z\rvert_{max}}$

Case 2. $R < Z_0$, $S = Z_0/R > 1$

$\lvert Z\rvert_{max}$[a]	$\lvert Z\rvert_{min}$[a]
Toward generator ← ⁞ ←——— $\lambda/4$ ———→ ⁞ → Toward load	
$\lvert Z\rvert_{max} = \dfrac{Z_0^2}{R}$	$\lvert Z\rvert_{min} = R$
$\lvert V\rvert_{max}$	$\lvert V\rvert_{min} = \dfrac{\lvert V\rvert_{max}}{S}$
$\lvert I\rvert_{min} = \dfrac{\lvert V\rvert_{max}}{\lvert Z\rvert_{max}}$	$\lvert I\rvert_{max} = \dfrac{\lvert V\rvert_{min}}{\lvert Z\rvert_{min}}$

[a]At these locations Z is purely resistive.

Example 4.4

Show that at the values of l where $|Z|$ is either maximum or minimum, Z is purely resistive.

Ans. Rewriting the impedance $Z(l)$ as given by (4.9) in terms of the reflection coefficient (4.17), we obtain

$$Z(l) = Z_0\left[\frac{1 + |\rho_L|e^{j(\theta - 2\beta l)}}{1 - |\rho_L|e^{j(\theta - 2\beta l)}}\right]$$

Now $|Z|$ is maximum at the same value of l where $|V|$ is maximum. As shown above (4.20), this occurs at $\theta - 2\beta l = 0$. For these values of l, $Z(l)$ becomes

$$Z = Z_0\left[\frac{1 + |\rho_L|}{1 - |\rho_L|}\right]$$

which is purely real. A similar construction shows that at the values of l where $|Z|$ is minimum, Z is purely resistive and given by

$$Z = Z_0\left[\frac{1 - |\rho_L|}{1 + |\rho_L|}\right]$$

Example 4.5

The amplitude of the forward voltage wave at 50 MHz traveling on a 100-Ω air-dielectric transmission line is 200 volts. The standing-wave ratio on the line is 1.5.

a. What is the value of $|V_-|$?
b. What is the time average power delivered to the load?
c. What are Z_{max} and Z_{min} on the line?
d. How far apart are the locations of Z_{max} and Z_{min}?

Ans.

a. From (4.29) we obtain

$$|\rho_L| = 0.20$$

so that

$$|V_-| = 0.20|V_+| = 40 \text{ volts}$$

b. From (2.32)

$$\langle P_L \rangle = \langle P_+ \rangle (1 - |\rho_L|^2)$$

where

$$\langle P_+ \rangle = \tfrac{1}{2}\mathrm{Re}\, V_+ I_+^* = \tfrac{1}{2}|V_+|\,|I_+|$$
$$= \frac{1}{2Z_0}|V_+|^2 = 200 \text{ watts}$$

so that

$$\langle P_L \rangle = 192 \text{ watts}$$

c. From (4.23) and (4.24) we obtain

$$Z_{\max} = SZ_0 = 150 \text{ ohms}$$

$$Z_{\min} = \frac{Z_0}{S} = 66.7 \text{ ohms}$$

d. $Z_{\max}$ and $Z_{\min}$ are $\lambda/4$ meters apart, where

$$\frac{\lambda}{4} = \frac{c}{4f} = 1.5 \text{ m}$$

4.4 VOLTAGE AND CURRENT DISTRIBUTIONS

With the apparatus developed above, it is possible to obtain the standing-wave voltage and current patterns along the line for a few fundamental, albeit idealized, configurations. For example consider the open-ended quarter-wavelength line shown in Fig. 4.19. The source voltage operates at angular frequency ω and has amplitude V_S. Let us first construct the current along the line. The general solution is given by (4.6). To determine the amplitudes I_+ and I_- we note the following two conditions. First, at the load ($l = 0$) the current vanishes:

$$I(0) = I_+ + I_- = 0 \tag{4.35}$$

Our second condition stems from the fact that the line is $\lambda/4$ meters long, so that $Z = 0$ at the generator. That is, the generator "sees" a short circuit. (See Fig. 4.20.) It follows that the current at the generator ($\beta l = \pi/2$) has the amplitude

$$\frac{V_S}{R_S} = I_+ e^{j\pi/2} + I_- e^{-j\pi/2}$$

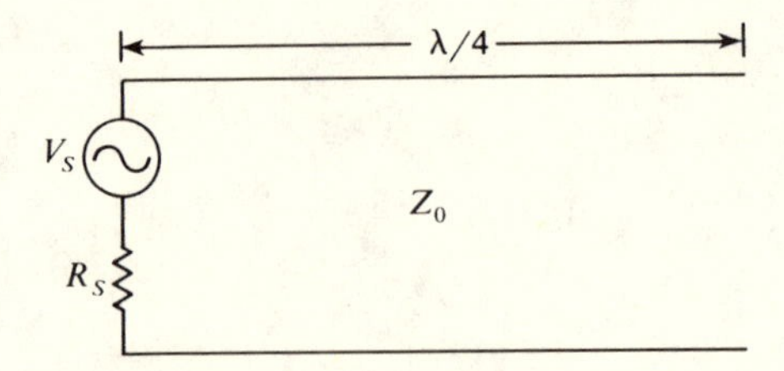

Figure 4.19 An open-ended circuit.

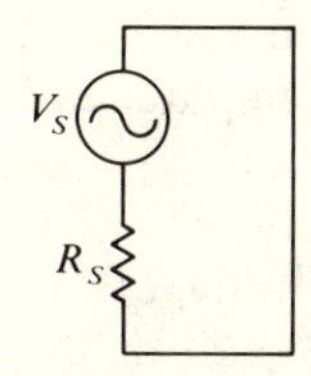

Figure 4.20 The generator "sees" a short circuit.

or equivalently

$$\frac{V_S}{R_S} = j(I_+ - I_-) \tag{4.36}$$

This equation together with (4.35) serve to determine the amplitudes

$$I_+ = -I_- = \frac{V_S}{2jR_S} \tag{4.37}$$

Substituting these values into (4.6) gives the desired current distribution

$$I(l) = \frac{V_S}{R_S}\left(\frac{e^{j\beta l} - e^{-j\beta l}}{2j}\right)$$

or equivalently

$$I(l) = \frac{V_S}{R_S}\sin\beta l \tag{4.38}$$

Having constructed I_+ and I_-, the voltage amplitudes V_+ and V_- are simply obtained from (2.12):

$$V_{\pm} = \pm Z_0 I_{\pm}$$

which gives the voltage standing-wave pattern

$$V(l) = \frac{Z_0 V_S}{jR_S}\cos\beta l \tag{4.39}$$

Writing $-j = e^{-j\pi/2}$ and inserting the time dependence $e^{j\omega t}$ into (4.38) and (4.39) gives

$$V(l) = \frac{Z_0 V_S}{R_S}\cos\beta l\, e^{j(\omega t - \pi/2)}$$

$$I(l) = \frac{V_S}{R_S}\sin\beta l\, e^{j\omega t} \tag{4.40}$$

Taking the real parts of both equations, we obtain

$$\operatorname{Re} V(l) = \frac{Z_0 V_S}{R_S}\cos\beta l \sin\omega t$$

$$\operatorname{Re} I(l) = \frac{V_S}{R_S}\sin\beta l \cos\omega t \tag{4.41}$$

This solution represents the standing-wave voltage and current oscillation along the open-ended line drawn in Fig. 4.19. A graph of the amplitude of these oscillations is shown in Fig. 4.21.

The second fundamental configuration we examine has a short circuit as in Fig. 4.22. For this case it is simpler to first construct the voltage. The general solution is given by (4.5). The amplitudes V_+ and V_- are found from the following conditions. At the load, $V(0) = 0$ which gives

$$0 = V_+ + V_- \tag{4.42}$$

The fact that the line is $\lambda/4$ meters long indicates that the generator "sees" an open circuit. (See Fig. 4.23.) It follows that the voltage at the generator has the amplitude V_S so that

$$V_S = V_+ e^{j\pi/2} + V_- e^{-j\pi/2}$$

or equivalently

$$V_S = j(V_+ - V_-) \tag{4.43}$$

Combining (4.42) and (4.43) gives

$$V_+ = -V_- = V_S/2j$$

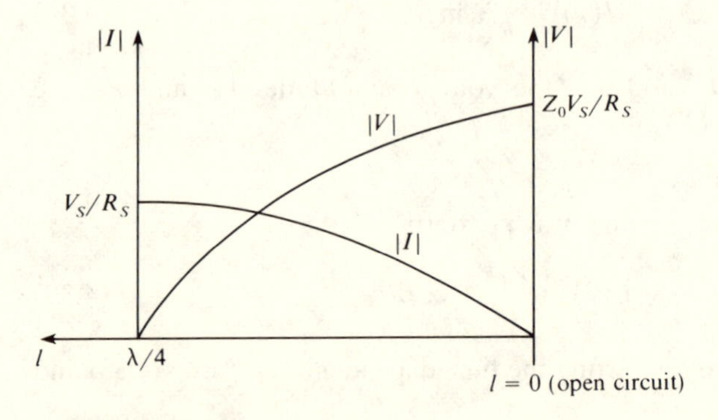

Figure 4.21 Current and voltage amplitudes.

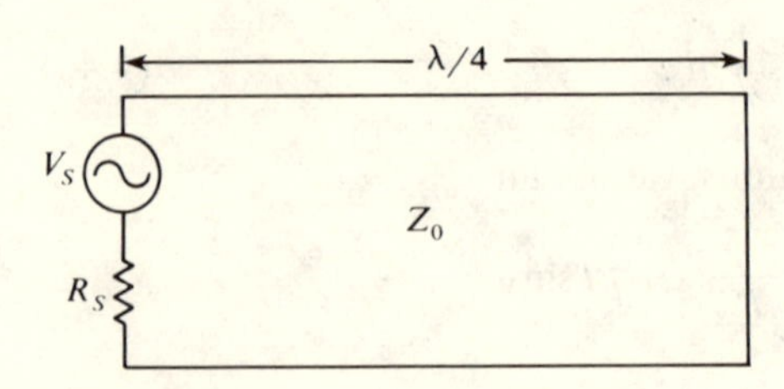

Figure 4.22 A short-circuited line.

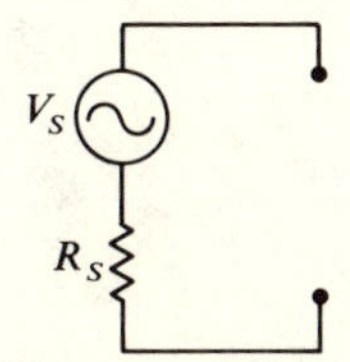

Figure 4.23 The generator circuit.

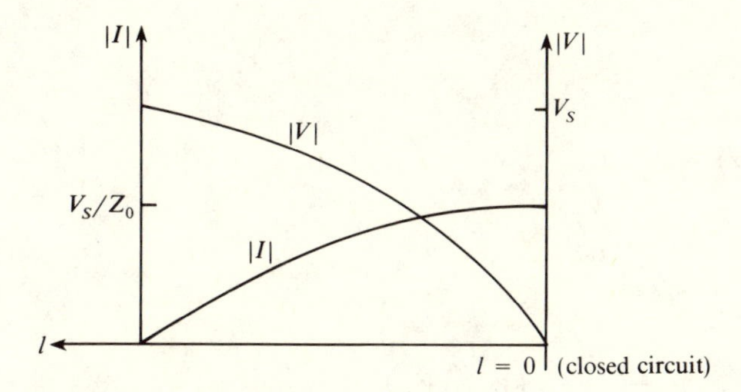

Figure 4.24 Voltages and current amplitudes.

Substituting these values into (4.5) gives the solution

$$V(l) = V_S \sin \beta l \tag{4.44}$$

The current amplitudes are again obtained from (2.12):

$$I_+ = I_- = \frac{V_+}{Z_0} = \frac{V_S}{2jZ_0}$$

Inserting these values into the general solution (4.6) yields the current distribution

$$I(l) = \frac{V_S}{jZ_0} \cos \beta l \tag{4.45}$$

A sketch of the oscillations (4.44) and (4.45) appears in Fig. 4.24.

Example 4.6

What are the voltage and current distributions for the configuration shown in Fig. 4.25?

Ans. Since the period of $Z(l)$ is $\lambda/2$, it follows that the circuit presented to the generator is a short circuit. (See Fig. 4.20.) The amplitude of current at the generator is

$$I\left(\frac{3\lambda}{2}\right) = \frac{V_S}{R_S}$$

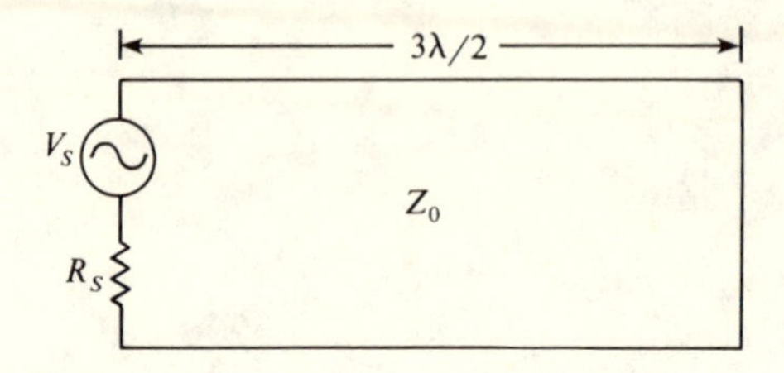

Figure 4.25 See Example 4.6.

so that (with $\beta l = 3\pi$)

$$\frac{V_S}{R_S} = I_+ + I_-$$

Equivalently we may write

$$\frac{V_S}{R_S} = \frac{1}{Z_0}(V_+ - V_-)$$

At the load we obtain

$$V(0) = V_+ + V_- = 0.$$

These latter two equations give

$$V_+ = -V_- = V_S Z_0 / 2R_S$$

from which we obtain

$$I_+ = I_- = V_S / 2R_S$$

The solutions sought are

$$V(l) = j\frac{V_S Z_0}{R_S}\sin\beta l$$

$$I(l) = \frac{V_S}{R_S}\cos\beta l$$

The voltage across the load and across the generator is zero. The current amplitude at the load and at the generator is V_S/R_S.

Example 4.7

a. Obtain an expression for the voltage along the transmission line for the configuration shown in Fig. 4.26, given that $R_L > Z_0$.
b. If $R_L = 100\ \Omega$ and $Z_0 = 50$ and $V_S = 150$ volts, what is the average power delivered to the load?

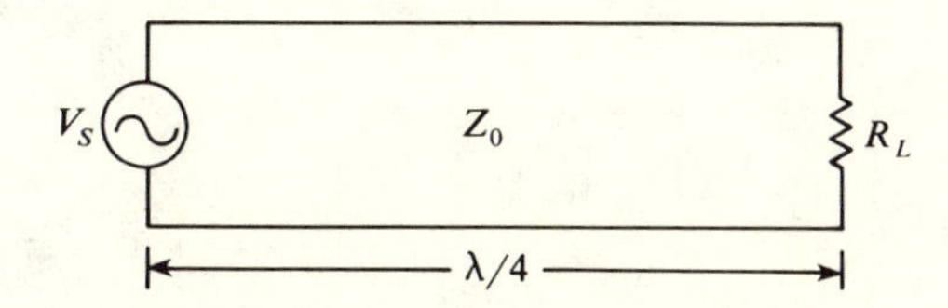

Figure 4.26 See Example 4.7.

Ans.

a. The voltage along the line is

$$V(l) = V_+ e^{i\beta l} + V_- e^{-i\beta l}$$

At the generator, the voltage amplitude is V_S, so that

$$V_S = j(V_+ - V_-)$$

Since $R_L > Z_0$, the load represents an impedance and voltage maximum. From (4.31), $S = R_L/Z_0$ and the voltage maximum at the load has the value

$$V(0) = S|V_S| = \frac{R_L|V_S|}{Z_0} = \frac{R_L V_S}{Z_0}$$

The latter equality follows from the fact that V_S is a real amplitude. With this information we may write

$$\frac{R_L V_S}{Z_0} = V_+ + V_-$$

Combining this with the previous equation relating V_S to V_+ and V_- gives

$$V_+ = \frac{V_S(1 + jR_L/Z_0)}{2j}$$

$$V_- = \frac{V_S(-1 + jR_L/Z_0)}{2j}$$

There results

$$V(l) = V_S \sin\beta l + \frac{V_S R_L}{Z_0}\cos\beta l$$

The voltage amplitude at the source is the minimum value V_S, whereas at the load it is the maximum value $V_S R_L/Z_0$.

b. From (2.26) and (2.32) we obtain

$$\langle P_L \rangle = \frac{1}{2}\frac{|V_+|^2}{Z_0}(1 - |\rho|)$$

With

$$|\rho| = \left|\frac{R_L - Z_0}{R_L + Z_0}\right| = \frac{1}{3}$$

and

$$|V_+|^2 = \frac{V_S^2}{4}\left[1 + \left(\frac{R_L}{Z_0}\right)^2\right]$$

there results

$$\langle P_L \rangle = 187.5 \text{ watts}$$

4.5 THE LOSSY TRANSMISSION LINE

Transmission lines in practical use are lossy. In frequency ranges where such loss is minimal, it is valuable to examine the transmission problem in the so-called low-loss limit. We will first obtain results applicable to the lossy-line problem, and from these general results obtain information relevant to the low-loss approximation.

The decrement in electrical power suffered in a lossy transmission line may be attributed to a series resistance per unit length and a shunt conductance per unit length. The unit-length equivalent circuit is as shown in Fig. 4.27. Equations (2.1) and (2.2) for the purely reactive line become

$$\frac{\partial V}{\partial z}\,dz = -L\,dz\frac{\partial I}{\partial t} - R\,dz\,I \tag{4.46}$$

$$\frac{\partial I}{\partial z}\,dz = -C\,dz\,\frac{\partial V}{\partial t} - G\,dz\,V \tag{4.47}$$

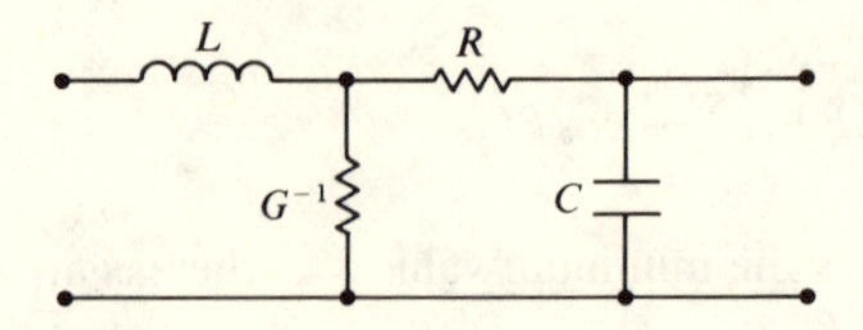

Figure 4.27 Unit-length equivalent circuit for a lossy transmission line.

These incremental equations yield the coupled differential equations

$$\frac{\partial V}{\partial z} = -\left(RI + L\frac{\partial I}{\partial t}\right) \tag{4.48}$$

$$\frac{\partial I}{\partial z} = -\left(GV + C\frac{\partial V}{\partial t}\right) \tag{4.49}$$

Let us assume that the time dependence of V and I goes as $e^{j\omega t}$. Then the preceding two equations become

$$-\frac{dV}{dz} = RI + j\omega LI \tag{4.50}$$

$$-\frac{dI}{dz} = GV + j\omega CV \tag{4.51}$$

Differentiating and combining these equations gives the separated equations

$$\frac{d^2V}{dz^2} = \gamma^2 V \tag{4.52}$$

$$\frac{d^2I}{dz^2} = \gamma^2 I \tag{4.53}$$

where the propagation constant γ is given by

$$\gamma^2 \equiv (R + j\omega L)(G + j\omega C) = \tilde{Z}\tilde{Y} \tag{4.54}$$

The series inductance per unit length is $\tilde{Z}$, and the shunt admittance per unit length is $\tilde{Y}$.

The second-order ordinary differential equation (4.52) has the general solution

$$V(z) = V_+e^{-\gamma z} + V_-e^{+\gamma z} \tag{4.55}$$

where V_+ and V_- are arbitrary constant amplitudes. Likewise (4.53) has the general solution

$$I(z) = I_+e^{-\gamma z} + I_-e^{+\gamma z} \tag{4.56}$$

where again, I_+ and I_- are constant amplitudes.

The propagation constant γ as given by (4.54) is complex, and we may write

$$\gamma = \alpha + j\beta \tag{4.57}$$

The decay constant α has units of m^{-1}, commonly termed *nepers/meter*.

Consider the term $V_+e^{-\gamma z}$ in (4.55) with the time factor included. This term becomes

$$V_+(z,t) = V_+e^{-\alpha z}e^{j(\omega t - \beta z)}$$

This function represents a wave propagating to larger z and attenuating as $e^{-\alpha z}$. Similarly $V_-(z,t)$ represents a modulated wave propagating to smaller

z. So again, as in the idealized lossless case, we may term V_+, I_+ forward waves and V_-, I_- reflected waves.

The characteristic impedance

$$Z_0 = \frac{V_+}{I_+} = -\frac{V_-}{I_-} \tag{4.58}$$

becomes, for the lossy line,

$$Z_0 = \sqrt{\frac{R + j\omega L}{G + j\omega C}} = \sqrt{\tilde{Z}\tilde{Y}} \tag{4.59}$$

In the limit that R and G go to zero, (4.59) is seen to return the relation (2.12) appropriate to the lossless line.

Reverting to the choice of origin which puts the load at $z = 0$ and the generator at $z = -l$, dividing (4.55) by (4.56), and using (4.58) gives the impedance l meters from the load:

$$Z(l) = Z_0\left(\frac{e^{\gamma l} + \rho_L e^{-\gamma l}}{e^{\gamma l} - \rho_L e^{-\gamma l}}\right) \tag{4.60}$$

This is the appropriate generalization of (4.9).

At the load, $l = 0$, and (4.60) returns the relation (4.10):

$$Z_L = Z_0\left(\frac{1 + \rho_L}{1 - \rho_L}\right)$$

Substituting this relation into (4.60) and reexpressing the exponentials in terms of their hyperbolic equivalents, we obtain

$$Z(l) = Z_0\left(\frac{Z_L \cosh \gamma l + Z_0 \sinh \gamma l}{Z_0 \cosh \gamma l + Z_L \sinh \gamma l}\right) \tag{4.61}$$

which is the generalization of (4.11). In the limit of no loss, $\alpha = 0$, $\gamma = j\beta$, and the above expression for $Z(l)$ is seen to reduce to the expression for the lossless line (4.11).

4.6 THE LOW-LOSS APPROXIMATION

In the so-called low-loss approximation

$$\begin{gathered} R \ll \omega L \\ G \ll \omega C \end{gathered} \tag{4.62}$$

Rewriting the propagation constant γ (4.45) in the form

$$\gamma = j\omega\sqrt{LC}\left(1 + \frac{R}{j\omega L}\right)^{1/2}\left(1 + \frac{G}{j\omega C}\right)^{1/2} \tag{4.63}$$

permits the Taylor series expansion

$$\gamma = j\omega\sqrt{LC}\left(1 + \frac{1}{2}\frac{R}{j\omega L} + \cdots\right)\left(1 + \frac{1}{2}\frac{G}{j\omega C} + \cdots\right)$$

It follows that in the low-loss limit (4.62), the propagation constant is well approximated by the expression

$$\gamma \simeq j\omega\sqrt{LC}\left(1 + \frac{R}{2j\omega L} + \frac{G}{2j\omega C}\right)$$

which permits the identification

$$\alpha = \frac{1}{2}\left(\frac{R}{Z_0} + GZ_0\right) \tag{4.64}$$

$$\beta = \frac{\omega}{v} = \frac{2\pi}{\lambda} \tag{4.65}$$

In these expressions Z_0 is the ideal characteristic impedance $\sqrt{L/C}$, and v is the wave speed (2.6), namely $1/\sqrt{LC}$.

In the same approximation (4.62), expansion of the expression for the characteristic impedance (4.59) gives the expression

$$Z_0 \simeq \frac{R}{G}\left(1 + \frac{j\omega L}{R} - \frac{j\omega C}{G}\right) \tag{4.66}$$

Finally we turn to the input impedance $Z(l)$ in the low-loss limit. A convenient expression results if the length of the line further satisfies the inequality $\alpha l \ll 1$. Expansion of (4.61) in this approximation gives

$$Z(l) \simeq Z_0\left[\frac{Z_L\cos\beta l + jZ_0\sin\beta l + \alpha l(Z_0\cos\beta l + jZ_L\sin\beta l)}{Z_0\cos\beta l + jZ_L\sin\beta l + \alpha l(Z_L\cos\beta l + jZ_0\sin\beta l)}\right] \tag{4.67}$$

Let us apply this result to find the input impedance to an open line $\lambda/2$ meters long with $G = 0$ and small series loss R (ohms/meter). For this case $|Z_L| = \infty$ and (4.67) gives

$$Z(\lambda/2) \simeq Z_0\left(\frac{\cos\beta l + j\alpha l\sin\beta l}{j\sin\beta l + \alpha l\cos\beta l}\right) \tag{4.68}$$

At $\beta l = \pi$, with α given by (4.64), we obtain

$$Z\left(\frac{\lambda}{2}\right) \simeq \frac{Z_0}{\alpha l} = \frac{2Z_0^2}{Rl} \gg Z_0 \tag{4.69}$$

This large impedance is due to the fact that the equivalent circuit for this line is the parallel network shown in Fig. 4.28. With $R = 0$ at resonance the

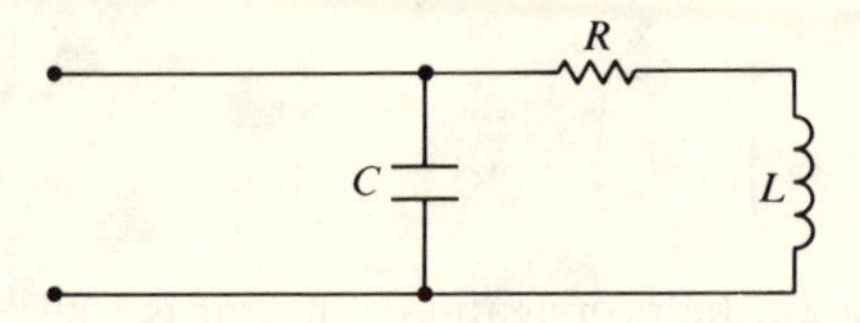

Figure 4.28

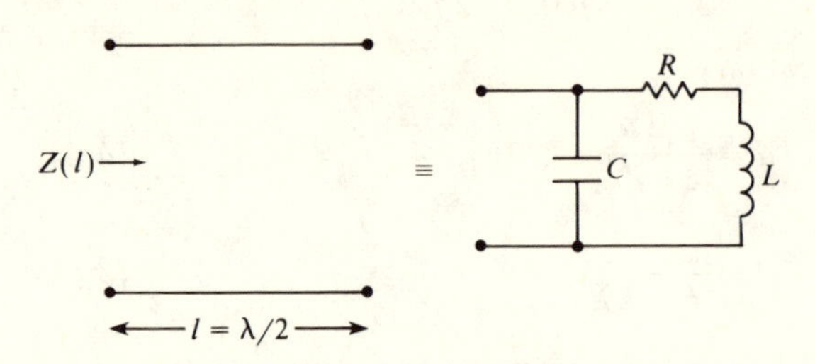

Figure 4.29 Equivalent circuit for $l = \lambda/2$.

circuit becomes an open circuit with $Z = \infty$. (See Table 1.1.) The insertion of a negligible resistance does not destroy the high impedance. So we have the equivalent circuit shown in Fig. 4.29.

Example 4.8

What is the minimum length of a short-circuited line in the limit $\alpha l \ll 1$ which gives the impedance of a slightly resistive parallel LC network?

Ans. For the short-circuited line, $|Z_L| = 0$ and (4.67) gives

$$Z\left(\frac{\lambda}{4}\right) \simeq Z_0\left(\frac{j\sin\beta l + \alpha l\cos\beta l}{\cos\beta l + j\alpha l\sin\beta l}\right)$$

At $\beta l = \pi/2$, we obtain (4.69):

$$Z = \frac{Z_0}{\alpha l} = \frac{2Z_0^2}{Rl} \gg Z_0$$

which is an appropriately large impedance. The value $\beta l = \pi/2$ corresponds to the length $l = \lambda/4$ (Fig. 4.30).

Note: The lengths of these high-input-impedance lines are easily inferred from the periodicity of $Z(l)$. Thus, as described previously, the large impedance of an open-circuit load appears again at $\lambda/2$ meters toward the

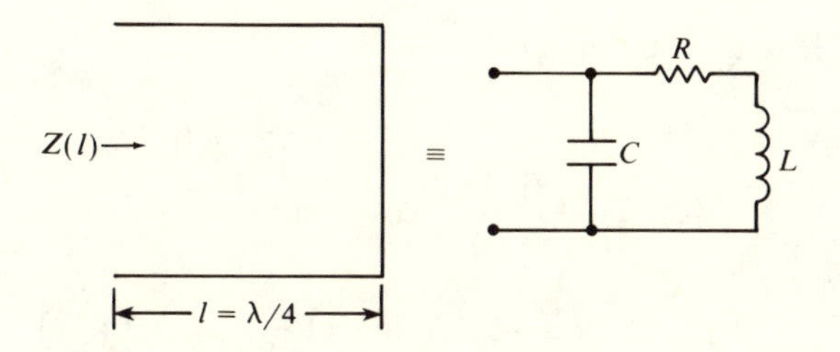

Figure 4.30 Equivalent circuit for $l = \lambda/2$.

generator. An open-circuit impedance also appears $\lambda/4$ meters toward the generator away from a closed-circuit load.

4.7 THE QUARTER-WAVELENGTH-LINE CAVITY EQUIVALENT

A resonant cavity is used to store electromagnetic energy at a given frequency.[3] As with a resonant circuit (Section 1.6), the efficiency of this process is measured by the Q of the device. In this section we shall see how a quarter-wavelength line terminated in a short circuit may serve as a resonant energy-storage device.

Let us first demonstrate that the total electromagnetic energy of the line is constant in the lossless quarter-wavelength section. Toward this end we recall the expressions (4.44) and (4.45) for the voltage and current respectively on the quarter-wavelength shorted line (with the time dependence reinserted):

$$V(l) = V_S \sin \beta l \cos \omega t \tag{4.70}$$

$$I(l) = -\frac{V_S}{Z_0} \cos \beta l \sin \omega t \tag{4.71}$$

The instantaneous electric and magnetic energy per unit length, W_E and W_M respectively, are given by

$$W_E = \tfrac{1}{2} C V^2 \tag{4.72}$$

$$W_M = \tfrac{1}{2} L I^2 \tag{4.73}$$

[3] Resonant cavities are discussed further in Appendix D.

The total energy per unit length on the line is

$$\begin{aligned} W(l,t) &= W_E + W_M \\ &= \tfrac{1}{2}CV_S^2\sin^2\beta l\cos^2\omega t + \frac{1}{2}\frac{LV_S^2}{Z_0^2}\cos^2\beta l\sin^2\omega t \\ &= \tfrac{1}{2}CV_S^2(\sin^2\beta l\cos^2\omega t + \cos^2\beta l\sin^2\omega t). \end{aligned} \tag{4.74}$$

Integrating this expression over the quarter-wavelength section gives the total energy

$$W(t) = \int_{-\lambda/4}^{0} W(l)\,dl = \tfrac{1}{2}CV_S^2\frac{\lambda}{8}(\cos^2\omega t + \sin^2\omega t) \tag{4.75}$$

$$W(t) = \frac{\lambda CV_S^2}{16} \tag{4.76}$$

At any instant in time the total energy in the quarter-wavelength line is constant. From (4.74) we see that the energy oscillates between being totally stored in the magnetic field and totally stored in the electric field. The result (4.75) that the total energy on the line is constant is not surprising in light of the fact, discovered in Example 4.8, that this section of line is equivalent to a parallel LC network at resonance. At this value of frequency the network becomes isolated from the source, and the current and voltage oscillate unabated with constant amplitude (see Table 1.1). We may conclude that the lossless quarter-wavelength line terminated in a short circuit acts as an ideal resonant cavity with infinite Q.

We wish now to obtain an expression for the Q of a slightly lossy quarter-wavelength line. As obtained in Example 4.8, the input impedance to this line is given by (4.69):

$$Z = Z_0/\alpha l \gg Z_0 \tag{4.77}$$

This is the impedance at the resonant frequency $f_0 = v/\lambda$, at which value the line has its maximum Q-value. An equivalent expression for Q to that given in Section 1.6, which is more relevant to cavities, is

$$Q = \frac{f_0}{\Delta f} \tag{4.78}$$

For the case at hand $\Delta f/2$ represents the shift in frequency away from f_0 required to reduce the power flow from a constant-current generator to the line by a factor of one-half. The equivalent circuit is shown in Fig. 4.31. The power flow to the circuit is given by (1.28):

$$\langle P \rangle = \tfrac{1}{2}\mathrm{Re}\, IV^*$$

which for the case at hand becomes

$$\langle P \rangle = \tfrac{1}{2}|I_S|^2\mathrm{Re}\, Z_{\mathrm{in}} \tag{4.79}$$

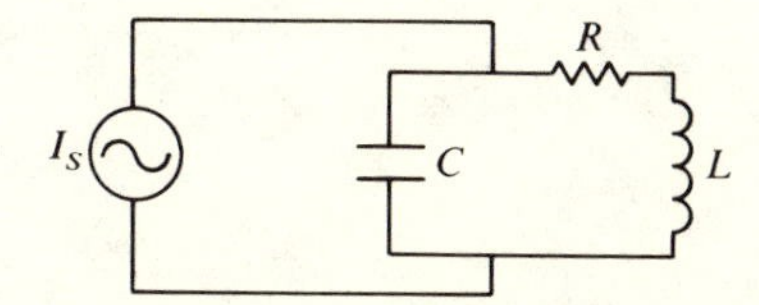

Figure 4.31

At resonance $\langle P \rangle$ has its maximum value,

$$\langle P \rangle = |I_S|^2 Z_0/2\alpha l \tag{4.80}$$

For a slight frequency shift to the frequency

$$f = f_0 + \delta f, \qquad f_0 = \frac{v}{4l}, \qquad \left|\frac{\delta f}{f_0}\right| \ll 1$$

the parameter βl becomes

$$\beta l = \frac{2\pi f l}{v} = \frac{2\pi l(f_0 + \delta f)}{v} = \frac{\pi}{2}\left(1 + \frac{\delta f}{f_0}\right)$$

It follows that

$$\cos \beta l = -\sin\left(\frac{\pi}{2}\frac{\delta f}{f_0}\right) \simeq -\frac{\pi}{2}\frac{\delta f}{f_0}$$

$$\sin \beta l = \cos\left(\frac{2\pi l \delta f}{v}\right) \simeq 1$$

In the limit $\alpha l \ll 1$, we may write (see Example 4.8)

$$Z_{\text{in}} \simeq Z_0\left(\frac{j\sin\beta l + \alpha l \cos\beta l}{\cos\beta l + j\alpha l \sin\beta l}\right)$$

which with the preceding approximations, and neglecting terms like $(\alpha l)(\delta f/f_0)$, gives

$$Z_{\text{in}} = \frac{Z_0}{\alpha l + j\left(\frac{\pi}{2}\frac{\delta f}{f_0}\right)}$$

It follows that

$$\operatorname{Re} Z_{\text{in}} = \frac{Z_0 \alpha l}{(\alpha l)^2 + \left(\frac{\pi}{2}\frac{\delta f}{f_0}\right)^2} \tag{4.81}$$

The power as given by (4.79) will be reduced from its maximum as given by (4.80) by a factor of $\frac{1}{2}$ when the two terms in the denominator of (4.81) are equal. A sketch of $\langle P \rangle / \langle P \rangle_{\text{max}}$ is shown in Fig. 4.32. We obtain, at

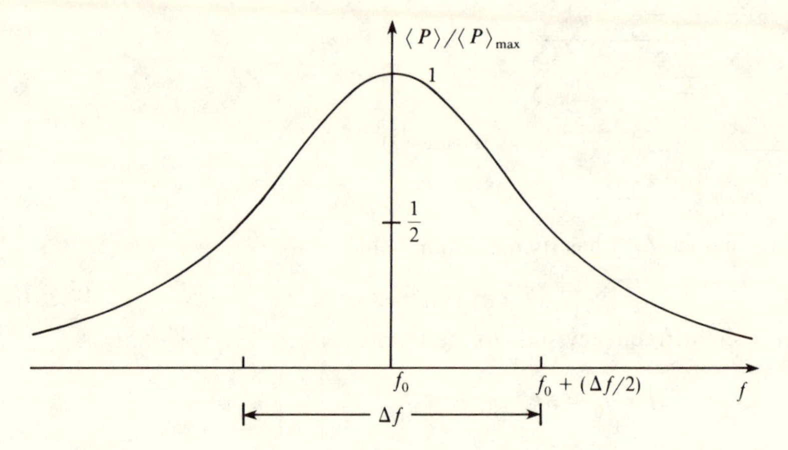

Figure 4.32 Average power flow as a function of frequency.

$\delta f = \Delta f/2$,

$$\alpha l = \frac{\pi}{4}\frac{\Delta f}{f_0}$$

which gives, with $\lambda = 4l$,

$$\frac{\Delta f}{f_0} = \frac{2\alpha}{\beta}$$

Substituting this value into (4.78) gives the expression sought:

$$Q = \frac{\beta}{2\alpha} \tag{4.82}$$

Equivalently we may write

$$Q = \frac{2\pi f_0 Z_0}{Rv} = \frac{\omega L}{R} \tag{4.83}$$

The latter expression is seen to agree with the form found for the Q of a dissipative LC circuit in Example 1.4. Since the input impedance to an open-ended half-wavelength line is the same as that of a closed quarter-wavelength line, the Q evaluated for the half-wavelength open line is also given by (4.82), (4.83).

In the limit of no dissipation, $\alpha \to 0$ and Q as given by (4.82) becomes infinite. In this same limit the stored energy in the LC network at resonance increases indefinitely at the expense of power delivered by the constant-current generator. (See Example 1.6.)

4.8 TRANSPORT ALONG A NERVE CELL[4]

The Nerve Cell

We conclude this chapter with a brief introductory description of the properties related to the propagation of a signal along a nerve cell, or neuron.

Neurons are composed of *dendrites*, *axons*, and a *cell body*. Typically the dendrites are a treelike aggregate of fibers which carry impulses to the cell body, whereas the axon is a fiber which carries impulses away from the cell body. We may represent a neuron schematically as shown in Fig. 4.33.

The larger nerve fibers are myelinated, that is, surrounded by longitudinal sections of fatty insulation called myelin. Myelin is formed by Schwann cells, which wrap themselves around axons during development. Unmyelinated fibers have a reasonably uniform cross section and carry a membrane of fatty molecules which is approximately 100 Å thick.

The terminus of a neuron may be implanted in the dendrites of a second neuron. This junction between two neurons is called a *synapse*. Impulses move along the axon of one neuron and across the synapse, exciting the dendrites of the second neuron and thereby passing the impulse from one neuron to another. The diameters of some different types of neurons are listed in Table 4.3.[5] The transmission across the synapse is primarily chemical, whereas that along a nerve fiber is primarily electrical.

Electrochemical properties. In most cells the internal concentration of sodium ions (Na^+) is less than that in the extracellular fluid. Measurement on both sides of the membrane of a neuron in the squid reveals a Na^+ concentration ratio of 1 : 9. On the other hand, the concentration of potassium ions (K^+) is greater in the interior of the cell than the exterior of the cell by a ratio of 20 : 1.[6] Furthermore, there is a higher concentration of large negative organic ions inside the cell than in the extracellular fluid. An electric potential across the cell membrane accompanies these ion concentration jumps, which is of the order of 70 mV negative with respect to the outside of the cell. For a membrane of thickness 100 Å this corresponds to an electric field of 70 kV/cm.[7]

[4] Neuron transport is not a standing-wave phenomenon as its inclusion in this chapter might imply. However, the equations developed in Section 4.5 are sufficient to render a rough description of the passive properties of a nerve fiber.

[5] A. Giese, "Cell Physiology," 3rd ed. (W. B. Saunders, Philadelphia, 1968).

[6] S. Kuffler and J. Nichols, "From Neuron to Brain" (Sinauer, Sunderland, Massachusetts 1976).

[7] The breakdown strength of air is 30 kV/cm.

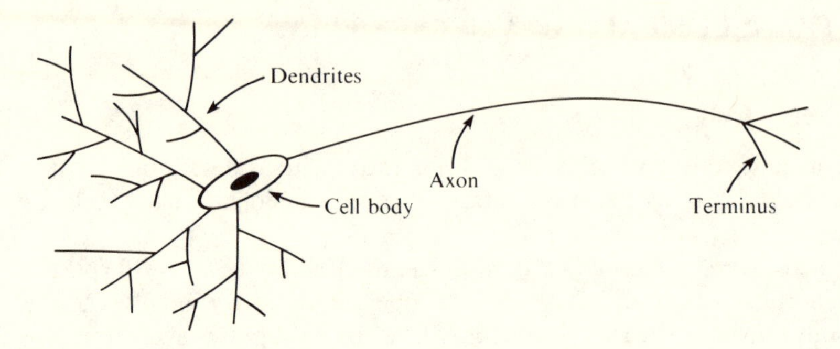

Figure 4.33 Components of a neuron.

Table 4.3 Axon diameters

Organism	Diameter (μm)
Giant squid	637
Giant earthworm	90
Giant shrimp	43
Catfish	6.6
Man	0.1–10

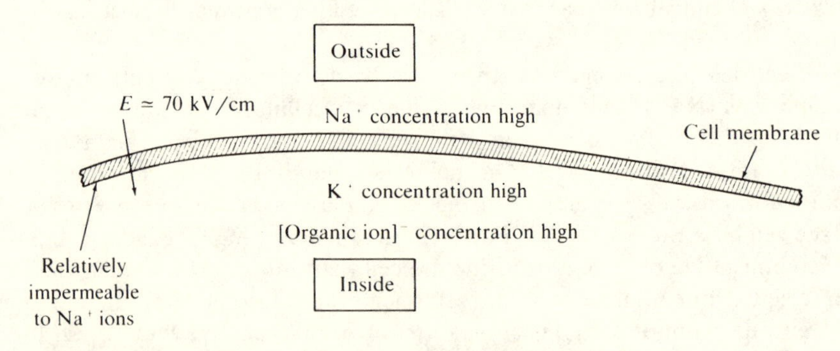

Figure 4.34 The resting state across the cell membrane.

In the resting state the permeability of the membrane to Na^+ is $\frac{1}{50}$ the permeability to K^+. (See Fig. 4.34.)

Generation of the action potential. The response of a nerve fiber to a stimulus divides into two categories depending on the amount by which the stimulus changes the voltage jump across the cell membrane. When the inside of the fiber is increased to a threshold voltage of about -50 mV or more (i.e., or less negative), a voltage spike called the *action potential* is

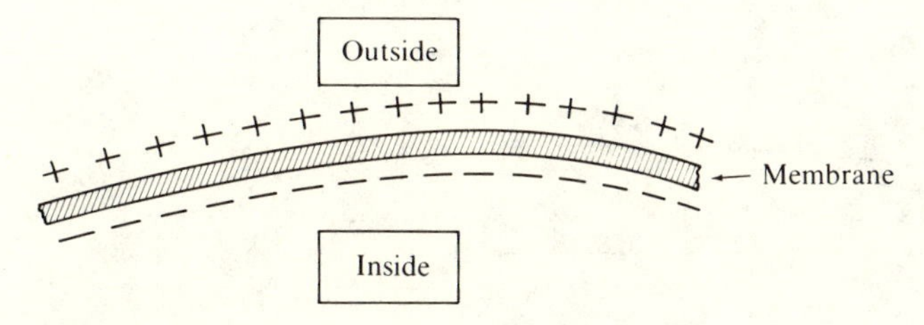

Figure 4.35 Resting state.

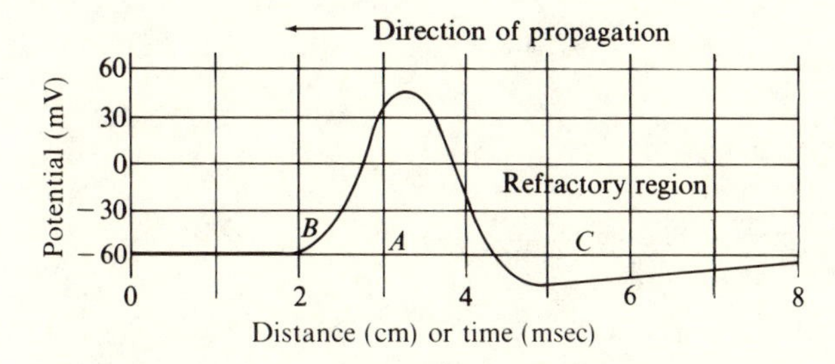

Figure 4.36 The action potential.

triggered, which then travels without distortion along the length of the fiber. The response of the fiber to stimuli of less strength is passive and remains localized.

The mechanism for the generation of the action potential in the unmyelinated nerve fiber is as follows. In the resting state the cell membrane is polarized as shown in Fig. 4.35. When the membrane is stimulated at a given point, it suffers a great increase in the permeability to sodium ions, thereby allowing these ions to fall through the membrane potential difference into the cell. This influx momentarily causes the inside of the fiber to become positively charged relative to the outside in the vicinity of the disturbance. Potassium ions then rush out of the cell near the excitation point to restore the resting-state polarity. How does this disturbance propagate?

Let us suppose that an action potential is propagating in a specific direction. A sketch of this potential at any instant of time is shown in Fig. 4.36.[8] At point A the potential is high due to the large influx of Na^+ ions. This local charge excess causes positive charge carriers to move away from A laterally in both directions. Ahead of the wave at B, this lateral influx of positive charge causes an increase (i.e., toward zero) of the inside potential, and threshold is once again reached, with accompanying increase in perme-

[8] R. D. Keynes, Scientific American, Dec. 1958.

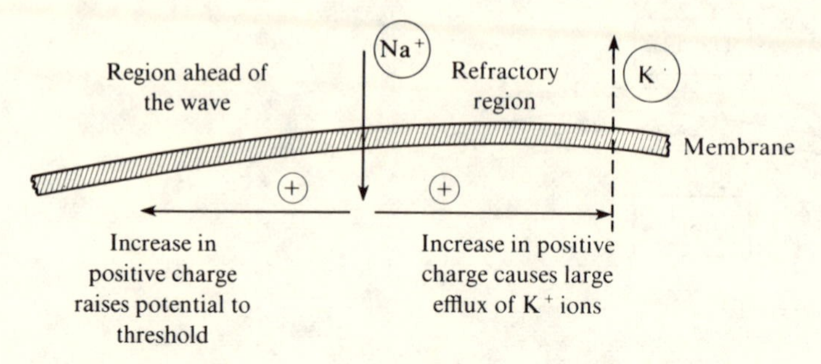

Figure 4.37

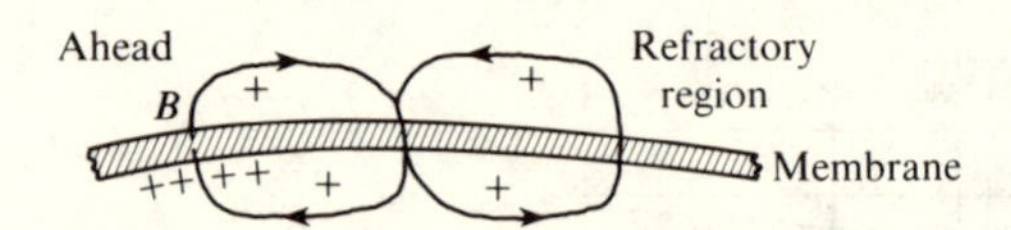

Figure 4.38 Current flow in the action potential.

ability to Na^+. At point C in the refractory region of the action potential, due to the fact that the wave has just passed this point, the membrane is left in a state which permits a large egress of K^+ ions, thereby restoring the resting potential and polarity. (See Fig. 4.37.)

Current loops accompany the action potential of the form shown in Fig. 4.38. Ahead of the wave at B the current flow is largely capacitive. Positive charge builds up on the inside of the membrane and is carried away from the outside of the membrane until threshold is reached with accompanying reversal of current flow. In this manner, the original disturbance propagates away from the stimulation point in the form of a wave of reversed polarity. This model of the action potential is due in large part to A. L. Hodgkin and A. F. Huxley.[9]

Passive properties. In the so-called *cable model*, a distributed network is adopted to represent the passive properties of a nerve fiber and has the structure shown in Fig. 4.39. The resting conductance per unit length of the membrane is G_m and C_m is the capacitance per unit length of the membrane. The resting potential across the membrane is maintained by the source voltage $\mathscr{E}_r$. The resistance per unit length in the interior of the cell is

[9] J. Physiology, vol. 116, pp. 449, 473, 497, 1952. See also, R. L. Liboff, J. Theoretical Biol., vol. 83, p. 427, 1980.

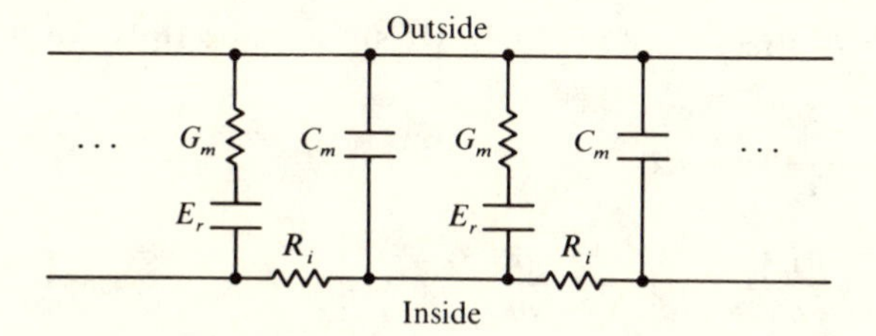

Figure 4.39 The equivalent network.

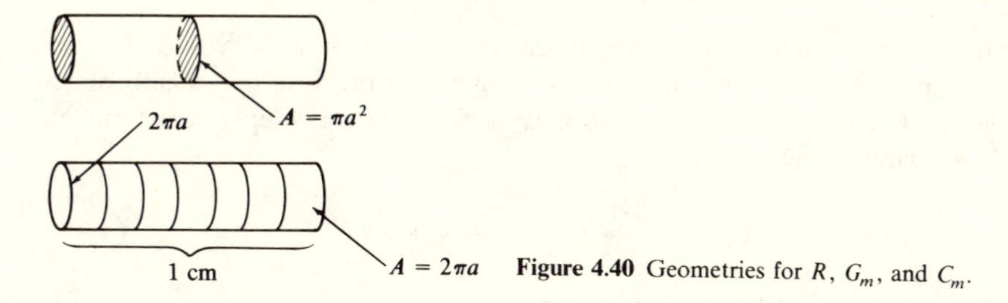

Figure 4.40 Geometries for R, G_m, and C_m.

R_i. If the radius of the fiber (in cm) is a, then typical values for these parameters are given by the following expressions:[10] (see Fig. 4.40)

$$R_i \simeq \frac{200}{\pi a^2} \text{ ohm/cm}$$

$$G_m^{-1} \simeq \frac{2000}{2\pi a} \text{ ohm/cm}$$

$$C_m \simeq 2\pi a \times 10^{-6} \text{ F/cm}$$

The axon surfaces relevant to these expressions are shown on the right.

In the resting state there is a capacitive charge of

$$Q = C\mathscr{E}_r \text{ C/cm}$$

and potential jump of $\mathscr{E}_r$ volts across the line. Since the capacitance effects an open circuit, there is no current flow in the resting state. The dynamics of the distributed network may be examined neglecting the presence of the potential source $\mathscr{E}_r$.

[10] B. Katz, "Nerve Muscle and Synapse" (McGraw-Hill, New York, 1966). See also D. Aidley, "The Physiology of the Excitable Membranes" (Cambridge U.P., Cambridge, 1971).

The network is governed by (4.48) and (4.49). Dropping subscripts, we write

$$\frac{\partial V}{\partial z} = -RI$$

$$\frac{\partial I}{\partial z} = -GV - C\frac{\partial V}{\partial t} \tag{4.84}$$

Eliminating I gives the single equation for V

$$\frac{\partial^2 V}{\partial z^2} = RC\frac{\partial V}{\partial t} + RGV \tag{4.85}$$

An identical equation results for the current I. Suppose a DC voltage source V_0 is placed across the line at $z = 0$. What is the resulting steady-state voltage $V(z)$? In steady state we may neglect the time-dependent term in (4.84), which gives

$$\frac{\partial^2 V}{\partial z^2} = RGV \equiv \kappa^2 V \tag{4.86}$$

This equation has the general solution

$$V = Ae^{-\kappa z} + Be^{\kappa z}$$

where A and B are arbitrary constants.

As $z \to 0$, $V \to 0$, which indicates that $B = 0$. At $z = 0$, $V = V_0$, so we obtain

$$V = V_0 e^{-\kappa z} \tag{4.87}$$

(See Fig. 4.41.) For a 1-μm-diameter fiber, from the values of R and G given above we find

$$\kappa = 4 \times 10^3 \text{ cm}^{-1}$$

which corresponds to a decay length

$$\kappa^{-1} = 2.5\ \mu\text{m}$$

In slightly more than 2 diameters the impressed voltage is attenuated to e^{-1} times its starting value.

For rapidly varying voltage, $C\omega \gg G$ and the last term in (4.85) may be neglected. This yields the so-called *diffusion equation*. Written for the

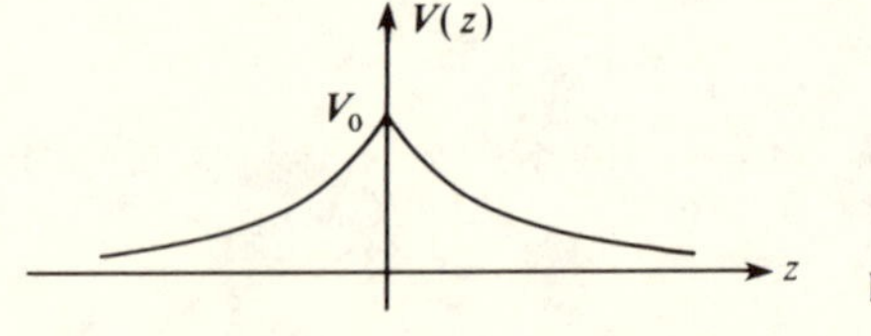

Figure 4.41

current I, it appears as

$$\frac{\partial^2 I}{\partial z^2} = RC\frac{\partial I}{\partial t} \tag{4.88}$$

To uncover the behavior implied by the equation we consider the case that a large pulse of current is injected into the line at the point $z = 0$ and $t = 0$. Such a current pulse may be conveniently represented by the *delta function*, $\delta(z)$:

$$I(z) = I_0 a\delta(z) \qquad (t = 0) \tag{4.89}$$

The parameter a represents a characteristic length. The delta function has the following defining properties:

$$\delta(z) = 0, \qquad z \neq 0$$

$$\int_{-\infty}^{\infty} \delta(z)\, dz = 1 \tag{4.90}$$

The solution to the diffusion equation (4.88) corresponding to the initial data (4.89) is the so-called normal distribution,

$$I(z,t) = \frac{aI_0}{(4\pi t/RC)^{1/2}} \exp\left(-\frac{z^2 RC}{4t}\right) \tag{4.91}$$

At $t = 0$ the current $I(z) = 0$ everywhere except at $z = 0$, thereby effecting the initial distribution (4.89). At later times the current spreads away from the source and approaches a flat distribution. As with the response to subthreshold stimuli, the original signal does not propagate away from the disturbance but remains essentially localized.

Example 4.9

A subthreshold harmonic voltage source with angular frequency ω is placed across the distributed RC line described above. In the limit of low conductance, $G \ll \omega C$:

a. Show that

$$\alpha = \beta = \sqrt{\frac{\omega RC}{2}}$$

b. Show that the propagation speed is

$$v = \sqrt{\frac{2\omega}{RC}} \simeq 50\sqrt{\omega D}$$

where D is the diameter $2a$.

c. What is the value of the decay constant α for a 1-μm-radius fiber at $\omega = 2\pi \times 1$ krad/sec?

Ans.

a. The propagation constant for the stated conditions follows from (4.54), and we obtain

$$\gamma = \sqrt{\frac{\omega RC}{2}}(1 + j) \equiv \alpha + j\beta$$

which gives the desired result.

b. The propagation speed is given by ω/β. The expression in terms of the fiber diameter D follows from values of R and C given in the text:

$$v = \frac{\omega}{\beta} = \sqrt{\frac{2\omega}{RC}} \simeq 50\sqrt{\omega D}$$

c. From part a and the values of R and C given in the text we find

$$\alpha = 112 \text{ nepers/cm}$$

This result indicates that without regeneration of the signal, it would suffer rapid decay. Note also that for the values given,

$$\omega C \simeq 13G$$

and our assumption that $\omega C \gg G$ is at best borderline.

PROBLEMS

4.1. An ideal transmission line terminated in a resistive load of 50 Ω has a standing-wave ratio of 3 : 1. A voltage minimum is measured $\lambda/4$ meters from the load.

a. What is Z_0?
b. If in steady state $|V_+| = 50$, what is $|I_+|$?

4.2. A generator is to be connected to a transmission line 20 meters long terminated in the load $Z_L = 50 + j60\ \Omega$. If the characteristic impedance of the line is 70 Ω, what must the internal impedance of the generator be in order for it to be matched to the line?

4.3. A quarter-wavelength line is terminated in an open circuit as shown in Fig. 4.42.

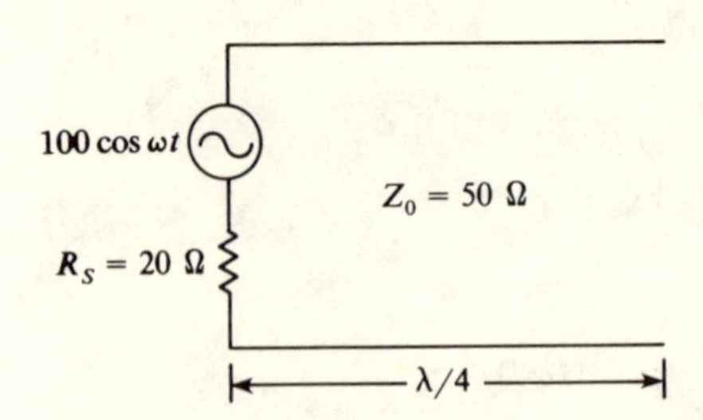

Figure 4.42 Configuration for Problem 4.3.

a. If the length of line is 0.60 m, what is the frequency f?
b. Obtain an expression for the standing-wave voltage on the line.
c. What is the amplitude of the voltage at the open end?

4.4. A low-loss transmission line 0.5 m long, terminated in a short circuit, acts as a resonant cavity with $Q \simeq 700$. What is the minimum resonant frequency of the line?

4.5. A generator is connected to a purely resistive load of 100 Ω through an ideal transmission line 25 meters long. The line has permittivity $\epsilon = 4\epsilon_0$. At 1.5 MHz:

a. What is the standing-wave ratio of the line?
b. What internal impedance of the generator is necessary for it to be perfectly matched to the line?
c. If the line is slightly lossy, what is the range of decay constants α for which your answer to part b will still be reasonably correct?

4.6. Introducing the parameter

$$\tau \equiv t/RC$$

show by direct differentiation that the normal distribution (4.91),

$$I(z,\tau) = \frac{aI_0}{(4\pi\tau)^{1/2}} \exp\left(-\frac{z^2}{4\tau}\right)$$

is a solution to the diffusion equation (4.88),

$$\frac{\partial^2 I}{\partial z^2} = \frac{\partial I}{\partial \tau}$$

CHAPTER

FIVE

SMITH CHART AND MATCHING PROBLEMS

5.1 THE COMPLEX MAPPING

The input impedance l meters from a load with reflection coefficient ρ_L is given by (4.9):

$$Z(l) = Z_0 \frac{1 + \rho_L e^{-j2\beta l}}{1 - \rho_L e^{-j2\beta l}} \tag{5.1}$$

Let us introduce the variable

$$\rho = \rho_L e^{-j2\beta l} \tag{5.2}$$

This parameter represents the value of the reflection coefficient l meters from the load. Then (5.1) may be rewritten

$$Z(l) = Z_0 \frac{1 + \rho}{1 - \rho} \tag{5.3}$$

Introducing the *normalized impedance*

$$z \equiv \frac{Z(l)}{Z_0} \tag{5.4}$$

permits (5.3) to be further simplified:

$$z = \frac{1 + \rho}{1 - \rho} \tag{5.5}$$

This equation may be viewed as a relation which maps any given region of the complex ρ-plane onto a corresponding region of the complex z-plane.

To see this we write

$$z = r + jx$$
$$\rho = u + jv \tag{5.6}$$

The variable r represents normalized resistance, and x normalized reactance. Substituting these relations into (5.5) gives

$$r + jx = \frac{1 + u + jv}{1 - u - jv}$$

Equating real and imaginary parts of the two sides of this equation and performing some algebra, we obtain

$$\left(u - \frac{r}{1+r}\right)^2 + v^2 = \left(\frac{1}{1+r}\right)^2 \tag{5.7a}$$

$$(u-1)^2 + \left(v - \frac{1}{x}\right)^2 = \left(\frac{1}{x}\right)^2 \tag{5.7b}$$

From (5.7a) we see that the lines of constant r in the complex z-plane get mapped into circles in the complex ρ-plane. As the resistance of the line goes to infinity, the radii of these circles shrink to zero. For $r = 0$, (5.7a) represents a unit-radius circle in the ρ-plane centered at the origin. Furthermore all circles described by (5.7a) pass through the point $u = 1$, $v = 0$. A sketch of this part of the transformation appears in Fig. 5.1. We may conclude that the right half z-plane is mapped onto the unit disk in the ρ-plane.

Equation (5.7b) indicates that lines of constant reactance are also mapped onto circles of radius $1/x$ in the ρ-plane. All such circles also go through the point $u = 1$, $v = 0$. The line $x = 1$ is mapped onto a unit circle

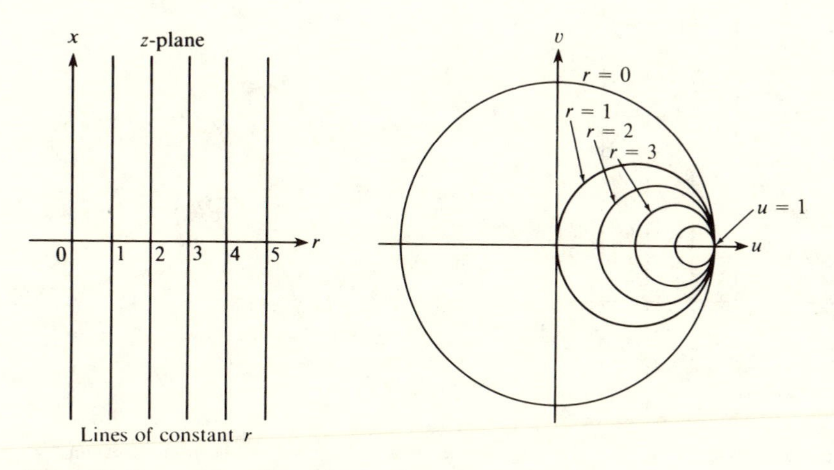

Figure 5.1 The z and ρ planes. Curves of constant r.

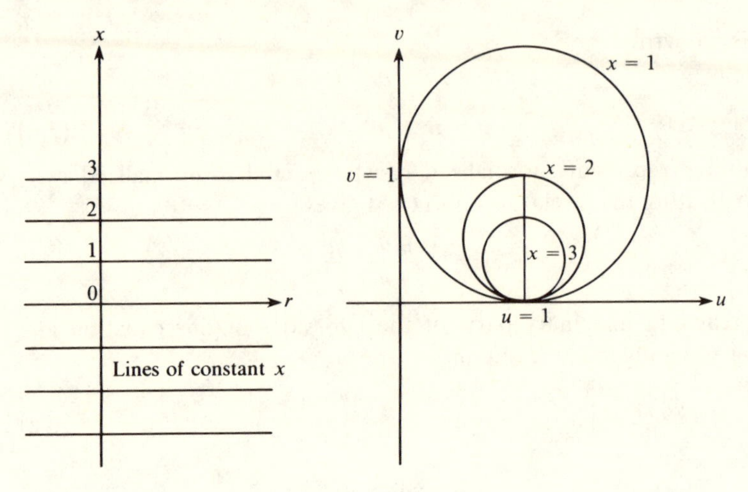

Figure 5.2 Curves of constant x.

in the ρ-plane centered at $u = 1$, $v = 1$. As $x \to 0$, the radius of the circles becomes infinite and the center moves to the point $u = 1$, $v = \infty$. (See Fig. 5.2.) We may conclude that the upper half z-plane is mapped onto a unit circle in the first quadrant of the ρ-plane by the transformation (5.5).

In the application of these results to transmission lines, $r \geq 0$, whereas x may be positive or negative. So we are concerned with the right half z-plane. As found above, the transformation (5.5) maps this domain onto the unit circle in the ρ-plane. Furthermore we see that the orthogonal grid of

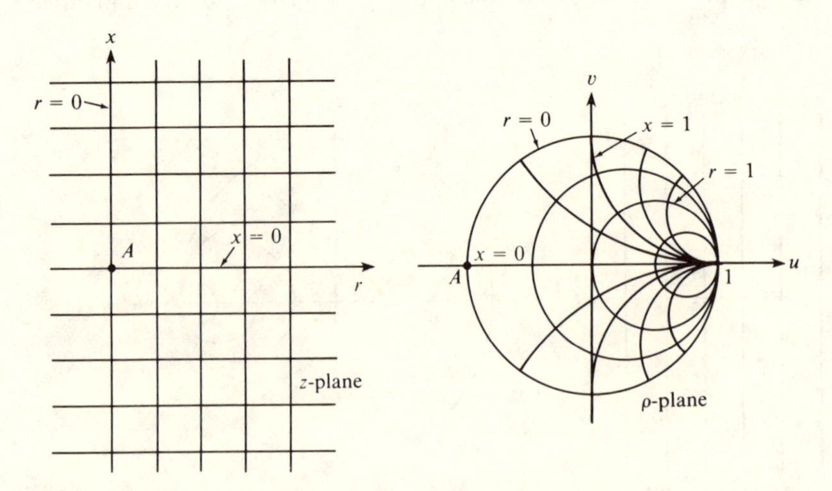

Figure 5.3 Mapping of the right half z-plane onto the ρ-plane by the transformation $z = (1 + \rho)/(1 - \rho)$.

constant-x and -r lines in the z-plane is mapped into a grid of orthogonal circles in the ρ-plane. Note particularly that the points $x = \pm\infty$, $r = 0$ in the z-plane are both mapped onto the point $\rho = 0$, $v = 1$ in the ρ-plane. (See Fig. 5.3.)

5.2 THE SMITH CHART[1]

The unit circle $|\rho| = 1$ together with its interior is called a Smith chart. How is this chart useful in the study of transmission lines?

To answer this question we note the following property of a homogeneous line: the magnitude of the reflection coefficient

$$|\rho| = |\rho_L e^{-j2\beta l}| = |\rho_L| \tag{5.8}$$

is constant along it. It follows that in the complex ρ-plane, the curve which represents the reflection coefficient of a line terminated in a given load is a circle, $|\rho| = |\rho_L|$. Now any point (u, v) in the unit circle corresponds to a point (r, x) in the right half z-plane. Suppose that the load has normalized impedance $z_L = r_L + jx_L$. This point corresponds to the point in the ρ-plane which is the intersection of the two circles generated by setting $r = r_L$ and $x = x_L$. At this point in the ρ-plane, $\rho = \rho_L$ and $l = 0$. Moving away from this point on the circle $|\rho| = |\rho_L|$ corresponds to moving away from the load, or equivalently, moving toward the generator. Since $\rho = \rho_L e^{-j2\beta l}$, increasing l (moving toward the generator) corresponds to clockwise rotation in the ρ-plane.

Suppose $R_L = 100\ \Omega$, $X_L = 100\ \Omega$, and $Z_0 = 100\ \Omega$. Then $z_L = 1 + j1.0$. This point and the corresponding circle of constant $|\rho|$ are depicted in Fig. 5.4. Progressing toward the generator, the circle of constant $|\rho|$ intersects points of different $z = r + jx$. These are the values of the input normalized impedance on the transmission line at different distances from the load. In moving through 2π radians on this circle one returns to the point z_L. What does this displacement on the Smith chart correspond to on the transmission line? To answer this question we recall that in Section 4.1 it was established that the impedance along the line is periodic with periodicity $\lambda/2$. Since one returns to the starting input impedance z_L after going around once on the constant $|\rho|$ circle, it follows that a complete turn through 2π radians on the Smith chart corresponds to displacement on the transmission line of $\lambda/2$ meters. A complete Smith chart is shown in Fig. 5.5.

---

[1] Named for P. H. Smith, Electronics, vol. 12, p. 29, 1939; vol. 17, p. 130, 1944.

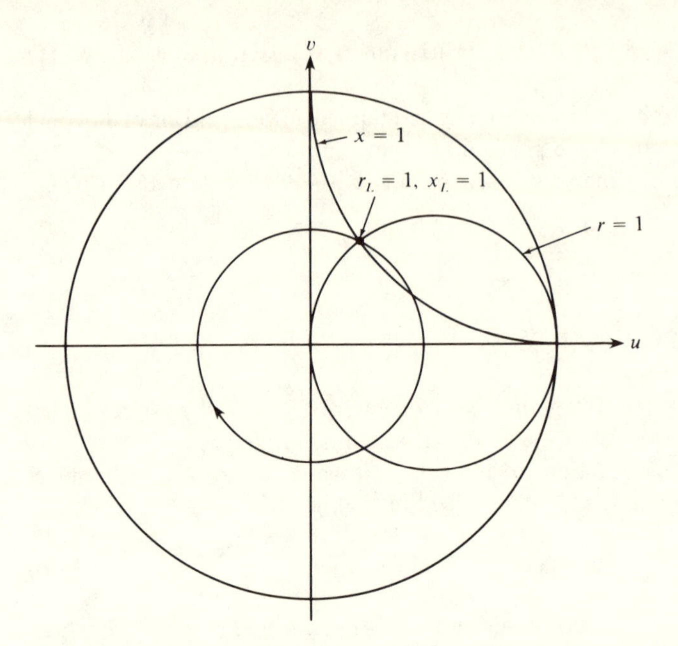

Figure 5.4 Moving toward generator on a transmission line terminated in $z_L = 1.0 + j1.0$.

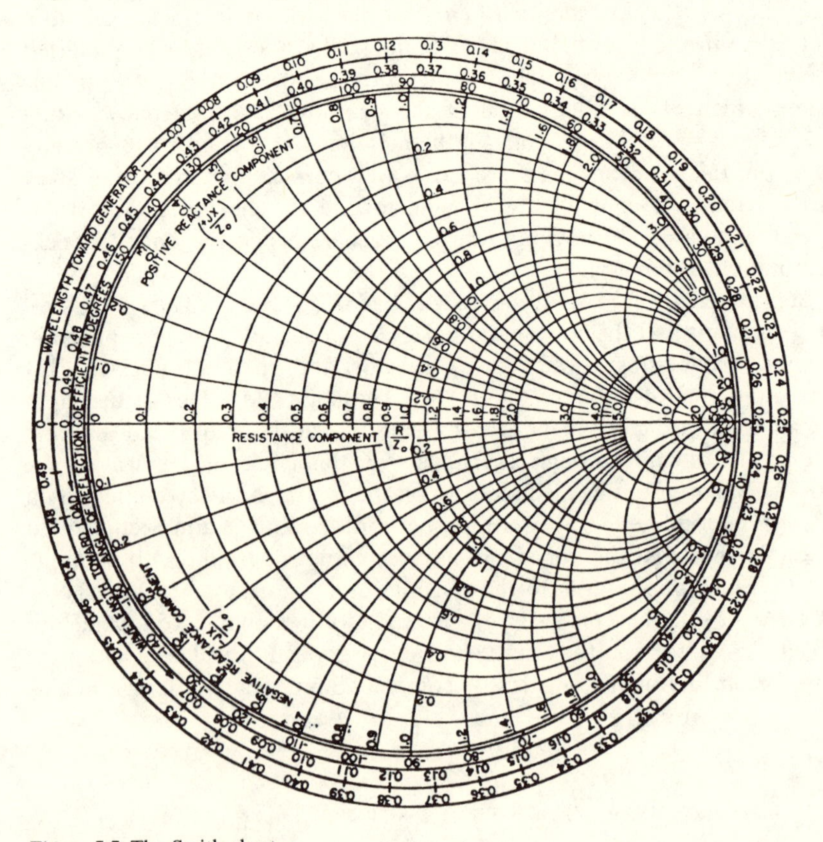

Figure 5.5 The Smith chart.

Example 5.1

A 300-Ω twin-lead transmission line with $\epsilon = 1.37\epsilon_0$ is connected to a load impedance $Z_L = 120 + j240$ at one end and 160-MHz source at the other end.

a. At what distance from the load is the impedance purely real?
b. What are the values of the impedances at these locations?

Ans. For $\epsilon = 1.37\epsilon_0$, we have $v = c/\sqrt{1.37} = 2.56 \times 10^8$ m/sec and

$$\lambda = \frac{v}{f} = \frac{2.56 \times 10^8}{160 \times 10^6} = 1.60 \text{ m}$$

At the load,

$$z_L = \frac{120}{300} + j\frac{240}{300} = 0.4 + j0.8$$

a. First we must find this point on the Smith chart (step 1) and draw a circle through it and about the origin (step 2). The outer dial on the chart registers wavelengths toward the generator. Drawing a line from the origin through the point z_L, we find that it intersects the outer dial at

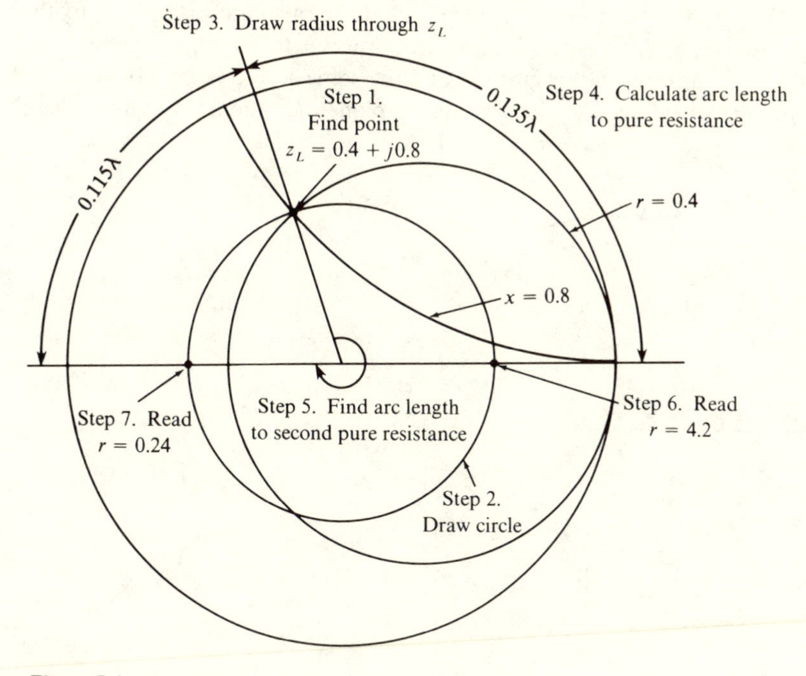

Figure 5.6

0.115λ (step 3). The impedance becomes purely resistive at 0.135λ and $0.135\lambda + (\lambda/4) = 0.385\lambda$ meters from the load (steps 4 and 5).

b. Values of resistance at these locations are read from the Smith chart (steps 6 and 7):

$$\text{at } 0.135\lambda = 0.22 \text{ m}, \qquad r = 4.2, \qquad Z = Z_0 r = 1.26 \text{ k}\Omega$$

$$\text{at } 0.385\lambda = 0.62 \text{ m}, \qquad r = 0.24, \qquad Z = Z_0 r = 72\ \Omega$$

These operations on the Smith chart are shown in Fig. 5.6.

5.3 EXTREME CURRENT AND VOLTAGE POINTS

Taking the absolute magnitudes of both sides of (4.16) gives

$$|V| = |V_+|\,|1 + \rho| \tag{5.9}$$

In like manner, from (4.22) we obtain

$$|I| = \frac{|V_+|}{Z_0}|1 - \rho| \tag{5.10}$$

Recall that we have set (5.2)

$$\rho = \rho_L e^{-j2\beta l}$$

From (5.9) and (5.10) we may conclude that the absolute magnitude of the voltage at any point on the line is proportional to $|1 + \rho|$, whereas the absolute magnitude of the current at that same point is proportional to $|1 - \rho|$. When referred to the complex ρ-plane, i.e., the Smith chart, these two lengths are as shown in Fig. 5.7. It is quite clear that the maximum

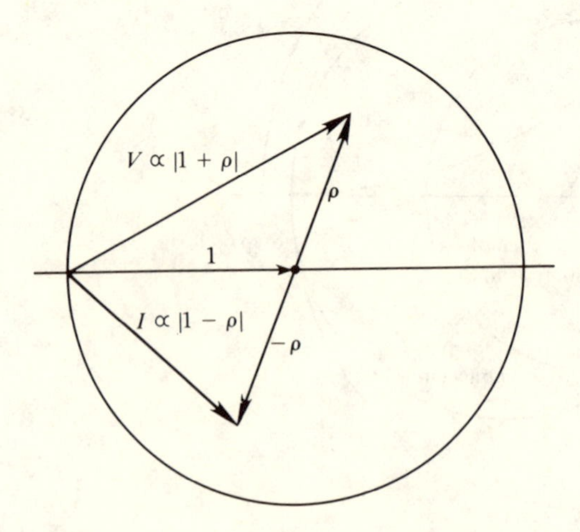

Figure 5.7 Current and voltage on the Smith chart.

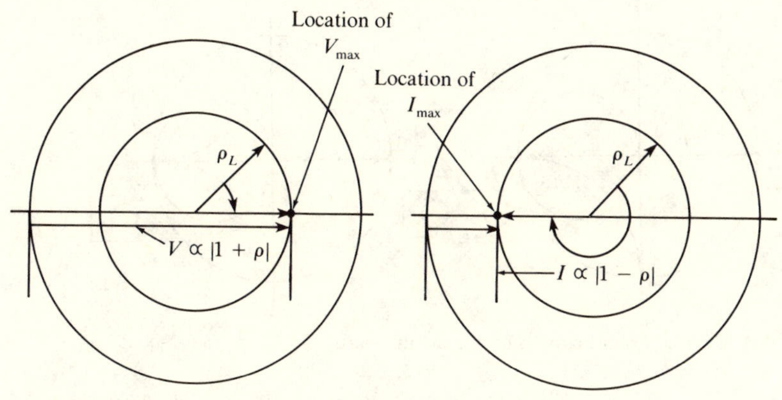

Figure 5.8 Voltage and current maxima on the Smith chart.

voltage occurs when ρ lies on the positive real axis. The maximum current occurs when ρ lies on the negative real axis. These points are shown in Fig. 5.8.

The Smith chart may equally be employed to obtain the admittance along a line. From (5.5) we see that

$$z(\rho) = \frac{Z(l)}{Z_0} = \frac{1 + \rho}{1 - \rho}$$

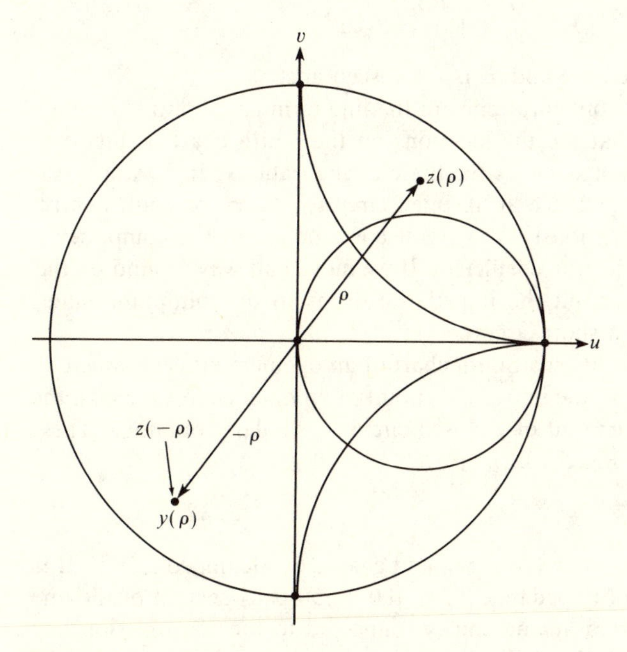

Figure 5.9 Impedance and admittance on the Smith chart.

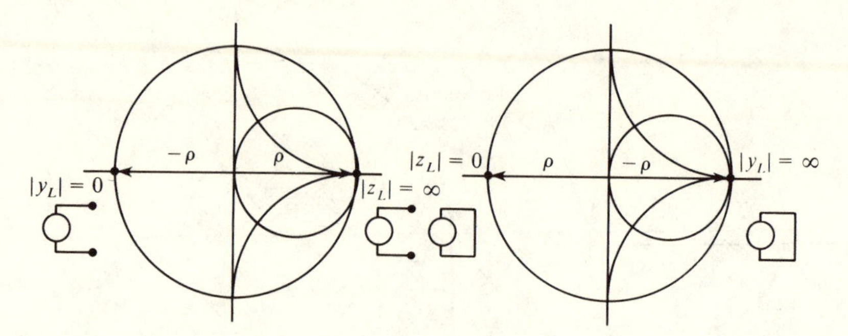

Figure 5.10 Open and short circuits on the Smith Chart.

The normalized admittance is then given by

$$y(\rho) = \frac{Y(l)}{Y_0} = \frac{Z_0}{Z(l)} = \frac{1-\rho}{1+\rho} = z(-\rho) \tag{5.11}$$

It follows that

$$y(\rho) = z(-\rho) \tag{5.12}$$

At a given value of the complex vector ρ with corresponding impedance z, the admittance is obtained by reflecting the vector ρ through the origin. (See Fig. 5.9.) The real and imaginary values of admittance are

$$y = \frac{G}{Y_0} + j\frac{B}{Y_0} \equiv g + jb$$

where G is the conductance and B is the susceptance.

Two fundamental configurations are the line terminated in a short and in an open circuit. What are the locations on the Smith chart of the load impedance and admittance values for these configurations? In Example 4.1 one found that the impedance at an open circuit is $-j\infty$, or equivalently, $x = -\infty$. As found previously, this point corresponds to the components $u = 1$, $v = 0$ of the reflection coefficient. If we move half way around on the Smith chart from this point the impedance drops to the minimum value, zero corresponding to a short circuit.

The admittance point on a Smith chart of an open circuit is obtained by reflecting the ρ-vector to the impedance point of an open circuit through the origin. The admittance point of a closed circuit is similarly obtained. These values are depicted in Fig. 5.10.

Example 5.2

A transmission line with $\epsilon = 1.5\epsilon_0$ and characteristic impedance 50 Ω is terminated in a load of impedance $Z_L = 100 + j50$ Ω. A section of the line 0.30 m long and shorted at one end is connected to the transmission line 1.53 m from the load. At 100 MHz, what is the input impedance at the stub?

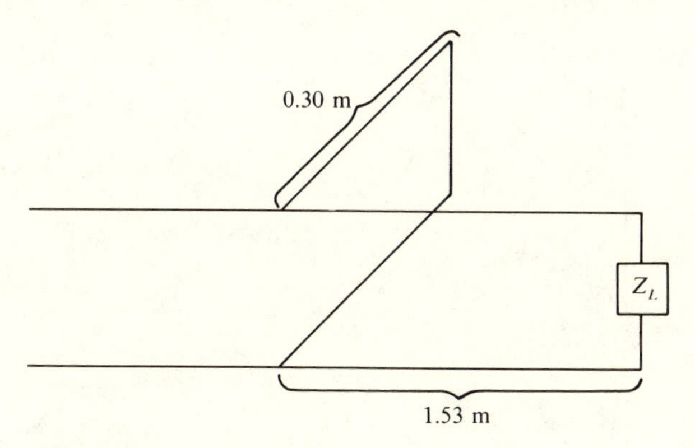

Figure 5.11

Ans. The configuration is depicted in Fig. 5.11. The wavelength of the waves is

$$\lambda = \frac{v}{f} = \frac{3}{\sqrt{1.5}} = 2.45 \text{ m}$$

The length in wavelengths of the distance to the stub is

$$l_1 = \frac{1.53}{2.45} = 0.625 = 0.5 + 0.125$$

and that of the stub is

$$l_2 = \frac{0.30}{2.45} = 0.123$$

The normalized impedance of the load is

$$z_L = \frac{100 + j50}{50} = 2 + j1$$

Because the stub is connected in parallel to the line, it will prove convenient to work with admittance. Reflecting the ρ-vector to z_L through the origin (step 1) gives the load admittance

$$y_L = 0.40 - j0.20$$

Moving toward the generator 0.125 wavelengths (step 2) from this point, we read the admittance

$$y(l_1) = 0.50 + j0.51$$

Next we find the admittance of the stub looking into it at the open end. Moving away from the point $|y_L| = \infty$ through $l_2 = 0.123$ wavelengths, we find (step 3)

$$y_{\text{st}}(l_2) = -j1.01$$

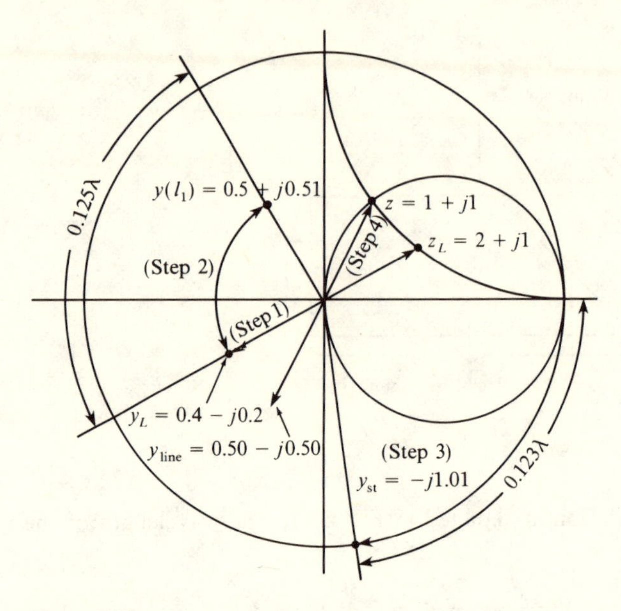

Figure 5.12

Adding this value to the admittance $y(l_1)$, we find

$$y_{\text{line}} = y(l_1) + y_{\text{st}}(l_2) = 0.50 - j0.50$$

Locating this point on the Smith chart and reflecting through the origin (step 4), we find the normalized impedance

$$z = 1.0 + j1.0$$

This value corresponds to the impedance

$$Z = 50 + j50\ \Omega$$

These operations are shown in Fig. 5.12.

5.4 VOLTAGE STANDING-WAVE RATIO

As described in Section 4.3, the standing-wave ratio is directly related to the value of impedance where it is purely resistive. If this value of the impedance is R, then as listed in Table 4.2, the standing-wave ratio is

$$S = \begin{cases} R/Z_0 & \text{if} \quad R > Z_0 \\ Z_0/R & \text{if} \quad R < Z_0 \end{cases}$$

In either case, the purely real impedance values are R and Z_0^2/R and are separated by $\lambda/4$ meters. When normalized, these impedances become R/Z_0 and Z_0/R, respectively, and are seen to be equal to the standing-wave

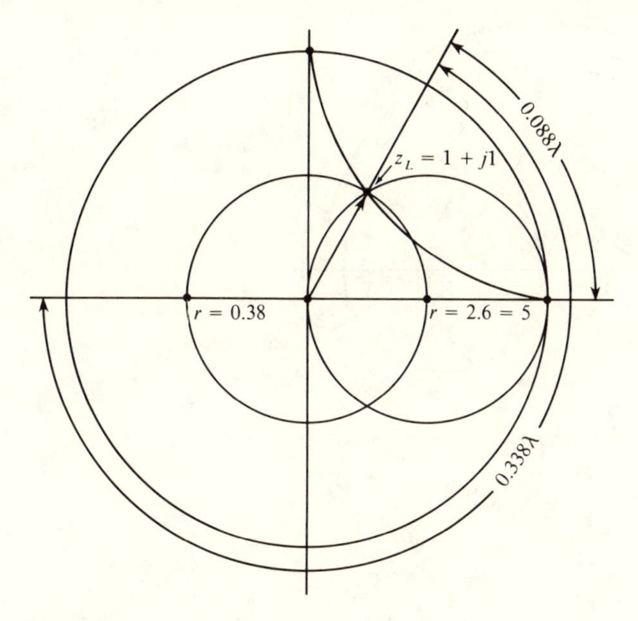

Figure 5.13

ratio. On the Smith chart such purely resistive values occur on the u-axis (i.e., $v = 0$).

So the Smith chart affords a simple technique for obtaining the standing-wave ratio and the locations along the transmission line where the impedance is real and the voltage and currents have their extreme values.

For example, consider the case that $z_L = 1.0 + j1.0$. Drawing the circle through this point on the Smith chart indicates that the purely resistive value of normalized impedance are $r = 2.6$ and $r = 0.38$. Since $S > 1$, the standing-wave ratio for the line is $S = 2.6$. Maximum voltage occurs 0.088λ meters from the load, and minimum voltage occurs 0.338λ meters from the load. These operations are shown in Fig. 5.13.

Example 5.3

For the configuration described in Example 5.1:

a. What is the standing-wave ratio for the line?
b. If it is known that $Z_0 > R$, what are the values of $Z_{\max}$, $Z_{\min}$, and R?
c. At which locations on the line is the voltage maximum and minimum, respectively?

Ans.

a. From the Smith chart we find the purely resistive values $r = 4.2$ and $r = 0.24$. Since $S > 1$, it follows that $S = 4.2$.

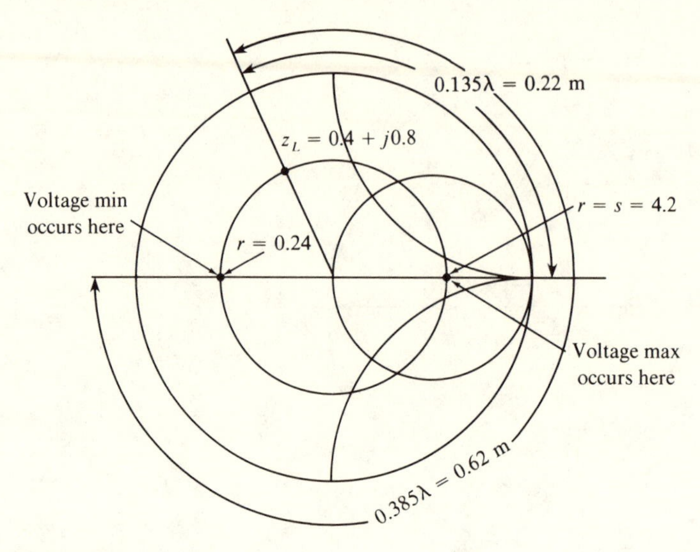

Figure 5.14

b. The purely resistive values of Z are $Z_{max} = 4.2Z_0 = 1.26$ kΩ and $Z_{min} = R = 0.24Z_0 = 72$ Ω.

c. Voltage maxima and minima occur where Z is purely resistive. Thus, the voltage maximum occurs 0.22 meters from the load and voltage minimum occurs 0.62 meters from the load. These operations on the Smith chart are shown in Fig. 5.14.

5.5 THE SINGLE-STUB TUNER

In *tuning* a transmission line, one matches the line to the load. When this condition is met, there is no reflected wave and $\rho = 0$. So the problem of tuning, when referred to the Smith chart, is concerned with reaching the origin of the chart. How is this accomplished?

To answer this question we note that at the origin of the Smith chart, $r = 1$ and $x = 0$. We note also that $g = G/Y_0 = 1$ and $b = B/Y_0 = 0$ at the origin. So tuning will be accomplished if we find a point on the line where $y = 1 + jb$ and then simply connect a shunt inductance $-jb$ [see Eq. (1.14)] across the line at that point (see Fig. 5.15). The locus of points on the Smith chart corresponding to $y = 1 + jb$ is the circle $g = r = 1$.

In shunt tuning the purely reactive element is effected by a length of transmission line which is either open or shorted at one end. (See Fig. 5.16.) Consider that a transmission line is terminated in the load z_L. At what location can a purely reactive shunt element be placed on the line to match

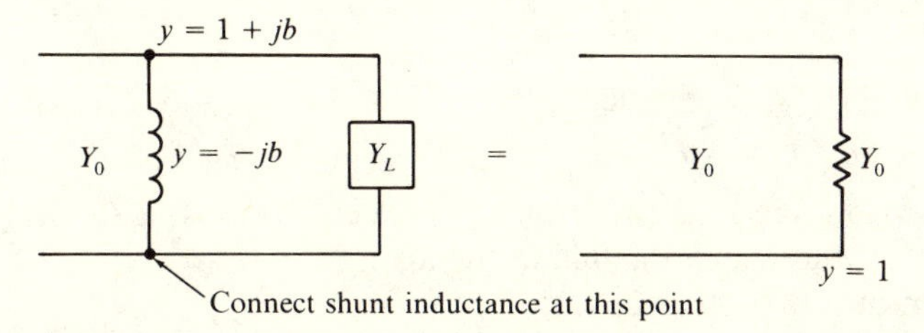

Figure 5.15 Matching with a shunt inductance.

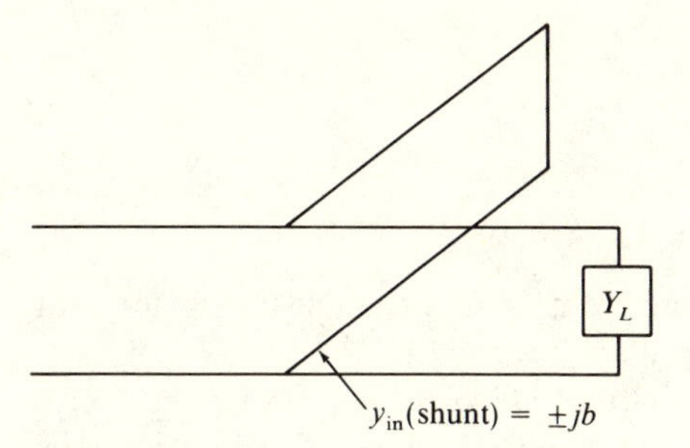

Figure 5.16

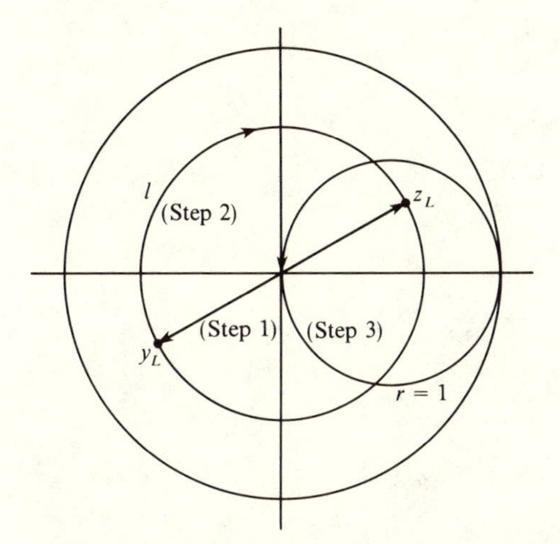

Figure 5.17 Steps in evaluating jb and l for the single-stub tuner on the Smith chart.

the line to the load? First we must find the load admittance. This value is determined by reflecting $\rho(z_L)$ through the origin (step 1). Then we draw a circle through y_L and observe where it intersects the $r = 1$ circle (step 2). At this point of intersection we read off the value of susceptance jb (step 3). If a shunt element with susceptance $-jb$ is placed l meters from the load, the line will be matched. (See Fig. 5.17.)

Example 5.4

A transmission line with characteristic impedance $Z_0 = 50\ \Omega$ is terminated in a load $Z_L = 100 + j45\ \Omega$.

a. At what distance from the load (in terms of wavelength) must a shunt susceptance be placed to match the line to the load?
b. What is the magnitude of the susceptance?
c. If a short-circuited piece of the same 50-Ω line is to be used as the shunt, how long should it be?

Ans.

a. From the Smith chart we find y_L (step 1)

$$Z_L = 2 + j0.90$$

$$y_L = 0.42 - j0.19$$

Drawing a circle through this point, we note that there is a displacement of 0.196λ to the $r = 1$ circle (step 2).

b. At the intersection of these two circles we read $b = 0.95$ (step 3). It follows that a shunt susceptance $y = -j0.95$ placed 0.196λ from the load will match the line to the load.

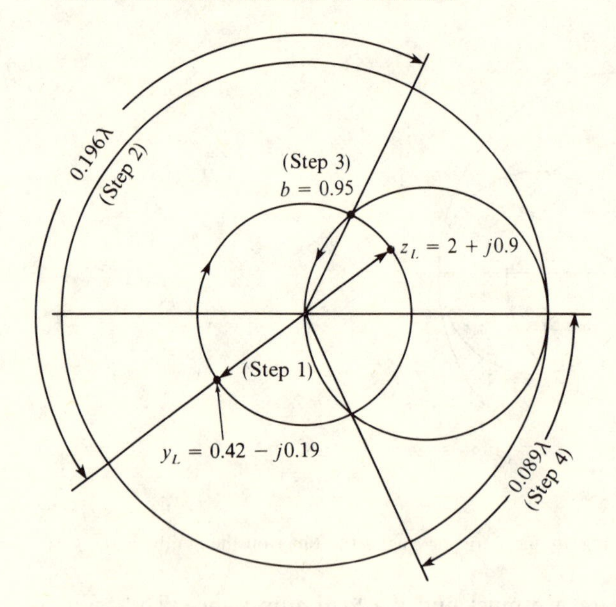

Figure 5.18 Steps in Example 5.4.

c. Starting at the $|y_L| = \infty$ point at the extreme right of the Smith chart, we note that a section of short-circuited line 0.089λ meters long (step 4) will have an input admittance of $-j0.95$. (See Fig. 5.18.)

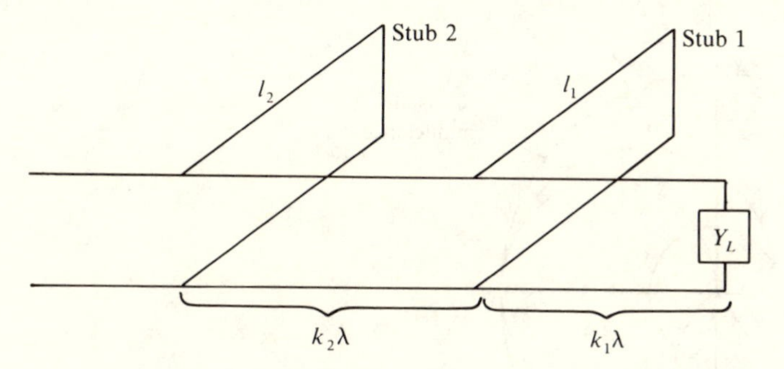

Figure 5.19 The double-stub tuner.

5.6 THE DOUBLE-STUB TUNER

In one variation of a double-stub tuner, two short-circuited shunt elements are attached to the line at different distances from the load, as depicted in Fig. 5.19. The stub lengths are then adjusted until matching is attained. The values of l_1 and l_2 may be found with the aid of the Smith chart. Consider the $r = 1$ circle. Let us call this circle C. The *conjugate* circle to C, which we shall label $\tilde{C}$, is displaced from the circle C through the arc length $k_2\lambda$ as depicted in Fig. 5.20. Any point $\tilde{P}$ on $\tilde{C}$ is displaced from its starting location P through the arc displacement $k_2\lambda$ toward the generator. This construction indicates the function of the two stubs. Namely, stub 1 serves to bring the line admittance to a value on the circle $\tilde{C}$. Suppose for example, the load admittance is y_L. Moving through the displacement $k_1\lambda$ brings the

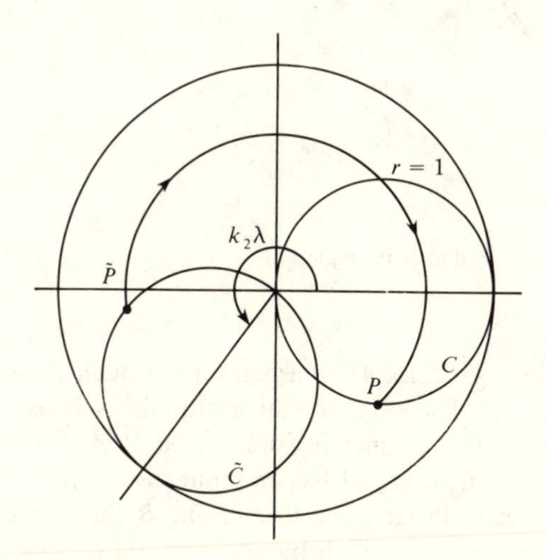

Figure 5.20

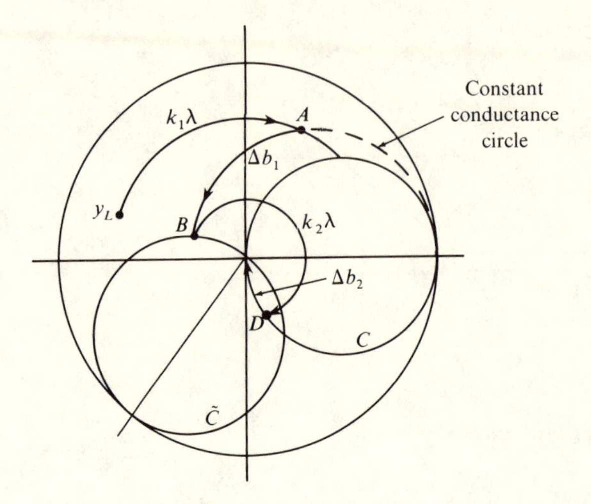

Figure 5.21a Elements of the double-stub tuner on the Smith chart.

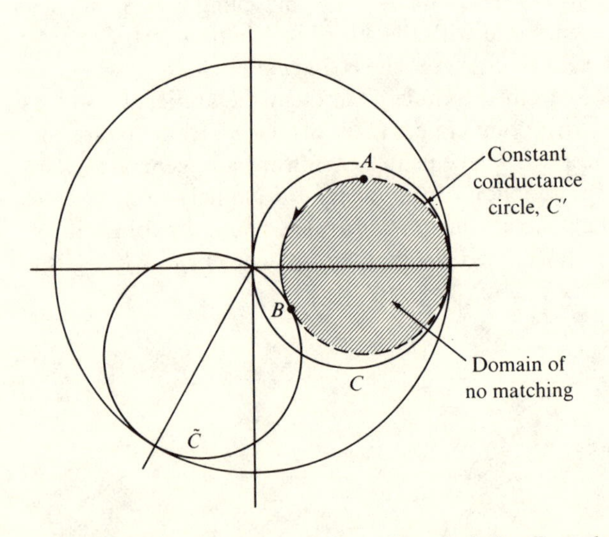

Figure 5.21b Domain of no matching. (Described after Example 5.5).

admittance to point A as shown in Fig. 5.21a. The dashed line through the point A is a constant-conductance line. The effect of stub 1 at point A is to carry the admittance along the constant-conductance line through A. We wish to reach the point B on the circle $\tilde{C}$. If stub 1 has an input susceptance $j\Delta b_1$, then this displacement is effected. Progressing from point B through the displacement $k_2\lambda$ brings us to point D which by construction of the conjugate circle, lies on the $r = 1$ circle, C. Once at D, a susceptance $j\Delta b_2$

at the input of stub 2 effects matching. This procedure is detailed in the following example.

Example 5.5

Consider again the configuration of Example 5.4. However, now let there be two parallel connected 50-Ω short-circuited lines of adjustable length located $\lambda/8$ and $\lambda/2$ from the load. What lengths must the stubs be to effect matching?

Ans. The configuration is shown in Fig. 5.22. So the circles C and $\tilde{C}$ are separated by $\frac{3}{8}\lambda$. (See Fig. 5.23.) Moving toward the generator away from y_L to the location of the first stub brings us to point A. (See Fig. 5.24.)

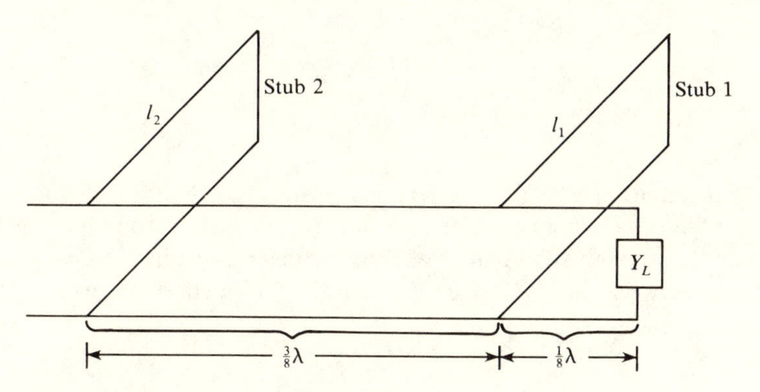

Figure 5.22 Parameters for Example 5.5.

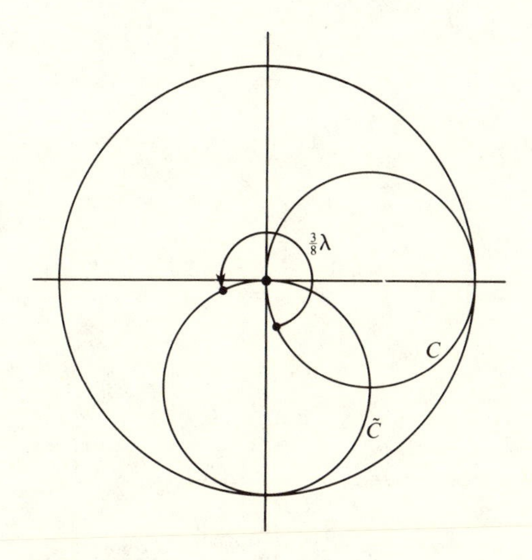

Figure 5.23

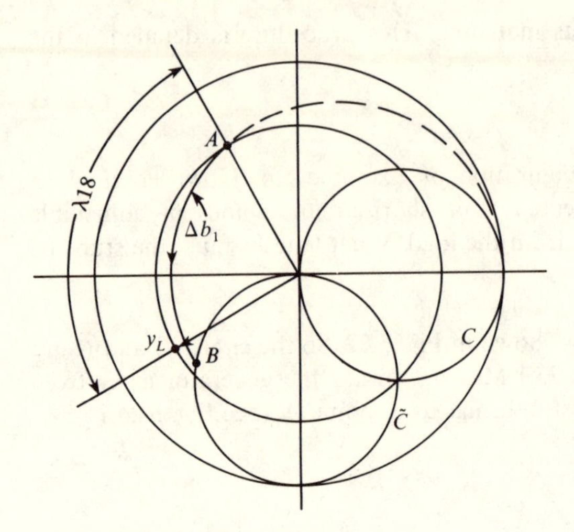

Figure 5.24

The input admittance of this stub 1 must bring the admittance of the line to point B on $\tilde{C}$. Points A and B lie on a constant-conductance ($g = 0.51$) line. The input susceptance of stub 1 is determined by the change in susceptance between points B and A. From the Smith chart we read

$$\Delta b_1 = -0.12 - (0.51) = -0.63$$

It follows that an input susceptance of stub 1 equal to $-j0.63$ will bring the line admittance to point B.

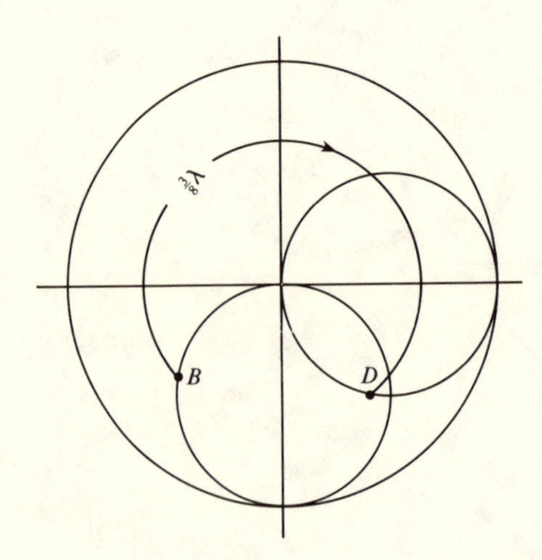

Figure 5.25

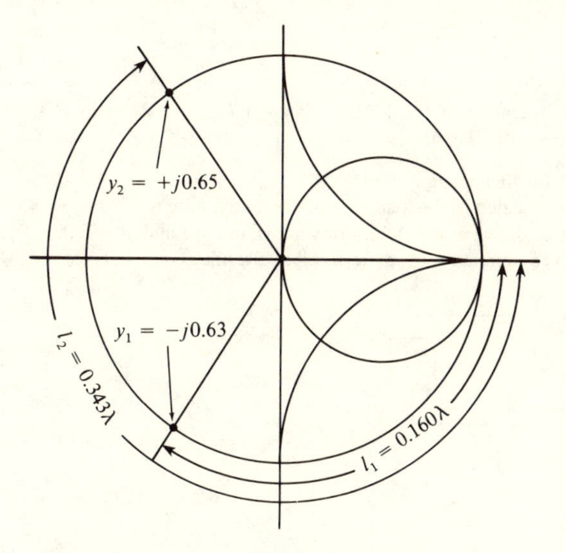

Figure 5.26 Lengths of stubs for Example 5.5.

Moving through the arc $\frac{3}{8}\lambda$ from point B brings the admittance to point D on the circle C. (See Fig. 5.25.) From the Smith chart we see that the susceptance at D is $-j0.65$. So a susceptance of $+j0.65$ at the input of stub 2 will cause the line to be matched to the load.

The lengths of the stubs are read from the Smith chart. (See Fig. 5.26.) In summary, for stub 1:

$$y_1 = -j0.63, \qquad Y_1 = -j31.5\ \Omega^{-1}, \qquad l_1 = 0.160\lambda$$

For stub 2:

$$y_2 = +j0.65, \qquad Y_2 = +j32.5\ \Omega^{-1}, \qquad l_2 = 0.343\lambda$$

While the double-stub tuner is an effective impedance matching device, it should be noted that not all loads can be matched for a given stub spacing. As can be seen from an inspection of Figure 5.21b, the loads within the circle C' cannot be matched since none of the corresponding constant conductance circles intersect circle $\hat{C}$. Clearly, the limiting conductance circle for matching is circle C' which is tangent to circle $\tilde{C}$. Any point on this circle, e.g., point A, can be transformed to the point of tangency B, on circle $\tilde{C}$ and, as discussed above, matching can then be achieved.

Application of the Smith chart is again encountered in Chapter 7, where the theory of waveguides is developed.

PROBLEMS

5.1. A twin-lead transmission line operating at 180 MHz with $\epsilon = 1.2\epsilon_0$ and $Z_0 = 75\ \Omega$ is terminated in the complex load $Z_L = 50 + j30\ \Omega$.

a. What is the standing wave ratio for this configuration?
b. A section of line l meters long connected to the load is removed, and a pure resistance R is connected to the line in its place as shown in Fig. 5.27. For what values of l and R will there be no change in the current-voltage standing-wave patterns along the line?

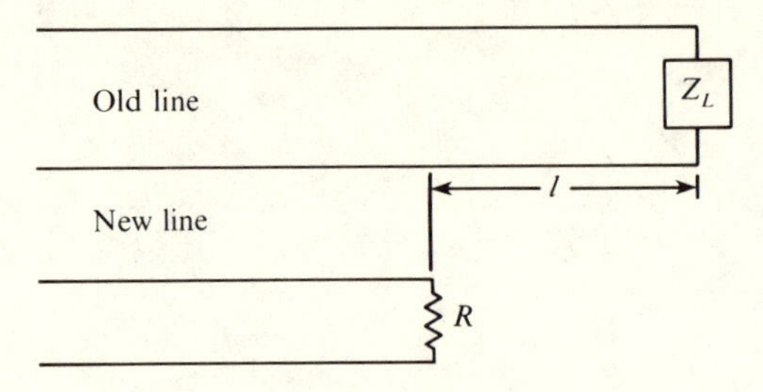

Figure 5.27 Configuration for Problem 5.1.

5.2. A 40-Ω transmission line with $\epsilon = 1.3\epsilon_0$ operating at 30 MHz is terminated in a short circuit.

a. Construct an elementary RC network which when placed across the line 0.78 meters from the load will effect matching.
b. How do the values of R and C which effect matching change as the connection point is moved further toward the generator?

5.3. The load of a 50-Ω line consists of a 30-Ω resistance shunted across one end. The other end is open. The section is $\lambda/16$ meters long. The line is matched to this load through a double-stub tuner. The first stub is placed $\lambda/16$ meters from the load, and the second stub is placed $3\lambda/8$ meters toward the generator from the first stub. The configuration is shown in Fig. 5.28. What are l_1 and l_2 in terms of the wavelength λ?

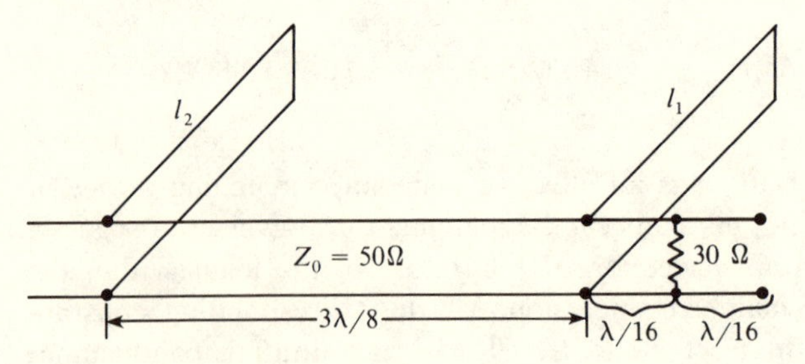

Figure 5.28 Configuration for Problem 5.3.

5.4. Show that the two purely resistive impedance values that lie on the same circle about the origin of a Smith chart obey the relation

$$r_1 r_2 = 1$$

Hint: Recall (4.33).

5.5. Power is led to a 300-Ω load from a generator oscillating at 1.5 MHz through an 80-Ω line. Using properties of the Smith chart, obtain the characteristic impedance $\tilde{Z}_0$, and length l, of a section of line which will match the 80-Ω line to the 300-Ω load. The propagation speed in the matching section at the given frequency is $v = 0.45c$. (See Fig. 5.29.) Hint: For matching to occur, the purely resistive normalized impedance at A must have the value $z = 80/\tilde{Z}_0$. Recall also Problem 5.4.

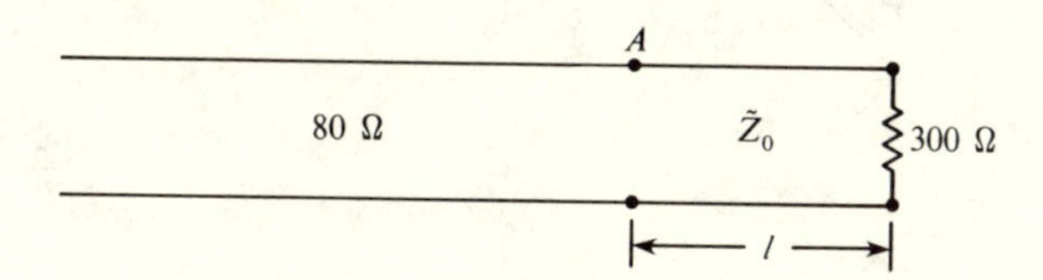

Figure 5.29 Configuration for Problem 5.5.

5.6. Student A is given a box of resistors, student B is given a box of capacitors, and student D is given a box of inductors. Each is told to match a 100-Ω line with $\epsilon = 1.3\epsilon_0$ to a purely resistive 50-Ω load at 1.7 MHz using only one element from his collection. What are the values R_A, C_B, L_D which must be used, and at what distances from the load and in what manner must these elements be inserted in the line?

5.7. The output voltage of a receiver at 15 MHz is matched to a 150-Ω air dielectric transmission line. This line is to be connected to a 300-Ω air-dielectric transmission line of length 30 m which is terminated in a receiver to which it is matched. (See Fig. 5.30.) Describe quantitatively how this matching may be accomplished with:

a. A pure resistance.
b. A quarter-wave transformer.
c. A stub tuner.

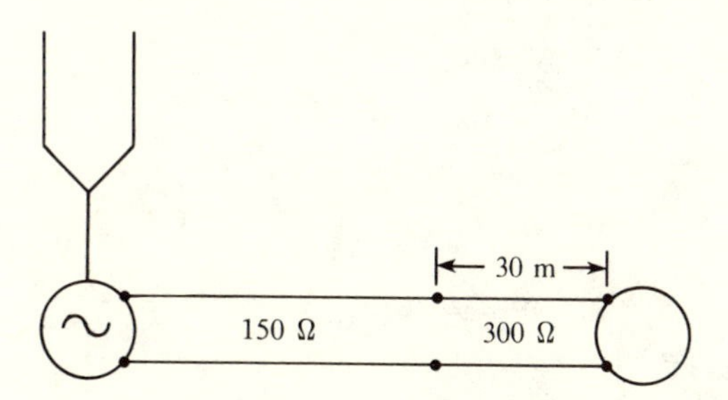

Figure 5.30 Configuration for Problem 5.7.

5.8. A generator supplies power to two receiving stations L_1 and L_2 as shown in Fig. 5.31.

a. Given that $Z_0 = 100\ \Omega$, $Z_{L_1} = 100 + j200$, and $Z_{L_2} = 300 + j100$, obtain the impedance Z_{P_2} at P_2 looking into the load L_2.
b. What value of the generator impedance will insure maximum power delivery to the loads?

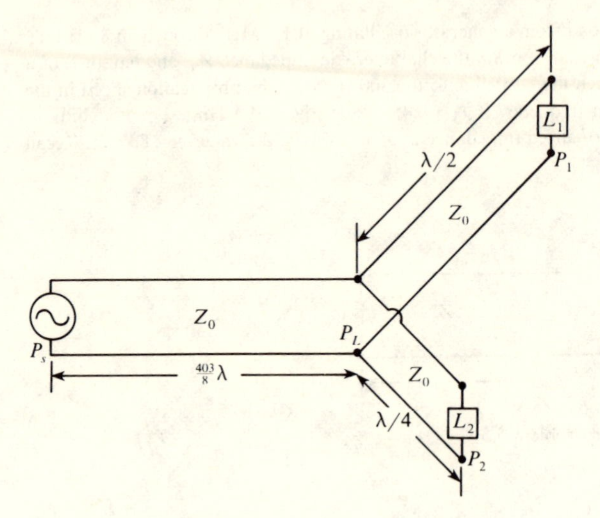

Figure 5.31 Configuration for Problem 5.8.

5.9. The transmission line shown in Fig. 5.32a has $Z_0 = 50\ \Omega$. Waves on the line have speed $v = 2 \times 10^8$ m/sec. Operating at 50 MHz, as the length of the stub is varied, the locus of $z(d)$ describes a circle as shown in Fig. 5.32b. If the length of the line between source and load is $d = 5.0$ m, find the values of the load resistances R_A and R_B.

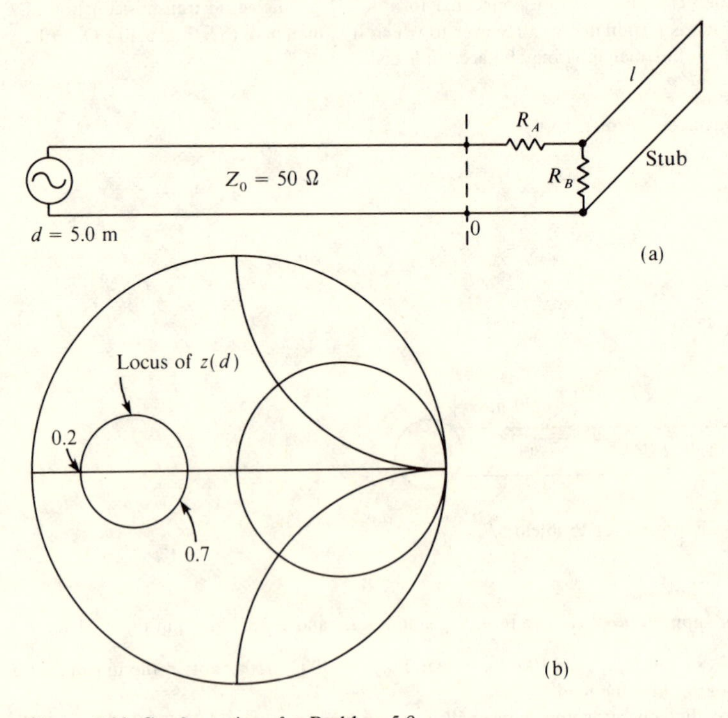

Figure 5.32 Configurations for Problem 5.9.

PART TWO

ELECTROMAGNETIC WAVES

CHAPTER

SIX

ELEMENTS OF STATICS AND MAXWELL'S EQUATIONS

6.1 MATHEMATICAL PRELIMINARIES

In discussing electromagnetic theory it proves convenient if the student has a working familiarity with some basic notions of vector calculus. This section serves as a short review of these concepts.[1]

Vector Fields and the Divergence

A vector field is a set of three scalar functions. The values of these functions define a vector at any point in space. The electric field due to a point charge at the origin is the field

$$E_x = K\frac{qx}{\left(x^2+y^2+z^2\right)^{3/2}}$$

$$E_y = K\frac{qy}{\left(x^2+y^2+z^2\right)^{3/2}}$$

$$E_z = K\frac{qz}{\left(x^2+y^2+z^2\right)^{3/2}} \tag{6.1}$$

where K is the constant

$$K = \frac{1}{4\pi\epsilon_0} = 8.89 \times 10^9 \text{ m/F} \simeq 9 \times 10^9 \text{ m/F}$$

[1]Additional formulas of vector calculus may be found in Appendices A and B.

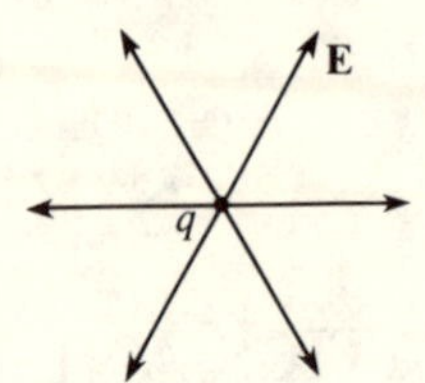

Figure 6.1 Vector field of a point charge.

and ϵ_0 is the permittivity of free space. (See Fig. 6.1.) The electric vector corresponding to the field (6.1), at any point (x, y, z), is

$$\mathbf{E} = \mathbf{a}_x E_x + \mathbf{a}_y E_y + \mathbf{a}_z E_z$$
$$= K\frac{q(\mathbf{a}_x x + \mathbf{a}_y y + \mathbf{a}_z z)}{(x^2 + y^2 + z^2)^{3/2}}$$

where vectors $(\mathbf{a}_x, \mathbf{a}_y, \mathbf{a}_z)$ are parallel to the three cartesian axes. Introducing the radius vector $\mathbf{r}$ from the origin to the point (x, y, z)

$$\mathbf{r} = \mathbf{a}_x x + \mathbf{a}_y y + \mathbf{a}_z z$$

permits the last equation to be written

$$\mathbf{E} = K\frac{q\mathbf{r}}{r^3} = K\frac{q\hat{\mathbf{r}}}{r^2} \qquad (\hat{\mathbf{r}} \equiv \mathbf{r}/r) \tag{6.2}$$

Eq. (6.2) is more easily recognized to be the Coulomb field due to a point charge.

The *divergence* is an operation on a vector field which yields a scalar function. The divergence of a field $\mathbf{F}$ in Cartesian coordinates is written

$$\operatorname{div}\mathbf{F} = \frac{\partial}{\partial x}F_x + \frac{\partial}{\partial y}F_y + \frac{\partial}{\partial z}F_z \tag{6.3}$$

This operation may be conveniently written in terms of the del operator, ∇, which has components

$$\nabla = \mathbf{a}_x\frac{\partial}{\partial x} + \mathbf{a}_y\frac{\partial}{\partial y} + \mathbf{a}_z\frac{\partial}{\partial z}$$

So we may write

$$\operatorname{div}\mathbf{F} = \nabla\cdot\mathbf{F}$$

For example, consider the divergence of the Coulomb field (6.1):

$$\nabla\cdot\mathbf{E} = kq\left[\left(\frac{1}{r^3} - \frac{3x^2}{r^5}\right) + \left(\frac{1}{r^3} - \frac{3y^2}{r^5}\right) + \left(\frac{1}{r^3} - \frac{3z^2}{r^5}\right)\right]$$
$$= kq\left(\frac{3}{r^3} - \frac{3}{r^3}\right) = 0$$

So for the Coulomb field of a point charge at the origin we find $\nabla\cdot\mathbf{E} = 0$ for all $\mathbf{r} \neq 0$.

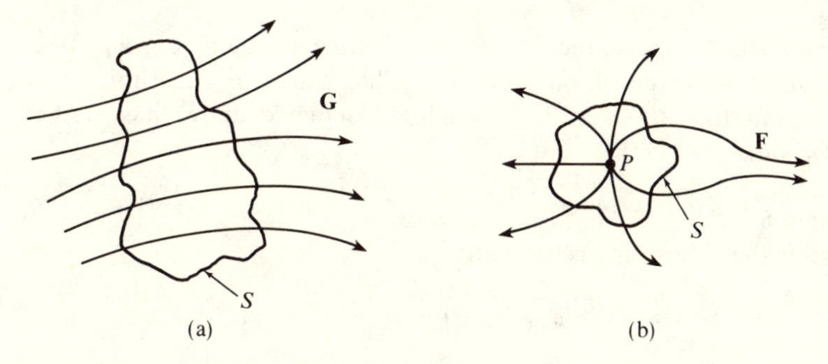

Figure 6.2 Vector fields passing through a closed surface.

Gauss's Theorem. Sources and Sinks

Gauss's theorem states that if a closed surface encloses the volume V, then

$$\int \operatorname{div} \mathbf{F} \, d\mathbf{r} = \oint \mathbf{F} \cdot d\mathbf{S} \tag{6.4}$$

Here we are writing $d\mathbf{r}$ for the differential of volume, $dx\,dy\,dz$. The right-hand side of the above equality is a surface integral which involves the inner (or "dot") product between the vector $\mathbf{F}$ and the surface element $d\mathbf{S}$ at a specific point on the surface. (See Example 6.2.)

If the volume of integration in (6.4) is shrunk to zero about a given point, then by the mean-value theorem we may write

$$\lim_{V \to 0} \int \operatorname{div} \mathbf{F} \, d\mathbf{r} = \lim_{V \to 0} \operatorname{div} \mathbf{F} \int d\mathbf{r}$$

and from (6.4) we obtain

$$\operatorname{div} \mathbf{F} = \lim_{V \to 0} \frac{1}{V} \oint \mathbf{F} \cdot d\mathbf{S} \tag{6.5}$$

The surface integral on the right-hand side of this equation represents the net flux of $\mathbf{E}$ field lines[2] which emanate from the closed surface. Consider the two cases shown in Fig. 6.2. In case (a), $\oint \mathbf{G} \cdot d\mathbf{S} = 0$, whereas in case (b), $\oint \mathbf{F} \cdot d\mathbf{S} > 0$. From our integral expression for the divergence (6.5) we may conclude that for case (a), $\nabla \cdot \mathbf{G} = 0$, whereas for case (b), $\nabla \cdot \mathbf{F} \neq 0$ at P.

The physical significance of the divergence of a field is that its value at any point indicates whether or not the field has a *source* at the point. A field has a source at a point if there is a net efflux of field lines from that point.

[2]A field line at any point in space is tangent to the field at that point. The flux of field lines (lines/m^2) is proportional to the magnitude of the field.

For the fields shown above, the field **F** has a source at P and $\nabla \cdot \mathbf{F} > 0$. If the **F** field lines are reversed, one says that **F** has a *sink* at P. In this event one would find that $\nabla \cdot \mathbf{F} < 0$. The field **G** depicted above has neither sources nor sinks.

Example 6.1

Consider the following vector field:

$$\mathbf{F} = \mathbf{a}_x xy + \mathbf{a}_y\left(\frac{-y^2}{2}\right) + \mathbf{a}_z\left(\frac{b^4}{x^2 + y^2}\right)$$

where b is a constant.

a. What is the divergence of this field?
b. What is the value of the integral

$$\oint \mathbf{F} \cdot d\mathbf{S}$$

where the surface of integration is a sphere of radius r about the origin?
c. Does this field have any sources or sinks?

Ans.

a.

$$\nabla \cdot \mathbf{F} = \frac{\partial}{\partial x}F_x + \frac{\partial}{\partial y}F_y + \frac{\partial}{\partial z}F_z = y - y + 0 = 0$$

So **F** is divergenceless at all points in space.
b. By (6.4)

$$\oint \mathbf{F} \cdot d\mathbf{S} = \int \operatorname{div} \mathbf{F}\, d\mathbf{r} = 0$$

So *any* closed surface integral of **F** vanishes.
c. Since $\nabla \cdot \mathbf{F} = 0$ at all points in space, **F** has neither sources nor sinks.

Example 6.2

To show that the divergence of a constant field is zero, illustrate that the surface integral of the field about the origin is zero.

Ans. The configuration is shown in Fig. 6.3. At the location on the surface where **dS** is shown we see that

$$\mathbf{F} \cdot d\mathbf{S} = F\cos\theta\, dS$$

The differential element of surface in spherical coordinates is given by

$$dS = r^2\, d\Omega = r^2 d\phi \sin\theta\, d\theta$$

The *solid angle* subtended by dS about the origin is $d\Omega$. We may therefore write

$$\mathbf{F} \cdot d\mathbf{S} = Fr^2 \cos\theta\, d\phi \sin\theta\, d\theta$$

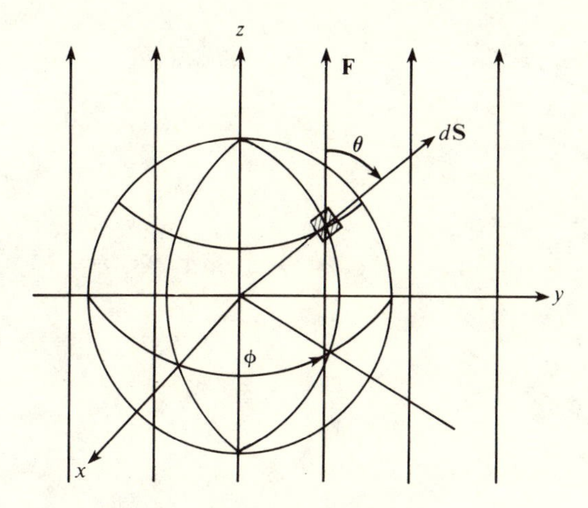

Figure 6.3 Configuration for Example 6.2.

Integrating over the surface of the sphere gives

$$\oint \mathbf{F} \cdot d\mathbf{S} = Fr^2 2\pi \int_0^{2\pi} \cos\theta \sin\theta \, d\theta = 0$$

It follows by (6.5) that $\nabla \cdot \mathbf{F} = 0$ about the origin. But for a constant force field any point in space may serve as an origin. So $\nabla \cdot \mathbf{F} = 0$ everywhere.

Stokes's Theorem and the Curl

The Cartesian components of the curl of a vector field F are given by the determinant

$$\operatorname{curl} \mathbf{F} = \begin{vmatrix} a_x & a_y & a_z \\ \dfrac{\partial}{\partial x} & \dfrac{\partial}{\partial y} & \dfrac{\partial}{\partial z} \\ F_x & F_y & F_z \end{vmatrix} \tag{6.6}$$

It follows that

$$(\operatorname{curl} F)_x = \frac{\partial}{\partial y} F_z - \frac{\partial}{\partial z} F_y$$

$$(\operatorname{curl} F)_y = \frac{\partial}{\partial z} F_x - \frac{\partial}{\partial x} F_z$$

$$(\operatorname{curl} F)_z = \frac{\partial}{\partial x} F_y - \frac{\partial}{\partial y} F_x \tag{6.7}$$

In terms of the del operator we may write

$$\operatorname{curl} \mathbf{F} = \nabla \times \mathbf{F}$$

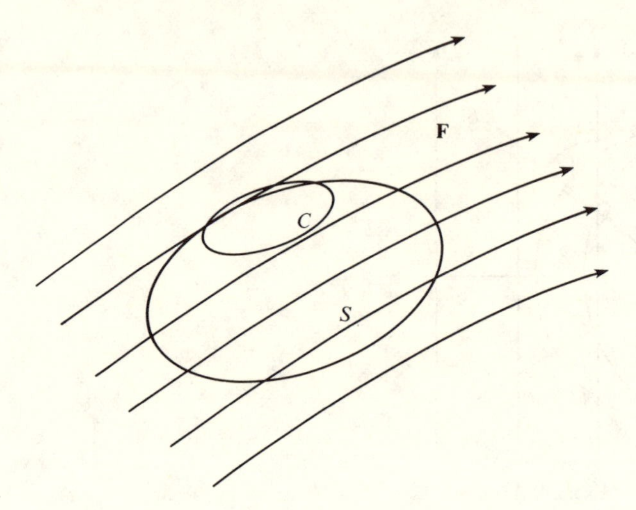

Figure 6.4 The open surface S and closed curve C.

The curl of the Coulomb field (6.1) has components

$$(\text{curl}\,\mathbf{E})_x = Kq\left[-\frac{3zy}{r^5} + \frac{3yz}{r^5}\right] = 0$$

$$(\text{curl}\,\mathbf{E})_y = (\text{curl}\,\mathbf{E})_z = 0$$

So the curl of the Coulomb field due to a point charge is zero.

An extremely useful integral relation for the curl of a vector field is afforded by Stokes's theorem. This theorem states the following. If S is a surface which is closed except for a portion cut out of it along a curve C, then

$$\int_S \nabla \times \mathbf{F} \cdot d\mathbf{S} = \oint_C \mathbf{F} \cdot d\mathbf{l} \tag{6.8}$$

(See Fig. 6.4.) The right-hand side of (6.8) is a line integral about the closed curve C. It involves the inner product between the vector $\mathbf{F}$ and the differential of line element $d\mathbf{l}$. Consider, for example, the circular line integral of a constant vector field $\mathbf{F}$. Let the normal to the plane of the circle of integration be perpendicular to $\mathbf{F}$. From Fig. 6.5 we see that for this orientation

$$\mathbf{F} \cdot d\mathbf{l} = F\cos\theta\, dl = F\cos\theta\, r\, d\theta$$

$$\oint \mathbf{F} \cdot d\mathbf{l} = Fr\int_0^{2\pi} \cos\theta\, d\theta = 0$$

More generally, it may be shown that the line integral of a constant force field vanishes for any closed curve in space. With Stokes's theorem we may then infer that $\nabla \times \mathbf{F} = 0$ for a constant vector field.

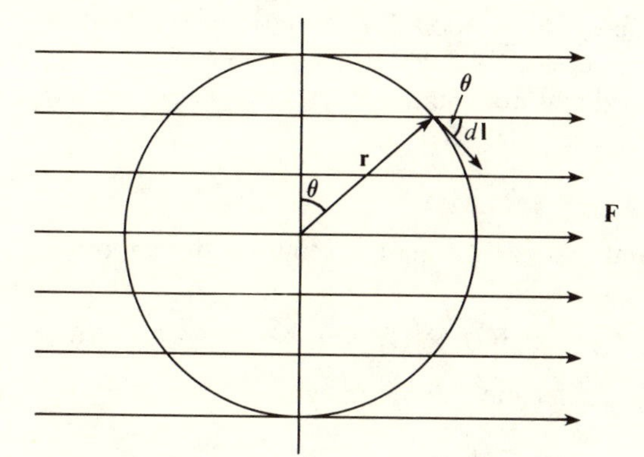

Figure 6.5

Geometrical Interpretation of the Curl

The geometrical significance of the curl may be seen in the following example. Consider the one-dimensional field

$$E_x = E_0\left(\frac{y}{d}\right)^2$$

where E_0 is a constant amplitude and d is a constant distance. Let a dielectric bar with charges q fixed to its ends be placed in this field so that it is free to turn in the x, y plane about a fixed center. (See Fig. 6.6.) Will the rod rotate due to the electric field? The answer is that it will rotate if the work done on it by the electric field in a rotation is greater than zero. An expression for this work is

$$W = 2q\oint \mathbf{E} \cdot d\mathbf{l}$$

By Stokes's theorem we may write

$$W = 2q\int(\nabla \times \mathbf{E}) \cdot d\mathbf{S} \simeq 2q|\nabla \times \mathbf{E}|A$$

where A is the area swept out by the bar in one rotation. So the rod will turn if $|\nabla \times \mathbf{E}| \neq 0$. This rotational aspect of the curl operation motivated

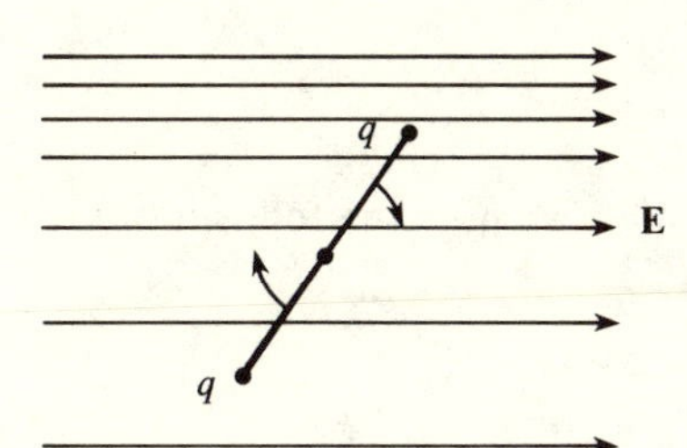

Figure 6.6 A charged rod in an inhomogeneous field **E**.

an earlier notation rot $\mathbf{E}$, where "rot" stood for "rotation." As will be shown, $\nabla \times \mathbf{E} = 0$ for any static electric field. It follows that for a static electric field, $W = 0$ and the rod will not rotate.

Divergence of the Curl

Let us use Stokes's theorem and Gauss's theorem to establish the important relation

$$\operatorname{div}(\operatorname{curl} \mathbf{G}) = \nabla \cdot (\nabla \times \mathbf{G}) = 0 \tag{6.9}$$

where $\mathbf{G}$ is any vector field. Consider the volume integral

$$\int_V \nabla \cdot (\nabla \times \mathbf{G})\, d\mathbf{r} = \int_s (\nabla \times \mathbf{G}) \cdot d\mathbf{S}$$

This equality follows from Gauss's theorem (6.4). The surface S which bounds the volume V is a closed surface. So Stokes's theorem (6.8) gives

$$\int_{S\ (\text{closed})} (\nabla \times \mathbf{G}) \cdot d\mathbf{S} = 0$$

It follows that

$$\int_V \nabla \cdot (\nabla \times \mathbf{G})\, d\mathbf{r} = 0$$

for any arbitrary volume V. We may conclude that (6.9) is correct.

The Gradient

So far we have considered vector differential operations on vector fields. An important vector differential operation on a scalar field $\Phi(x, y, z)$ is given by the gradient. In Cartesian coordinates this operator has the following representation:

$$\operatorname{grad} \Phi = \nabla \Phi = \mathbf{a}_x \frac{\partial \phi}{\partial x} + \mathbf{a}_y \frac{\partial \phi}{\partial y} + \mathbf{a}_z \frac{\partial \phi}{\partial z} \tag{6.10}$$

Let us use Stokes's theorem to show that

$$\operatorname{curl}(\operatorname{grad} \Phi) = \nabla \times (\nabla \Phi) = 0 \tag{6.11}$$

Again let S be a surface with a portion cut out of it along the curve C. Then Stokes's theorem (6.8) gives

$$\int_S \nabla \times (\nabla \Phi) \cdot d\mathbf{S} = \int_C \nabla \Phi \cdot d\mathbf{l} \tag{6.12}$$

To evaluate the right-hand side of this equation we note the following. The integral

$$\int_{\mathbf{r}_1}^{\mathbf{r}_2} \nabla \Phi \cdot d\mathbf{l} = \Phi(\mathbf{r}_2) - \Phi(\mathbf{r}_1) \tag{6.13}$$

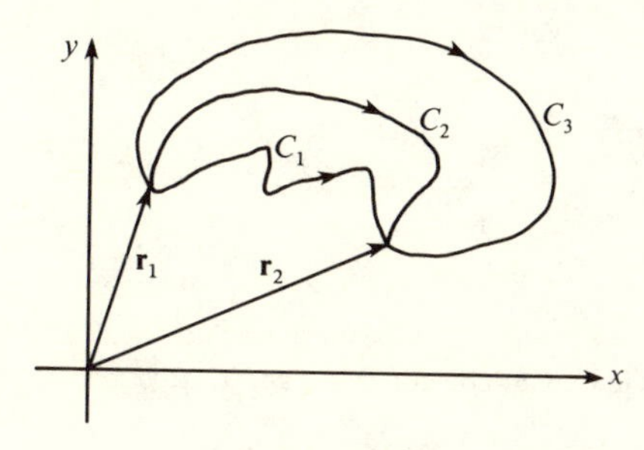

Figure 6.7 Arbitrary paths from $\mathbf{r}_1$ to $\mathbf{r}_2$.

for any arbitrary path connecting r_1 to r_2. (See Fig. 6.7.) It follows that

$$\oint \nabla\Phi \cdot d\mathbf{l} = \Phi(\mathbf{r}_1) - \Phi(\mathbf{r}_2) = 0 \tag{6.14}$$

So the right-hand side of (6.12) is zero and we may set

$$\int_S \nabla \times (\nabla\Phi) \cdot d\mathbf{S} = 0$$

for all arbitrary open surfaces S. We may conclude that (6.11) is valid.

Example 6.3

Show without explicit calculation that the curl of the static Coulomb field due to a point charge q at the origin is zero everywhere.

Ans. We may rewrite this field (6.1) in the form

$$E = -Kq\left[\mathbf{a}_x \frac{\partial}{\partial x}\left(\frac{1}{r}\right) + \mathbf{a}_y \frac{\partial}{\partial y}\left(\frac{1}{r}\right) + \mathbf{a}_z \frac{\partial}{\partial z}\left(\frac{1}{r}\right)\right]$$

so that

$$E = -Kq\nabla\left(\frac{1}{r}\right)$$

It follows by the property (6.11) that $\nabla \times \mathbf{E} = 0$.

Example 6.4

Evaluate the line integral along the curves C_1 and C_2 (Fig. 6.8), respectively, of the two-dimensional vector field

$$\mathbf{F} = \mathbf{a}_x \frac{x}{b^2} + \mathbf{a}_y \frac{y}{b^2}$$

Ans. Along C_1 we obtain

$$\int_{C_1} \mathbf{F} \cdot d\mathbf{l} = \int_0^b dx\, F_x(x,0) + \int_0^b dy\, F_y(b,y)$$

$$= \tfrac{1}{2} + \tfrac{1}{2} = 1$$

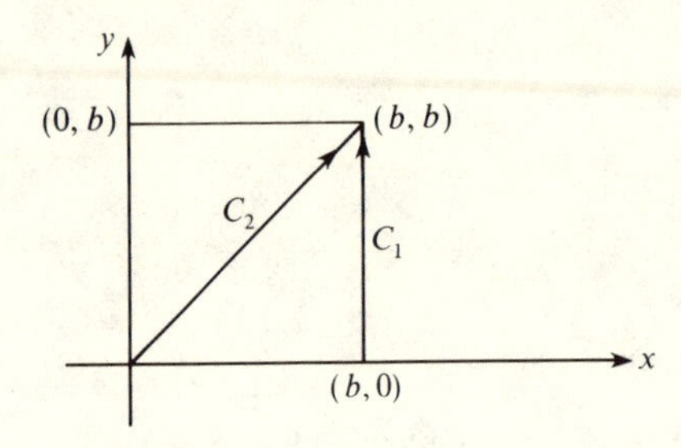

Figure 6.8 The curves C_1 and C_2.

Along C_2 we may write

$$d\mathbf{l} = \tfrac{1}{2}(\mathbf{a}_x + \mathbf{a}_y)\, dl, \qquad x = y = l$$

$$\int_{C_2} \mathbf{F} \cdot d\mathbf{l} = \int_0^b dl \left(\frac{2l}{b^2}\right) = 1$$

Thus for the case at hand $\int_{C_1} \mathbf{F} \cdot d\mathbf{l} = \int_{C_2} \mathbf{F} \cdot d\mathbf{l}$. We note that $\mathbf{F}$ may be put in the form

$$\mathbf{F} = \nabla \frac{1}{2}\left(\frac{x^2 + y^2}{b^2}\right)$$

and our result follows directly from (6.14).

Example 6.5
Using Stokes's theorem, find the direction of curl of the vector field

$$\mathbf{F} = \mathbf{a}_x F_0 \frac{y^2}{2b^2}$$

where F_0 and b are constants.

Ans. Let the closed curve C lie in a plane. Let $\mathbf{n}$ be the normal to the plane. (See Fig. 6.9.) Consider the loop integral

$$I = \oint_C \mathbf{F} \cdot d\mathbf{l}$$

in the limit that the area bounded by the curve C shrinks to zero. Suppose evaluation of I with $\mathbf{n}$ in the z-direction gives a finite value of I, whereas evaluation with $\mathbf{n}$ in either the x- or the y-direction gives $I = 0$. Stokes's

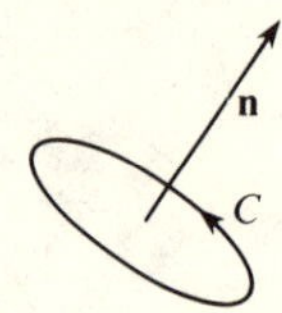

Figure 6.9 Configuration for Example 6.5.

theorem then implies that $\nabla \times \mathbf{F}$ lies in the z-direction. Let us show for the case at hand that I is nonvanishing only if the normal $\mathbf{n}$ is the z-direction. Suppose $\mathbf{n}$ is in the x-direction. Then $d\mathbf{l}$ has components $(0, \alpha\, dy, \beta\, dz)$. But $\mathbf{F}$ only has an x-component. If $\mathbf{n}$ is in the y-direction, $d\mathbf{l}$ has components $(\gamma dx, 0, \delta\, dz)$. Since $F_x = F_x(y)$, the closed dx integration would give no contribution. The parameters $\alpha, \beta, \gamma, \delta$ represent direction variables. It follows that $\nabla \times \mathbf{F}$ has only one component. Direct calculation of $\nabla \times \mathbf{F}$ gives

$$\nabla \times \mathbf{F} = \left(0, 0, -\frac{F_0 y}{b^2}\right)$$

in agreement with our geometrical finding.

Example 6.6
Establish the important vector relation

$$\nabla \cdot (\mathbf{F} \times \mathbf{G}) = \mathbf{G} \cdot \nabla \times \mathbf{F} - \mathbf{F} \cdot \nabla \times \mathbf{G}$$

Ans. To prove the validity of this relation we need only consider, say, the x-derivatives of both sides. For the left-hand side we obtain

$$\begin{aligned} \nabla \cdot (\mathbf{F} \times \mathbf{G})_x &= \frac{\partial}{\partial x}(F_y G_z - G_y F_z) \\ &= F_y \frac{\partial}{\partial x} G_z + G_z \frac{\partial}{\partial x} F_y - G_y \frac{\partial}{\partial x} F_z - F_z \frac{\partial}{\partial x} G_y \end{aligned}$$

The x-derivatives of $\nabla \times \mathbf{F}$ are included in its y- and z-components:

$$(\nabla \times \mathbf{F})_y = -\frac{\partial}{\partial x} F_z + \cdots$$

$$(\nabla \times \mathbf{F})_z = +\frac{\partial}{\partial x} F_y - \cdots$$

So for the right-hand side of our relation we obtain

$$-G_y \frac{\partial}{\partial x} F_z + G_z \frac{\partial}{\partial x} F_y + F_y \frac{\partial}{\partial x} G_z - F_z \frac{\partial}{\partial x} G_y$$

which is the same form as was obtained for the left-hand side.

6.2 GAUSS'S LAW

The electric field in the vicinity of a stationary aggregate of charge obeys the relation

$$\oiint \mathbf{E} \cdot d\mathbf{S} = \frac{1}{\epsilon_0} Q \tag{6.15}$$

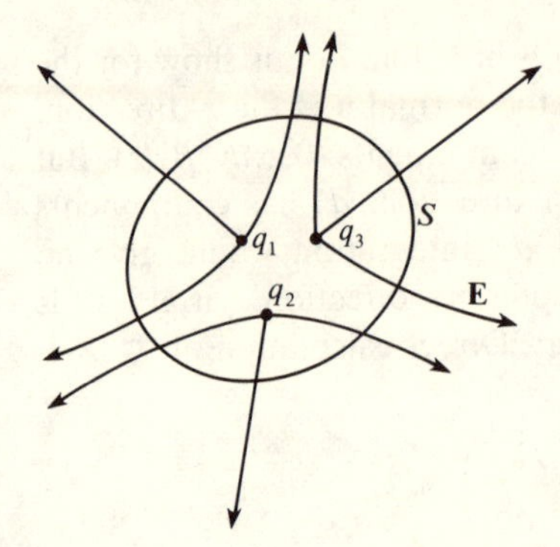

Figure 6.10 Three positive charges enclosed by the surface S.

In this equation, the closed surface S contains the total charge Q. (See Fig. 6.10.)

Equation (6.15) is called Gauss's law of electrostatics. Introducing the charge density ρ (C/m³) permits this equation to be written[3]

$$\oiint_S \mathbf{E} \cdot d\mathbf{S} = \frac{1}{\epsilon_0} \int_V \rho \, d\mathbf{r} \tag{6.16}$$

The closed surface S has volume V. With Gauss's theorem (6.4) the latter relation may be rewritten

$$\int_V \left(\nabla \cdot \mathbf{E} - \frac{1}{\epsilon_0} \rho \right) d\mathbf{r} = 0 \tag{6.17}$$

Shrinking the volume V to an arbitrarily small size and employing the mean-value theorem, we obtain

$$\nabla \cdot \mathbf{E} = \frac{1}{\epsilon_0} \rho \tag{6.18}$$

In a dielectric medium this equation becomes

$$\nabla \cdot \mathbf{E} = \frac{1}{\epsilon} \rho \tag{6.19}$$

where ϵ is the permittivity of the medium and ρ is the free charge density (as opposed to bound charge). Equation (6.18) is more conveniently written in terms of the displacement vector **D**, which for "linear" material is related to the electric field through

$$\mathbf{D} = \epsilon \mathbf{E} \tag{6.20}$$

Thus, (6.19) may be written

$$\nabla \cdot \mathbf{D} = \rho \tag{6.21}$$

[3] This use of ρ should in no way be confused with its previous use to denote the reflection coefficient.

In vacuum, (6.21) reduces to (6.18). These equations indicate that charge is the *source* of the electric field.

Example 6.7
Show that Gauss's law (6.18) follows from Coulomb's law (6.2).

Ans. Let a single point charge be enclosed in the closed surface S as shown in Fig. 6.11. We must show that

$$\oiint \mathbf{E} \cdot d\mathbf{S} = 4\pi K q$$

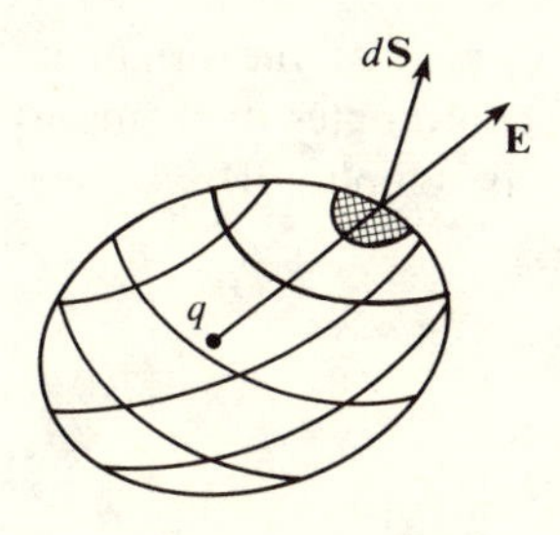

Figure 6.11 A single charge enclosed by a closed surface S.

Let q be at the origin. Then

$$\mathbf{E} \cdot d\mathbf{S} = Kq\frac{\hat{\mathbf{r}}}{r^2} \cdot d\mathbf{S}$$

The differential of solid angle (see Example 6.2) subtended by $d\mathbf{S}$ about the origin is

$$d\Omega = \frac{\hat{\mathbf{r}} \cdot d\mathbf{S}}{r^2}$$

so that

$$\mathbf{E} \cdot d\mathbf{S} = Kq\, d\Omega$$

Hence

$$\oiint \mathbf{E} \cdot d\mathbf{S} = 4\pi K q$$

6.3 THE SCALAR POTENTIAL AND LAPLACE'S EQUATION

In electrostatics, the electric field is *conservative*. This means that $\mathbf{E}$ may be written as the gradient of a potential function V:

$$\mathbf{E} = -\nabla V \tag{6.22}$$

Equivalently, with (6.11), we may say that

$$\nabla \times \mathbf{E} = 0 \tag{6.23}$$

It follows that in electrostatics, the integral

$$\int_{\mathbf{r}_1}^{\mathbf{r}_2} \mathbf{E} \cdot d\mathbf{l} = V(\mathbf{r}_1) - V(\mathbf{r}_2) \tag{6.24}$$

is independent of the path of integration [see Eq. (6.13)]. The integral

$$\Delta W = -q \int_{\mathbf{r}_1}^{\mathbf{r}_2} \mathbf{E} \cdot d\mathbf{l} = q \int_{\mathbf{r}_1}^{\mathbf{r}_2} dV \tag{6.25}$$

represents the work done by the charge q on the field in moving from $\mathbf{r}_1$ to $\mathbf{r}_2$.

The potential field due to a point charge q located at the origin is obtained from the Coulomb field (6.2). In spherical coordinates the component of the gradient of V in the direction of $\hat{\mathbf{r}}$ is simply $\partial V/\partial r$ (see Appendix A). It follows that

$$\mathbf{E} = Kq\frac{\hat{\mathbf{r}}}{r^2} = -\hat{\mathbf{r}}\frac{\partial}{\partial r}\left(\frac{Kq}{r}\right) = -\nabla V$$

and we obtain

$$V(r) = \frac{Kq}{r} \tag{6.26}$$

Superposition Principle

The potential field $V(\mathbf{r})$ obeys the following *superposition principle*. Let $\mathbf{E}_1$ and $\mathbf{E}_2$ be fields due to two separate charge configurations, respectively. The net field at any point $\mathbf{r}$ in space due to these charge configurations is

$$\mathbf{E}(\mathbf{r}) = \mathbf{E}_1(\mathbf{r}) + \mathbf{E}_2(\mathbf{r}) = -\nabla V_1(\mathbf{r}) - \nabla V_2(\mathbf{r})$$

Since the gradient is a linear operator, we may rewrite this latter expression in the form

$$\mathbf{E}(\mathbf{r}) = -\nabla(V_1(\mathbf{r}) + V_2(\mathbf{r})) = -\nabla V(\mathbf{r})$$

Thus

$$V(\mathbf{r}) = V_1(\mathbf{r}) + V_2(\mathbf{r})$$

The potential at any point in space due to separate charge configurations is the sum of the potentials of the individual charge configurations.

Let us use these results to obtain the potential function $V(z)$ along the axis of a disk of radius a which carries a uniform surface charge density σ C/m^2. (See Fig. 6.12.) An element of area on the surface of the disk is $r\,dr\,d\phi$, which in turn carries the charge $dq = \sigma r\,dr\,d\phi$ and is a distance $\sqrt{r^2 + z^2}$ from the point P. The potential from this element of surface at

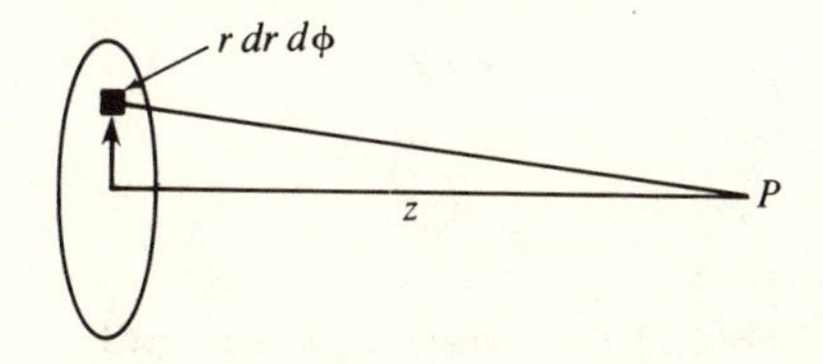

Figure 6.12 The charged disk.

the point P is

$$dV = \frac{K\,dq}{\sqrt{r^2 + z^2}} = \frac{K\sigma r\,dr\,d\phi}{\sqrt{r^2 + z^2}}$$

The superposition principle which the potential obeys allows us to write V at the point P as the integral of dV over the surface of the disk. Integrating over ϕ and r gives

$$V = 2\pi K\sigma \int_0^a \frac{r\,dr}{\sqrt{r^2 + z^2}} = 2\pi K\sigma \sqrt{r^2 + z^2}\,\Big|_0^a$$

$$V(z) = 2\pi K\sigma z\left(\sqrt{1 + \left(\frac{a}{z}\right)^2} - 1\right).$$

Far from the disk, $a/z \to 0$ and we obtain, letting $Q = \pi a^2\sigma$,

$$V(z) = \frac{2KQz}{a^2}\left[1 + \frac{1}{2}\left(\frac{a}{z}\right)^2 + \cdots - 1\right]$$

$$\sim \frac{KQ}{z}$$

which is the correct asymptotic result. The charge Q is the total charge on the disk:

$$Q = \int\!\!\int \sigma r\,dr\,d\phi$$

Laplace's Equation

If (6.22) is substituted into Gauss's law (6.18) we obtain

$$\nabla^2 V = \frac{\rho}{\epsilon_0} \tag{6.27a}$$

This equation is called Poisson's equation. In regions of space free of charge, (6.27a) reduces to *Laplace's equation*

$$\nabla^2 V = 0 \tag{6.27b}$$

This equation is the hallmark of the electrostatic field. An important property of it is that its solution interior to a closed surface on which the potential is prescribed is unique. There is only one solution to (6.27b) which

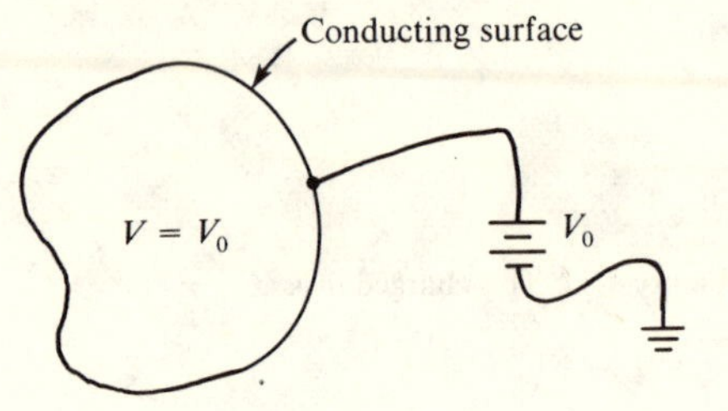

Figure 6.13 Grounded conducting closed surface.

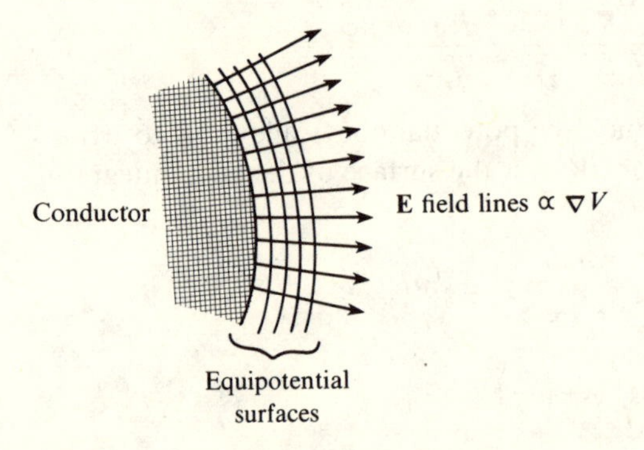

Figure 6.14 Equipotential surfaces are parallel to the surface of a conductor.

matches such boundary conditions. An important consequence of this property is as follows. The potential inside a closed charge-free domain bounded by a surface of constant potential is itself constant and equal to the potential on the boundary. (See Fig. 6.13.) It follows that the electric field in such a domain is zero. As indicated in the above sketch, a conducting surface is an equipotential surface. Since the gradient of V is normal to surfaces of constant V, it follows that E is normal to the surface of a conductor. (See Fig. 6.14.)

Example 6.8

A conducting surface carries surface charge density σ (C/m^2). Show that the electric field at the surface of the metal has the magnitude

$$E = \sigma/\epsilon_0$$

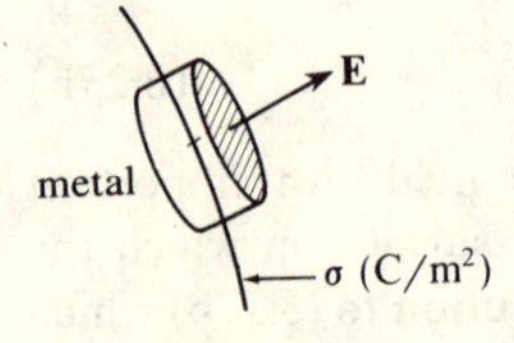

Figure 6.15 Pill-box configuration.

Ans. This result may be obtained from Gauss's law (6.16). To this end we construct an infinitesimal "pillbox" which straddles the surface, as shown in Fig. 6.15, and integrate (6.16) over the surface of this volume. Since $\mathbf{E} = 0$ inside the metal, the surface integral gives

$$\oint \mathbf{E} \cdot d\mathbf{S} = EA = \frac{1}{\epsilon_0} \int \rho \, d\mathbf{r} = \frac{1}{\epsilon_0} \sigma A$$

which yields the result.

Separation of Variables

A powerful technique for the solution of Laplace's equation and similar equations which emerge in the theory of waveguides is the method of separation of variables. This method is exhibited in the following example.

We wish to find the electric potential inside an infinitely long rectangular air-filled cylinder, three of whose sides are maintained at zero potential and whose fourth side is maintained at constant potential V_0. This configuration is shown in Fig. 6.16. The z-axis is out of the plane of the paper.

We must solve (6.27) subject to the boundary conditions

$$V(0, y) = V(a, y) = 0 \tag{6.28a}$$

$$V(x, 0) = 0 \tag{6.28b}$$

$$V(x, b) = V_0 \tag{6.28c}$$

In Cartesian space, Laplace's equation becomes

$$\frac{\partial^2 V}{\partial x^2} + \frac{\partial^2 V}{\partial y^2} + \frac{\partial^2 V}{\partial z^2} = 0 \tag{6.29}$$

From symmetry we see that we may assume that the potential V is independent of z. In this case (6.29) reduces to the simpler equation

$$\frac{\partial^2 V}{\partial x^2} + \frac{\partial^2 V}{\partial y^2} = 0 \tag{6.30}$$

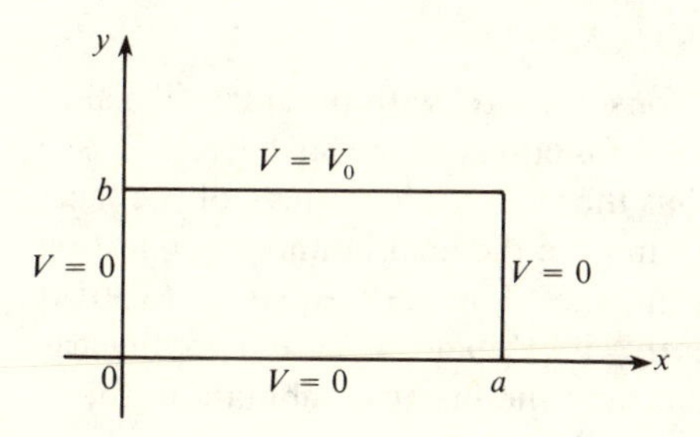

Figure 6.16 A rectangular cylinder with surfaces at constant potentials.

In the method of separation of variables, one assumes that $V(x, y)$ is in the product form

$$V(x, y) = X(x)Y(y) \tag{6.31}$$

Substituting this form into (6.30) and dividing by XY, we obtain

$$\frac{X_{xx}}{X} = -\frac{Y_{yy}}{Y} = -k^2 \tag{6.32}$$

where subscripts denote differentiation. The constant $-k^2$ stems from the fact that the left-hand member of the preceding equation is a function only of x, whereas the middle member is a function only of y. The only way these functions may be equal for all x, y is for them to be equal to the same constant, which we have labeled $-k^2$. Thus, Laplace's partial differential equation is reduced to two ordinary differential equations obtained from (6.32):

$$X_{xx} + k^2X = 0 \tag{6.33a}$$

$$Y_{yy} - k^2Y = 0 \tag{6.33b}$$

The general solution to (6.33a) is

$$X = A \sin kx + B \cos kx$$

The boundary condition (6.28a) gives

$$X(0) = B = 0$$

The condition

$$X(a) = A \sin ka = 0$$

indicates that the constant k is restricted to the values

$$k_n a = n\pi, \qquad n = 1, 2, \ldots \tag{6.34}$$

The general solution to (6.33b) is

$$Y = C \sinh ky + D \cosh ky$$

The boundary condition (6.28b) gives

$$Y(0) = D = 0$$

So the form our solution has assumed to this point is

$$V(x, y) = V_n \sin k_n x \sinh k_n y \tag{6.35}$$

Here we have replaced the product of constants AC with the single constant V_n. Any one of the infinite number of functions as given by (6.35) is a solution to Laplace's equation which has the value zero on three of the faces of the rectangular cylinder. In order to impose the final boundary condition (6.23c), we note the following important fact. Laplace's equation (6.30) is linear and homogeneous. Consequently, if V_1 and V_2 are two linearly independent solutions to this equation, then the linear combination $\alpha V_1 + \beta V_2$ is also a solution, where α and β are arbitrary constants. Since we have uncovered an infinite number of linearly independent solutions of Laplace's equation, it follows that the infinite sum

$$V(x, y) = \sum_{n=1}^{\infty} V_n \sin k_n x \sinh k_n y, \qquad k_n a = n\pi \tag{6.36}$$

is also a solution. We are now prepared to impose the final boundary condition (6.28c):

$$V_0 = V(x, b) = \sum_{n=1}^{\infty} V_n \sin k_n x \sinh k_n b \tag{6.37}$$

The student will recognize this summation to be a Fourier series. The constants V_n are determined by multiplying both sides of (6.37) by $\sin k_n x$ and integrating from $x = 0$ to $x = a$. There results

$$V_n = \begin{cases} \dfrac{4}{n\pi} \dfrac{V_0}{\sinh(n\pi a/b)}, & n \text{ odd} \\ 0, & n \text{ even} \end{cases}$$

So the solution to Laplace's equation satisfying all four boundary conditions (6.28) is

$$V(x, y) = \sum_{n \text{ odd}} \frac{4V_0}{n\pi \sinh(n\pi a/b)} \sin\frac{n\pi x}{a} \sinh\frac{n\pi y}{b}. \tag{6.38}$$

The electric field inside the cylinder follows from (6.22):

$$E_x = -\sum_{n \text{ odd}} \frac{4V_0}{a \sinh(n\pi a/b)} \cos\frac{n\pi x}{a} \sinh\frac{n\pi y}{b} \tag{6.39a}$$

$$E_y = -\sum_{n \text{ odd}} \frac{4V_0}{b \sinh(n\pi a/b)} \sin\frac{n\pi x}{a} \cosh\frac{n\pi y}{b} \tag{6.39b}$$

It follows that $E_x = 0$ on the midplane

$$x = a/2$$

and on the boundary

$$y = 0$$

The component $E_y = 0$ on the boundaries

$$x = 0 \quad \text{and} \quad x = a.$$

Furthermore, in obtaining the expression (6.38) we found that

$$V_0 = V(x, b) = \sum_{n \text{ odd}} \frac{4V_0}{n\pi \sinh(n\pi a/b)} \sin\frac{n\pi x}{a} \sinh n\pi$$

It follows that

$$\frac{dV_0}{dx} = 0 = -\sum_{n \text{ odd}} \frac{4V_0}{a \sinh(n\pi a/b)} \cos\frac{n\pi x}{a} \sinh n\pi$$

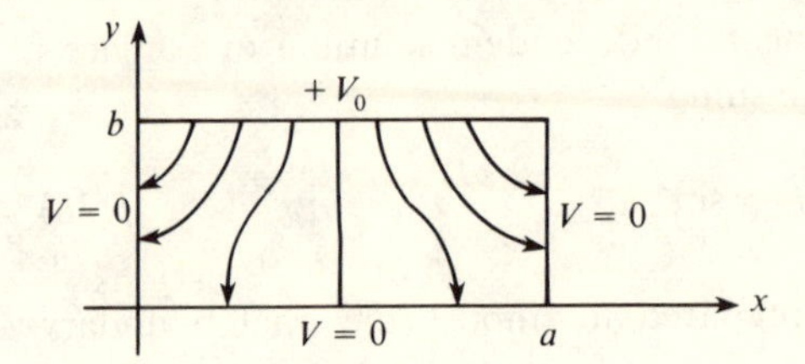

Figure 6.17 E-field lines inside cylinder.

Comparing this expression with (6.39a), we see that $E_x = 0$ on the boundary $y = b$ as well. So **E** is normal to all four surfaces of the rectangular prism. This of course is to be expected, since these surfaces are equipotential surfaces, and the gradient of V, as noted previously, is normal to surfaces of constant V.

A sketch of the field lines inside the cylinder is shown in Fig. 6.17.

Example 6.9

a. If the surfaces of the cylinder in the preceding configuration are conducting surfaces, obtain an expression for the surface charge density on the surface $y = b$.
b. Show that the dimensions of the expression you have obtained for σ are correct.

Ans.

a. The surface charge density is given by (see Example 6.8)

$$\sigma = \epsilon_0 E$$

It follows that

$$\sigma = \epsilon_0 E_y(x, b) = -\sum_{n \text{ odd}} \frac{4\epsilon_0 V_0}{b \sinh(n\pi a/b)} \sin\frac{n\pi x}{a} \cosh n\pi$$

b. The dimensions of our expression for σ are

$$[\sigma] = \left[\frac{\epsilon_0 V_0}{b}\right] = \frac{\mathrm{F-V}}{\mathrm{m}^2} = \frac{\mathrm{C}}{\mathrm{m}^2}$$

which are the correct dimensions for surface charge density.

The Method of Images

For certain symmetric charge distributions in the presence of metallic boundaries it is possible to solve for the static potential field without explicitly solving Laplace's equation. The technique involves replacing the metal surfaces with approximately located "image" charges which, with the

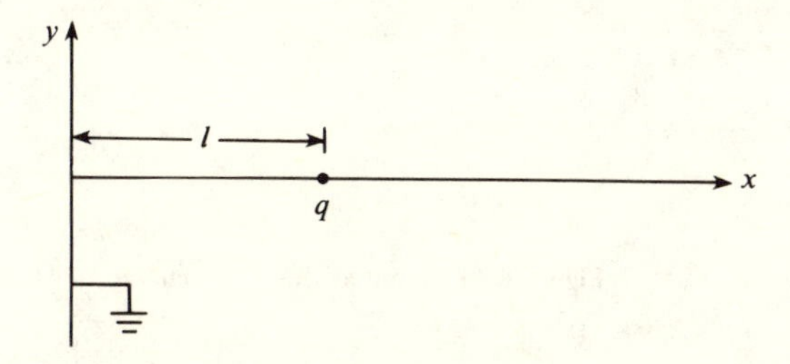

Figure 6.18 A charge q in front of a grounded plane.

original charge distribution, reproduces the same boundary conditions as the original configuration.

Suppose, for example, that a point charge q is placed l meters in front of a grounded conducting plane of infinite extent. The configuration is shown in Fig. 6.18. We must find the solution to Laplace's equation in the half space which includes the charge q, subject to the following boundary conditions:

a. $V(x, 0) = 0$.
b. V has a Coulomb singularity at the location of the charge q.
c. $V \to 0$ as $(x, y) \to (\infty, \pm\infty)$.

Exactly the same boundary conditions are met by the charge configuration consisting of a charge q at $(l, 0)$ and an *image* charge $-q$ at $(-l, 0)$. The **E**-field lines of this configuration are shown in Fig. 6.19.

So by our uniqueness theorem the potential due to $q, -q$ is the solution to Laplace's equation which we seek. This potential is simply given by

$$V = Kq\left[\frac{1}{|\mathbf{r} - \mathbf{l}|} - \frac{1}{|\mathbf{r} + \mathbf{l}|}\right]$$

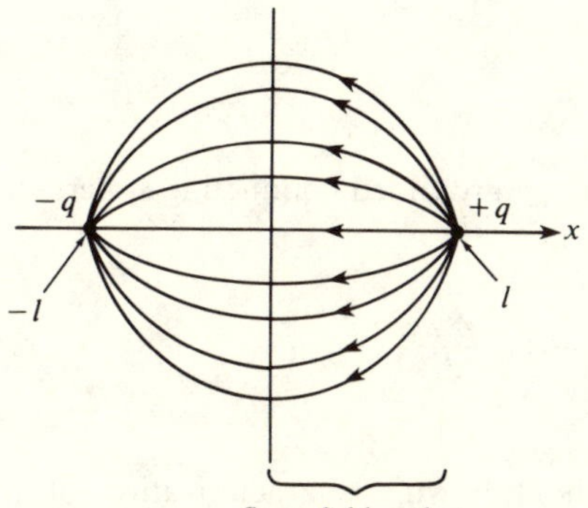

Figure 6.19 Field due to two charges of opposite sign and equal magnitude.

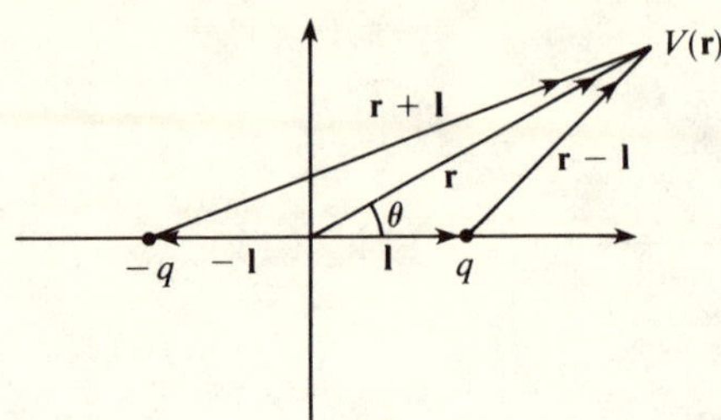

Figure 6.20 Potential due to a charge doublet.

(See Fig. 6.20.) This form may be written explicitly in terms of the angle θ and the radius r from the origin:

$$V = Kq\left(\frac{1}{\sqrt{r^2 + l^2 - 2rl\cos\theta}} - \frac{1}{\sqrt{r^2 + l^2 + 2rl\cos\theta}}\right)$$

This is the potential in the right half space which includes the charge. The potential in the half space to the left of the plane must match the boundary conditions: $V = 0$ on the surface $x = 0$, and V goes to zero at $(x, y) = (-\infty, \pm\infty)$. We conclude that the solution in the left half space ($x < 0$) is $V = 0$.

Example 6.10

A point charge $q = 10^{-14}$ C is placed $l = 1$ mm in front of a grounded conducting plane. What is the force on the charge due to the presence of the plane?

Ans. We have established in text that the charge q "sees" its image $-q$ a distance $2l$ meters from it. So the force on q is

$$F = qE = -\frac{Kq^2}{(2l)^2} = -\frac{9 \times 10^9 \times 10^{-28}}{(2 \times 10^{-6})^2} = -2.25 \times 10^{-7}\ \text{N}$$

This force is directed toward the plane.

Example 6.11

A point charge q is a distance d from a grounded conducting sphere of radius a. Using the method of images, find:

a. The potential V everywhere.
b. The force on the charge due to the presence of the sphere.

Ans.

a. To solve this problem we recall the following characterization of a sphere. If $\mathbf{r}_1$ is the vector to point P from the origin O_1 and $\mathbf{r}_2$ is the

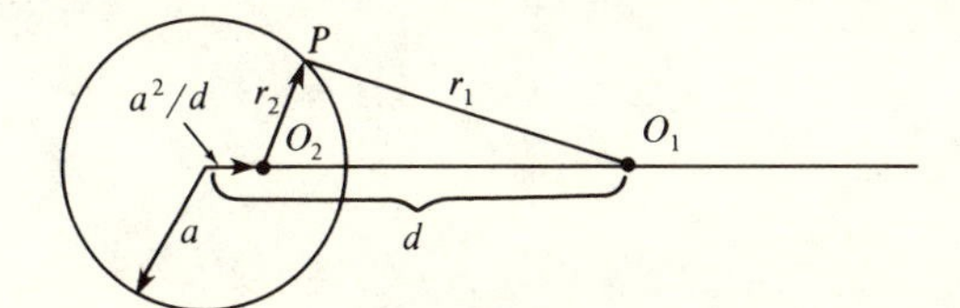

Figure 6.21 A conducting sphere and equivalent point-charge images.

vector to point P from the origin O_2, then the locus of points P such that

$$\frac{r_2}{r_1} = \text{constant}$$

is the surface of a sphere. Furthermore, if

$$\frac{r_2}{r_1} = \frac{a}{d}$$

and $d > a$, then O_2 is a distance $a(a/d)$ from the center of the sphere, O_1 is a distance d from the center of the sphere, and a is the radius of the sphere. (See Fig. 6.21.) Now let the charge q be at O_1. Let us introduce an image charge at O_2 of magnitude $-q'$. The potential on the surface of the sphere is

$$V(r = a) = \frac{q}{r_1} - \frac{q'}{r_2} = \frac{1}{r_2}\left(q\frac{r_2}{r_1} - q'\right)$$

But $r_2/r_1 = a/d$ on the surface of the sphere, so

$$V(r = a) = \frac{1}{r_2}\left(q\frac{a}{d} - q'\right)$$

Now in the problem at hand, the sphere is grounded and $V(r = a) = 0$. This condition may be satisfied if we choose the image charge q' to have the magnitude

$$q' = \frac{a}{d}q$$

This configuration is shown in Fig. 6.22. The potential outside the sphere

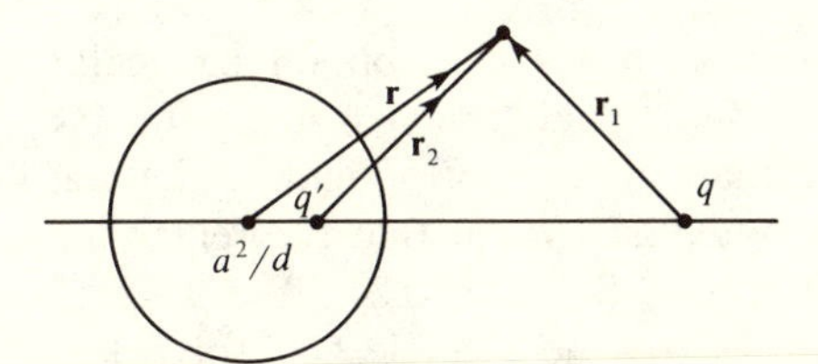

Figure 6.22 The vectors $\mathbf{r}$, $\mathbf{r}_1$, $\mathbf{r}_2$.

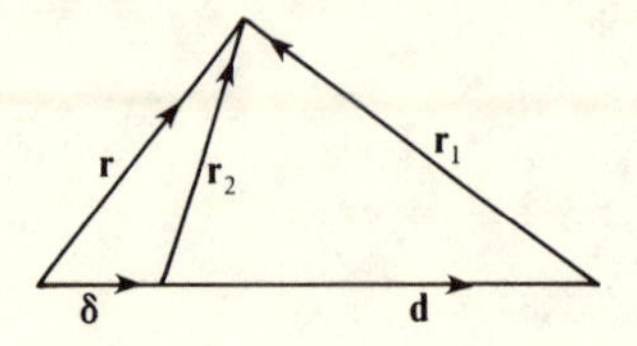

Figure 6.23 The vectors **δ** and **d**.

is

$$V(\mathbf{r}) = \frac{q}{r_1} - \frac{q(a/d)}{r_2}$$

Introducing the vectors

$$\boldsymbol{\delta} \equiv \frac{a^2}{d}\hat{\mathbf{a}}_z, \qquad \mathbf{d} \equiv d\hat{\mathbf{a}}_z$$

we see that

$$\mathbf{r}_2 + \boldsymbol{\delta} = \mathbf{r}$$

$$\mathbf{r}_1 + \mathbf{d} = \mathbf{r}$$

(Fig. 6.23). Thus the potential for $r > a$ becomes

$$V(r) = \frac{q}{|\mathbf{r} - \mathbf{d}|} - \frac{q(a/d)}{|\mathbf{r} - \boldsymbol{\delta}|}$$

This form may be more explicitly written in terms of the polar angle θ:

$$V(r, \theta) = \frac{q}{\sqrt{r^2 + d^2 - 2rd\cos\theta}} - \frac{q(a/d)}{\sqrt{r^2 + \delta^2 - 2r\delta\cos\theta}}$$

Since the sphere is at the potential $V = 0$, and there are no sources of potential inside the sphere, we obtain $V(r < a) = 0$.

b. The charge q "sees" its image q' a distance $d - a^2/d$ from itself. The force on q is therefore

$$F = \frac{Kqq'}{|d - a^2/d|^2} = \frac{Kq^2ad}{|d^2 - a^2|^2}$$

Potential Values by Symmetry Construction

For certain symmetric configurations, it is possible to obtain particular values of the potential (as opposed to functional forms) from symmetry considerations. For example, suppose that one side of a right equilateral triangular prism is maintained at the potential V_0 and that the remaining two sides are maintained at zero potential. (See Fig. 6.24.) We wish to ascertain the potential $V(C)$ at center C of the triangle. To solve this problem we first note that all three of the configurations in Fig. 6.25 give the

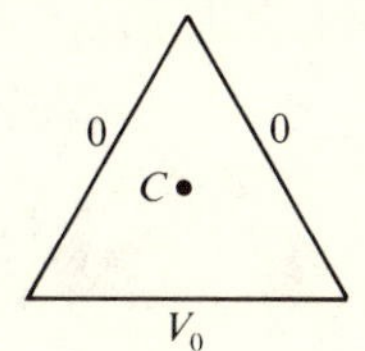

Figure 6.24 A triangular prism with sides maintained at constant potentials.

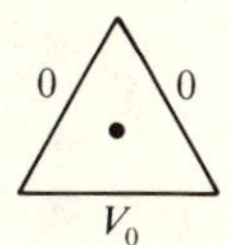

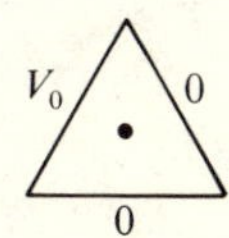

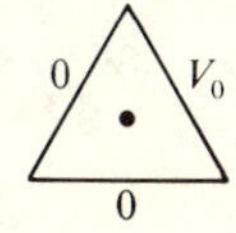

Figure 6.25 Three equivalent configurations.

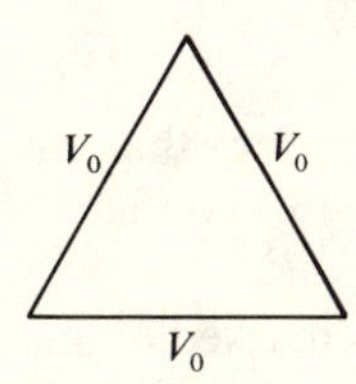

Figure 6.26

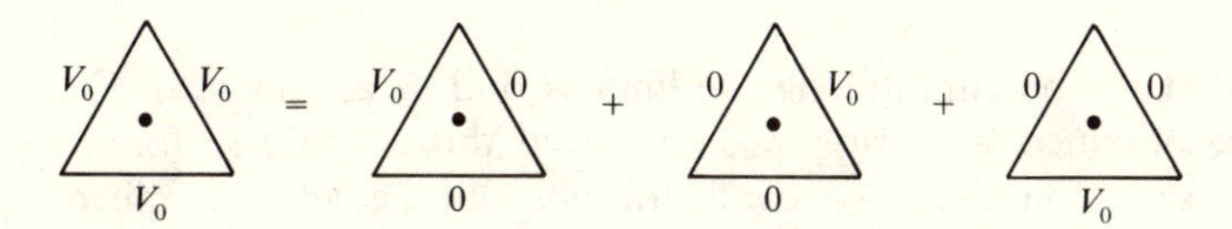

Figure 6.27

same potential at C. This is so because each configuration is obtained from the preceding one by a rotation about C through $\frac{2}{3}\pi$ radians. Secondly, by the superposition principle we note that the potential inside the equipotential prism shown in Fig. 6.26 is the sum of the potentials inside the preceding three rotated prisms. (See Fig. 6.27.) But the potential inside the equipotential prism is the constant V_0. It follows that

$$V_0 = 3V(C)$$

$$V(C) = \tfrac{1}{3}V_0$$

Example 6.12

The six sides of a hexagonal right prism are maintained at potentials of 60 volts and 0 volts as shown in Fig. 6.28.

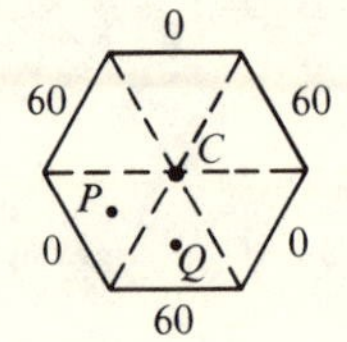

Figure 6.28 The hexagonal prism of Example 6.12.

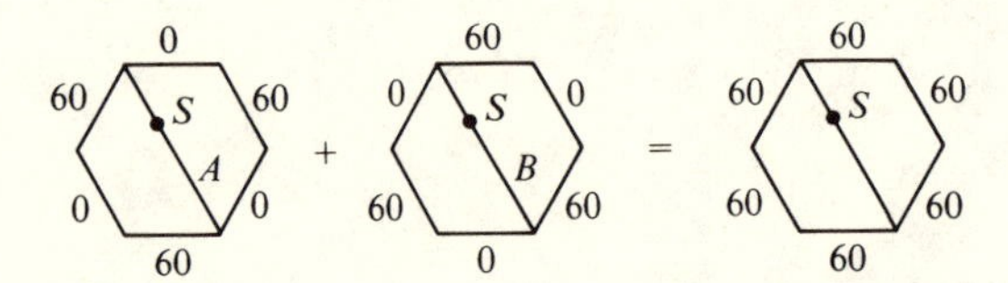

Figure 6.29

a. What is the potential at the center C of the hexagon? Show that the dashed surfaces in Fig. 6.28 are equipotential surfaces for this configuration.
b. What are the values of the potential at the triangle centers marked P and Q in the figure above?

Ans.

a. First note that the potential along the lines A and B as shown in Fig. 6.29 may be obtained from superposition. That this is the case follows from the observation that two configurations on the left are mirror images of each other through the line A (or B). So at any point along these lines the potentials of the two configurations are equal. By superposition we may conclude that

$$2V(S) = 60 \text{ volts}$$
$$V(S) = 30 \text{ volts}$$

and that the bisectors A and B are equipotential surfaces. Similarly for the remaining two bisectors. It follows that the potential at the center is $V(C) = 30$ volts.
b. First consider the triangle shown in Fig. 6.30. Then note the superposition depicted in Fig. 6.31. By construction in the text, the potential at the

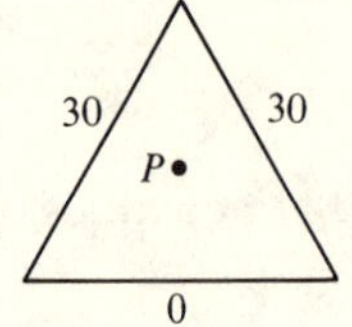

Figure 6.30

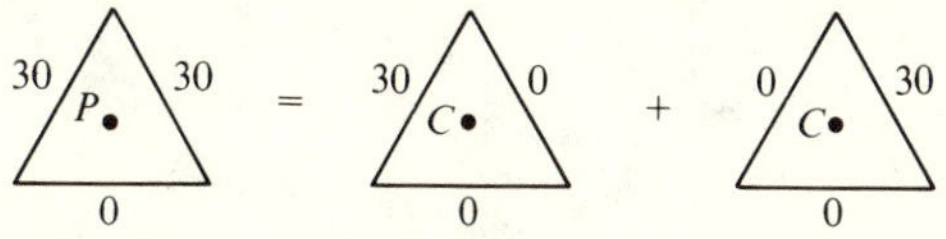

Figure 6.31

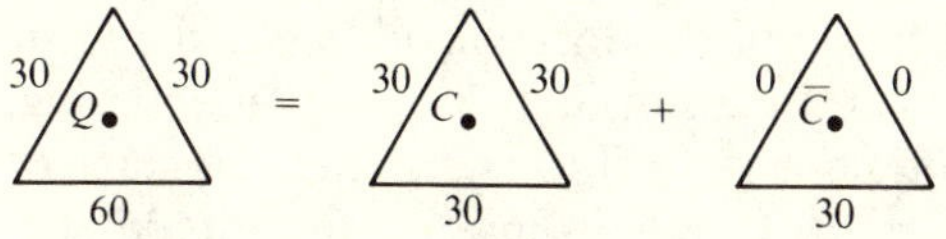

Figure 6.32 Completing the superposition in Example 6.12.

centers of the two triangles on the right is $V(C) = 10$ volts. It follows that $V(P) = 20$ volts. For the second configuration note the superposition shown in Fig. 6.32. Since $V(C) = 30$ and $V(\bar{C}) = 10$, we find that $V(Q) = 40$ volts.

6.4 THE BIOT-SAVART LAW AND AMPERE'S LAW

We turn next to the manner in which current acts as a source of magnetic field. Consider the configuration shown in Fig. 6.33. Current I (amperes) flows through a homogeneous wire. The vector differential length of wire is labeled $d\mathbf{l}$. What is the magnetic field at the vector displacement $\mathbf{r}$ from this source $d\mathbf{l}$? The magnetic field is given by the Biot-Savart law

$$d\mathbf{B} = \frac{\mu_0 I \, d\mathbf{l} \times \mathbf{r}}{4\pi r^3} \tag{6.40}$$

The constant μ_0 is the permeability of free space and has the value

$$\mu_0 = 4\pi \times 10^{-7} \text{ H/m} \simeq 1.26 \times 10^{-6} \text{ H/m}$$

Equation (6.40) indicates that $d\mathbf{B}$ is normal to both $d\mathbf{l}$ and $\mathbf{r}$.

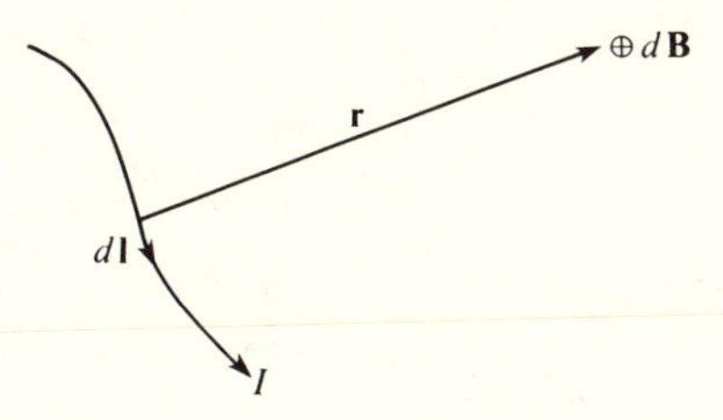

Figure 6.33 Vectors relevant to the Biot-Savart law.

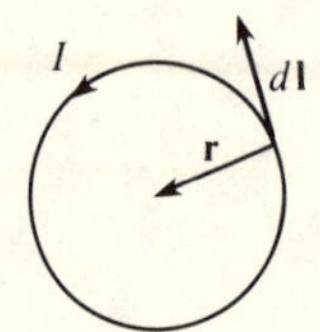

Figure 6.34 The vectors $d\mathbf{l}$ and $\mathbf{r}$.

A simple but important application of this formula addresses the value of the magnetic field at the center of a current loop of radius a. (See Fig. 6.34.) The vector $I\,d\mathbf{l} \times \mathbf{r}$ at the origin is normal to and out of the plane of the paper for all differential vectors $d\mathbf{l}$. It follows that the magnitude of $\mathbf{B}$ has the value

$$B = \frac{\mu_0 I}{4\pi}\int_0^{2\pi}\frac{d\theta}{a} = \frac{\mu_0 I}{2a}$$

An important vector relation which may be shown to follow from (6.40) is Ampere's law.[4] This law is a relation between magnetic field and current density $\mathbf{J}$ ($\mathrm{A/m^2}$):

$$\nabla \times \mathbf{B} = \frac{1}{\mu_0}\mathbf{J} \tag{6.41}$$

Let us see how this expression and the Biot-Savart law yield the same result for the magnetic field due to an infinitely long, uniform, homogeneous wire carrying current I.

To use (6.40) we introduce coordinates depicted in the sketch shown in Fig. 6.35. The direction of $\mathbf{B}$ at P is into the paper. Its magnitude is given by the integral

$$B = \frac{\mu_0 I}{4\pi}\int_{-\infty}^{\infty}\frac{R\,dl}{(R^2 + l^2)^{3/2}} = \frac{\mu_0 I}{2\pi R} \tag{6.42}$$

To employ Ampere's law in this problem we first suppose that the wire has radius a. Then the current density carried by the wire is

$$J = I/\pi a^2$$

Integrating (6.41) over a circular surface of radius $R > a$ which is oriented normal to the wire and applying Stokes's theorem, we obtain

$$\oint \mathbf{B}\cdot d\mathbf{l} = \mu_0\int \mathbf{J}\cdot d\mathbf{S} = \mu_0 I$$

$$B\,2\pi r = \mu_0 I$$

$$B = \frac{\mu_0 I}{2\pi R}$$

[4] This derivation may be found in J. D. Jackson, "Classical Electrodynamics," 2nd ed. (Wiley, New York, 1975).

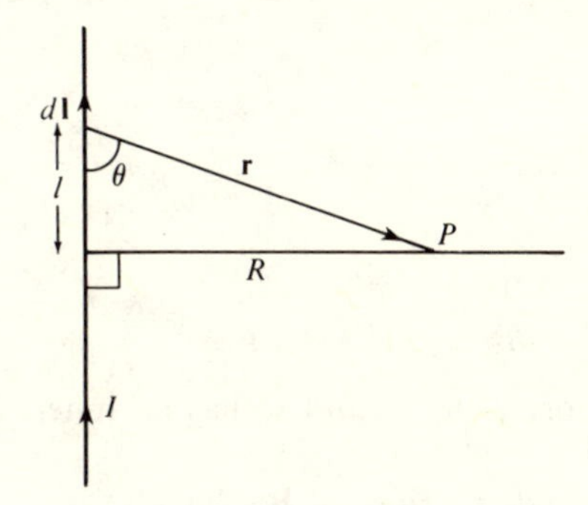

Figure 6.35 The infinite line current.

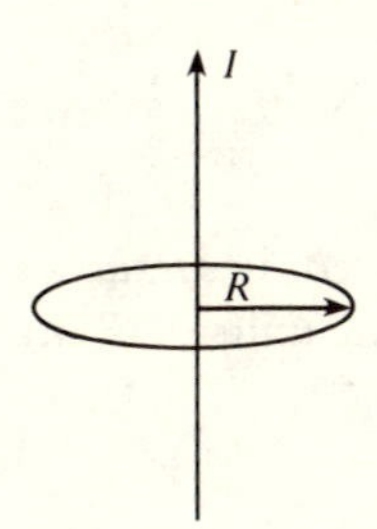

Figure 6.36 Surface of integration for the line current.

(See Fig. 6.36.) This result is the same as (6.42). However, it should be noted that in evaluating the loop integral in Stokes's theorem, it was assumed that **B** is in the direction given by the Biot-Savart law. [The result (6.42) was used previously in Example 2.2.]

Lorentz Force

Having described how a current is the source of magnetic field, let us now consider what effect a magnetic field has on a charge. A magnetic field exerts a force on a moving charge. Let a charge q move through a magnetic field **B**. Then there is a force on the charge which is given by the relation

$$\mathbf{F} = q\mathbf{v} \times \mathbf{B} \tag{6.43}$$

This force on the moving charge q is normal to the plane formed by the vectors **B** and **v**.

Let us demonstrate from the Lorentz force law (6.43) that a stationary magnetic field can do no work on a charged particle. Let the particle move between two points which we label 1 and 2. The work done on the particle is

$$W = \int_1^2 \mathbf{F} \cdot d\mathbf{l} = q\int_1^2 \mathbf{v} \times \mathbf{B} \cdot d\mathbf{l}$$

The velocity v may be written

$$\mathbf{v} = \frac{d\mathbf{l}}{dt}$$

It follows that

$$W = \int_1^2 q(d\mathbf{l} \times \mathbf{B}) \cdot \frac{d\mathbf{l}}{dt} = 0$$

The integral vanishes because $d\mathbf{l} \times \mathbf{B}$ is normal to $d\mathbf{l}$ and so has no inner product with $d\mathbf{l}$.

The Lorentz force law (6.43) reveals the dimensions of **B**. Namely

$$[B] = \left[\frac{F}{qv}\right] = \frac{\text{N}}{\text{C-m/sec}} = \frac{\text{N}}{\text{A-m}} \equiv \text{tesla}$$

$$1 \text{ tesla} = 1 \text{ weber/m}^2 = 10^4 \text{ gauss}$$

So magnetic field in the SI system of units is measured in *tesla* or equivalently *webers per square meter*. The cgs electromagnetic units of B are gauss.

Magnetic Intensity H

The magnetic intensity vector **H** is related to the magnetic field vector **B** through the permeability[5] μ (introduced previously in Section 2.5):

$$\mathbf{B} = \mu\mathbf{H} \tag{6.44}$$

In vacuum $\mu = \mu_0$. With the relation (6.44), Ampere's law (6.41) is more concisely written

$$\nabla \times \mathbf{H} = \mathbf{J} \tag{6.45}$$

Consider a very long cylinder with n turns/m of wire. The wire carries current I. (See Fig. 6.37.) Integrating Ampere's law (6.45) over the surface bounded by the curve C shown in the figure and using Stokes's theorem, we find that the magnetic intensity inside the cylinder is given by

$$H = nI \tag{6.45a}$$

In obtaining this result we assumed that the magnetic field outside the cylinder vanishes. This equation indicates that H has the dimensions of A/m. It is somewhat more conventional to use the longer phrase "ampere-turns per meter." Another popular unit of H is the *oersted*:

$$1 \text{ A/m} = 4\pi \times 10^{-3} \text{ oersted.}$$

[5] In anisotropic media **B** need not be parallel to **H**. Similarly **D** need not be parallel to **E** as expressed in (6.20).

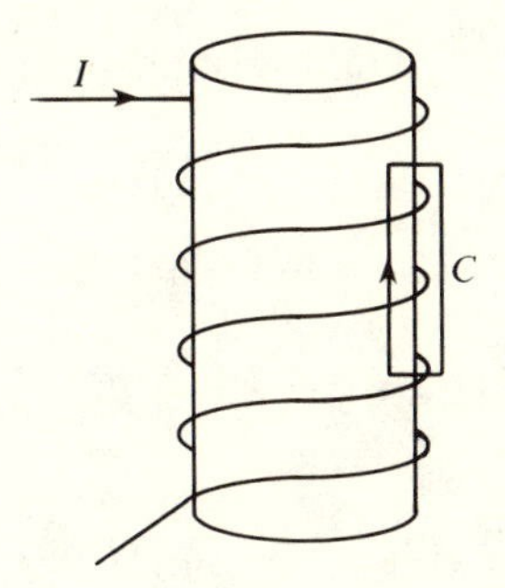

Figure 6.37 The long solenoid.

Example 6.13

What is the magnetic intensity vector field **H** inside a wire of radius a which carries a uniformly distributed line current of I amperes? The wire is straight, is infinitely long, and has permeability $\mu = \mu_0$.

Ans. The current density in the wire is

$$J = \frac{I}{\pi a^2}$$

Integrating (6.45) over a circular surface of radius $r < a$ and applying Stokes's theorem, we obtain

$$\oint \mathbf{H} \cdot d\mathbf{l} = 2\pi r H = \int_0^r J 2\pi r \, dr = J\pi r^2$$

$$H = \left(\frac{I}{2\pi a^2}\right) r$$

The Divergenceless of B

It is a universally observed fact that the magnetic field **B** has neither sources nor sinks. We recall from Section 6.1 that if a field has a source or sink at a given point in space, then the divergence of the field is nonvanishing at that point. So the fact that the field **B** has neither sources nor sinks may be expressed by the equation

$$\nabla \cdot \mathbf{B} = 0 \tag{6.46}$$

Thus in nature one never observes a net flux of **B**-field lines from a region of space. **B** field lines eventually close on themselves.

In terms of the magnetic intensity **H**, (6.46) is written

$$\nabla \cdot \mathbf{H} = 0 \tag{6.47}$$

Let us return to Ampere's law (6.45). In vacuum, $\mathbf{J} = 0$ and this equation becomes

$$\nabla \times \mathbf{H} = 0 \tag{6.48}$$

If we form the curl of both sides of this equation (see Appendix A), there results

$$\nabla \times (\nabla \times \mathbf{H}) = \nabla(\nabla \cdot \mathbf{H}) - \nabla^2 \mathbf{H} = 0$$

With (6.47), this expression reduces to

$$\nabla^2 \mathbf{H} = 0 \tag{6.49}$$

So we see that in magnetostatics the field obeys Laplace's equation. Note that Laplace's equation of a vector field in Cartesian coordinates comprises three Laplace equations for the three components of the field.

6.5 FARADAY'S LAW

Faraday's law states that an **E**-field is induced by a changing magnetic field. The electric field so generated obeys the relation

$$\nabla \times \mathbf{E} = -\frac{\partial \mathbf{B}}{\partial t} \tag{6.50}$$

This equation may be cast in more recognizable form with the aid of Stokes's theorem (6.8). Integrating both sides of (6.50) over a surface S bounded by the curve C, we obtain

$$\int_S \nabla \times \mathbf{E} \cdot d\mathbf{S} = \oint_C \mathbf{E} \cdot d\mathbf{l} = -\frac{\partial}{\partial t} \int_S \mathbf{B} \cdot d\mathbf{S} \tag{6.51}$$

The right-hand side of this equation represents the rate of change of magnetic flux

$$\Phi_B \equiv \int_S \mathbf{B} \cdot d\mathbf{S} \tag{6.52}$$

which passes through the surface S. The middle term in (6.51) represents the emf developed over the closed loop C due to this changing flux. The second equality in (6.51),

$$\oint_C \mathbf{E} \cdot d\mathbf{l} = -\frac{\partial}{\partial t} \Phi_B \tag{6.53}$$

is called Faraday's law. The flux Φ_B is measured in webers, and as noted previously

$$1 \text{ tesla} = 1 \text{ weber/m}^2$$

We note further that Faraday's law (6.50) indicates that if B is constant, $\partial B/\partial t = 0$ and we recapture (6.23):

$$\nabla \times \mathbf{E} = 0$$

appropriate to electrostatics. Taking the curl of this equation gives (see Appendix A)

$$\nabla \times (\nabla \times \mathbf{E}) = \nabla(\nabla \cdot \mathbf{E}) - \nabla^2 \mathbf{E} = 0$$

In charge-free space, Gauss's law (6.18) indicates that $\nabla \cdot \mathbf{E} = 0$ and the preceding equation gives

$$\nabla^2 \mathbf{E} = 0 \tag{6.54}$$

As noted previously, this equation represents three Laplace equations for the three components of **E** respectively and is satisfied by the electrostatic field in charge-free space. So we see that in electrostatics as well as in magnetostatics (6.49), fields obey Laplace's equation.

6.6 FIELD JUMP CONDITIONS

With the field equations obtained to this point, it is now possible for us to obtain the jump conditions which are obeyed by fields across surfaces of discontinuity. The pertinent equations are

$$\nabla \cdot \mathbf{D} = \rho \tag{6.21}$$

$$\nabla \times \mathbf{H} = \mathbf{J} \tag{6.45}$$

$$\nabla \cdot \mathbf{B} = 0 \tag{6.46}$$

$$\nabla \times \mathbf{E} = -\frac{\partial \mathbf{B}}{\partial t} \tag{6.50}$$

Consider first the jump undergone by the displacement field **D** in crossing a surface which separates two homogeneous media with respective dielectric constants ϵ_1 and ϵ_2. The configuration is shown in Fig. 6.38. Integrating (6.21) over the "pillbox" volume drawn above and applying Gauss's theorem (6.4), we obtain

$$\mathbf{D}_2 \cdot \mathbf{A} - \mathbf{D}_1 \cdot \mathbf{A} = \rho A \delta$$

The caps of the "pillbox" have area A and are separated by the distance δ. Let $D_\perp$ represent the component of **D** perpendicular to the surface of

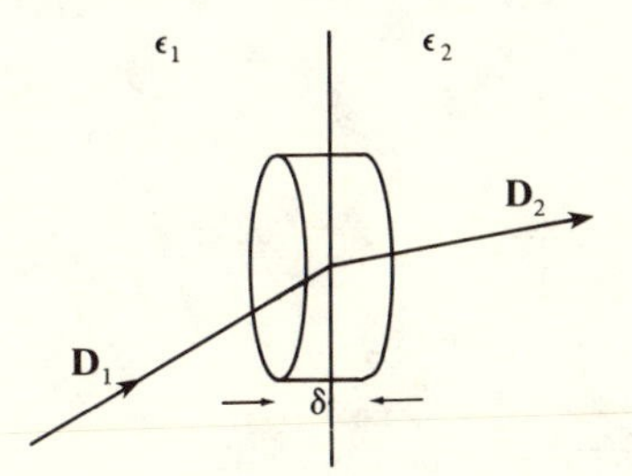

Figure 6.38 Integration volume for the jump in **D**.

discontinuity. Then the last equation may be rewritten

$$D_{2\perp} - D_{1\perp} = \rho\delta$$

In the limit that the thickness $\delta \to 0$, we obtain

$$D_{2\perp} - D_{1\perp} = \sigma$$

where σ is surface charge density with units C/m^2. This equation may be concisely written

$$[D_\perp] = \sigma \tag{6.55}$$

where [ ] denotes the jump. Introducing the dielectric constant ϵ permits (6.55) to be written in terms of electric field:

$$[\epsilon E_\perp] = \sigma \tag{6.56}$$

The jump conditions for the parallel components of the electric field follow from Faraday's law (6.50). Integrating this equation over the rectangular loop C shown in Fig. 6.39 and employing Stokes's theorem (6.8), we obtain

$$[E_{2\parallel} - E_{1\parallel}]l = l\delta \frac{\partial B}{\partial t}$$

The subscript $\parallel$ denotes the component of **E** parallel to the surface of discontinuity, and B represents the component of **B** normal to the loop C. Passing to the limit $\delta \to 0$, we obtain

$$[E_\parallel] = 0 \tag{6.57}$$

or

$$[D_\parallel/\epsilon] = 0 \tag{6.58}$$

The jump conditions for the normal components of the magnetic field **B** follow from (6.46). Integrating this equation over the volume of the "pill-box" shown above and applying Gauss's theorem (6.4), we find

$$[B_\perp] = 0 \tag{6.59}$$

or

$$[\mu H_\perp] = 0 \tag{6.60}$$

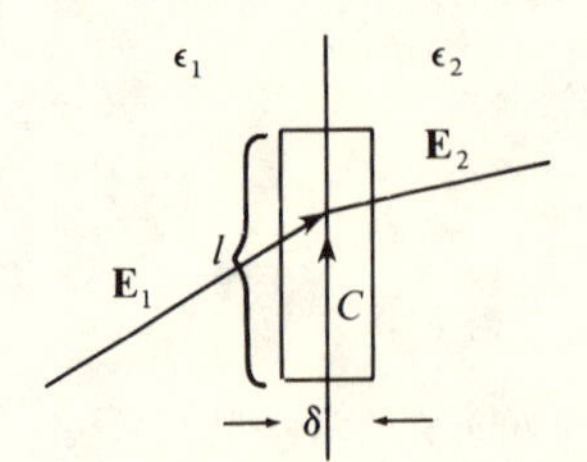

Figure 6.39 Integration loop for the jump in **E**.

The jump conditions for the tangential components of magnetic field follow from Ampere's law (6.45). A current density **J** flows through two homogeneous media with respective permeability constants μ_1 and μ_2. Integrating (6.45) over the rectangular surface shown above, we find

$$[H_{2\parallel} - H_{1\parallel}]l = J_\perp l\delta \tag{6.61}$$

where $J_\perp$ denotes the component of **J** normal to the rectangular loop. The *surface current* $K_\perp$ is defined as

$$K_\perp = \lim_{\delta \to 0} J_\perp \delta \tag{6.62}$$

The units of K are A/m. Passing to the limit $\delta \to 0$, (6.61) gives

$$[H_\parallel] = K_\perp \tag{6.63}$$

or

$$[B_\parallel/\mu] = K_\perp \tag{6.64}$$

Note that magnetic intensity **H** and surface current **K** have the same dimensions, viz., A/m.

Example 6.14

A uniform current $K_z = 6.2$ A/m flows in the $x = 0$ plane, in the $+z$ direction, as shown in Fig. 6.40. A magnetic field $B_1 = 1.3$ gauss is incident on the $x = 0$ plane at an angle $\theta_1 = \pi/6$ radians. If $\mu_1 = \mu_0$ and $\mu_2 = 1.3\mu_0$, what are the magnetic intensity vectors on both sides of the current sheet?

Ans. From (6.59) we obtain

$$B_{2x} = B_1 \cos\theta_1$$

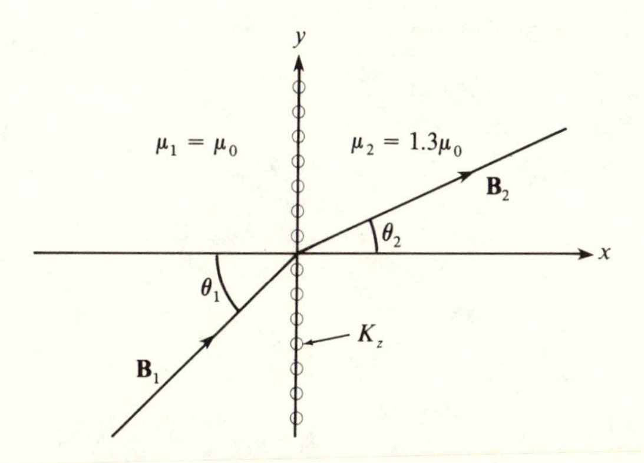

Figure 6.40 The surface current K_z.

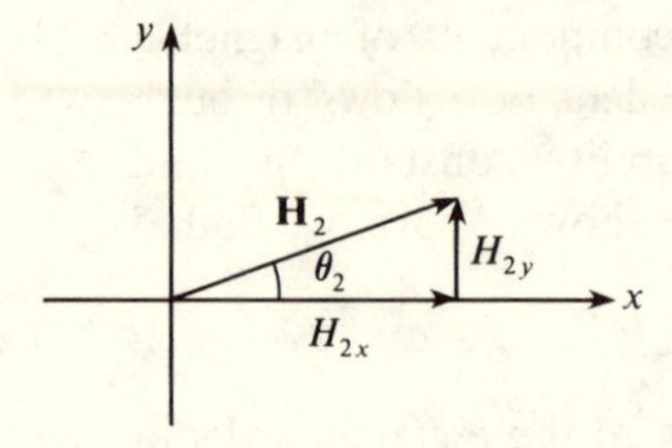

Figure 6.41 Components of $\mathbf{H}_2$.

so that

$$H_{2x} = \frac{B_1 \cos\theta_1}{\mu_2}$$

From (6.63) we obtain

$$H_{2y} = H_{1y} + K_z$$

or

$$H_{2y} = \frac{B_1 \sin\theta_1}{\mu_1} + K_z$$

It follows that the angle θ_2 that H_2 makes with the normal to the surface current K, as shown in Fig. 6.41, is given by

$$\tan\theta_2 = \frac{H_{2y}}{H_{2x}} = \frac{\dfrac{B_1 \sin\theta_1}{\mu_1} + K_z}{\dfrac{B_1 \cos\theta_1}{\mu_2}}$$

The SI value of B_1 is

$$B_1 = 1.3 \times 10^{-4} \text{ tesla}$$

so that

$$H_{2y} = 57.9 \text{ A/m}$$

$$H_{2x} = 68.9 \text{ A/m}$$

$$\tan\theta_2 = 0.84$$

$$\theta_2 = 0.70 \text{ rad} = 40.1°$$

$$\theta_1 = \pi/6 \text{ rad} = 30.0°$$

$$H_1 = \frac{B_1}{\mu_1} = 103 \text{ A/m}$$

$$H_2 = 90 \text{ A/m}$$

Example 6.15

A uniform surface current K_z flows in the $x = 0$ plane, which separates two homogeneous nonconducting media with respective permeabilities μ_1

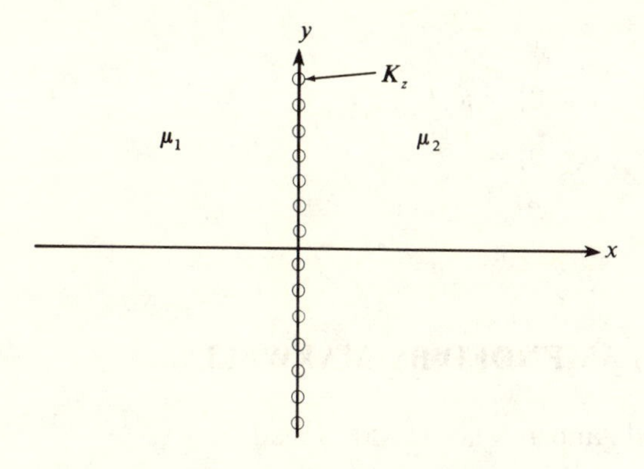

Figure 6.42 Configuration for Example 6.15.

and μ_2. (See Fig. 6.42.) It is known that **H** at $x = \pm\infty$ has only y-components, which are of equal magnitude.

a. What is **H** everywhere?
b. Show that your solution satisfies Laplace's equation (6.49) for **H**.
c. What is **B** everywhere?

Ans.

a. From the jump condition (6.63) we see that the y-components of **H** across the $x = 0$ plane satisfy the equation

$$H_y(x = 0_+) = H_y(x = 0_-) + K_z$$

Since the magnitudes of H_y are equal at $x = \pm\infty$ and H has no sources other than K_z, we see that the solution comprising the constant components

$$H_y(x > 0) = K_z/2$$
$$H_y(x < 0) = -K_z/2$$
$$H_x = H_z = 0$$

satisfies all boundary conditions.

b. Laplace's equation (6.49) is equivalent to the three equations

$$\nabla^2 H_x = 0$$
$$\nabla^2 H_y = 0$$
$$\nabla^2 H_z = 0$$

In the above solution H_x, H_y, and H_z are constants and therefore trivially solve these equations.

c. The components of B are

$$B_y(x > 0) = \mu_2 K_z/2$$
$$B_y(x < 0) = -\mu_1 K_z/2$$
$$B_x = B_z = 0$$

6.7 AMPERE'S LAW AS AMENDED BY MAXWELL

In 1864, Maxwell generalized Ampere's law (6.45) to read

$$\nabla \times \mathbf{H} = \mathbf{J} + \frac{\partial \mathbf{D}}{\partial t} \tag{6.65}$$

The time derivative of $\mathbf{D}$ on the right-hand side finds motivation in the *continuity equation* for the charge density ρ (C/m^3) and current density $\mathbf{J}$:

$$\frac{\partial \rho}{\partial t} + \nabla \cdot \mathbf{J} = 0 \tag{6.66}$$

Integrating this equation over a closed volume V and applying Gauss's theorem (6.4), we obtain

$$\frac{\partial}{\partial t} Q = -\oint \mathbf{J} \cdot d\mathbf{S}, \qquad Q = \int_V \rho \, d\mathbf{r}$$

The charge Q is contained in the volume V. This latter equation says that the only way the contained charge Q can change is by virtue of charge passing through the bounding surface. So (6.66) is a very fundamental equation. It says that charge is *conserved*.

Is the Ampere-Maxwell equation (6.65) consistent with the continuity equation (6.66)? Taking the divergence of both sides of (6.65) and recalling the rule (6.9), we obtain

$$\frac{\partial}{\partial t}(\nabla \cdot \mathbf{D}) + \nabla \cdot \mathbf{J} = 0 \tag{6.67}$$

Gauss's law (6.21) tells us that the divergence of the field $\mathbf{D}$ is the charge density ρ. So (6.67) returns the continuity equation (6.66). Note specifically that Ampere's equation (6.45), which contains only $\mathbf{J}$ on its right-hand side, contradicts the continuity equation.

Further physical significance of the Maxwell-Ampere equation (6.65) is revealed if we integrate it over a surface S bounded by the curve C. With Stokes's theorem (6.8), we obtain

$$\oint \mathbf{H} \cdot d\mathbf{l} = I + \frac{\partial}{\partial t} \int_S \mathbf{D} \cdot d\mathbf{S} \tag{6.68}$$

or equivalently

$$\oint \mathbf{H} \cdot d\mathbf{l} = I + I_D \tag{6.69}$$

where we have written I_D for the *displacement current*

$$I_D \equiv \frac{\partial}{\partial t} \int_S \mathbf{D} \cdot d\mathbf{S} \tag{6.70}$$

Equation (6.69) indicates that displacement current as well as true current will support a magnetic field.

Ohm's Law

It will prove convenient for later discussion to write the Maxwell-Ampere equation (6.65) entirely in terms of fields. This may be accomplished with the aid of Ohm's law.

This law is appropriate to homogeneous isotropic conducting media, for which it is found that the current is linearly proportional to the applied electric field:

$$\mathbf{J} = \sigma \mathbf{E} \tag{6.71}$$

The constant σ represents the *conductivity* of the medium. For current flowing in a uniform wire of cross-sectional area A and length l, the conductivity is inversely proportional to the resistance R (see Fig. 6.43):

$$R = \frac{l}{\sigma A} = \rho \frac{l}{A} \tag{6.72}$$

The resistivity ρ of the medium, like the conductivity σ, is a property of the substance, not of its geometry. The dimensions of these parameters are as follows.

$$[R] = \text{ohms}$$

$$[\sigma] = \frac{1}{\text{ohm-m}} = \text{mho/m}$$

$$[\rho] = \text{ohm-m}$$

A few typical values of resistivity are listed in Table 6.1.

With Ohm's law (6.71), the Ampere-Maxwell equation (6.65) may be written

$$\nabla \times \mathbf{H} = \sigma \mathbf{E} + \epsilon \frac{\partial \mathbf{E}}{\partial t} \tag{6.73}$$

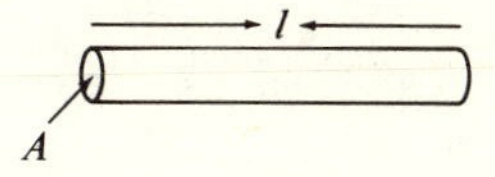

Figure 6.43 The parameters l and A.

Table 6.1

Material	Resistivity ρ (ohm-m at 20°C)	Density (g/cm^3)
Aluminum	2.8×10^{-8}	2.7
Copper	1.7×10^{-8}	8.9
Iron	1.0×10^{-7}	7.8
Nickel	7.8×10^{-8}	8.9
Silver	1.6×10^{-8}	10.5

Maxwell's Equations

In summary, the equations we have found which couple the fields **D**, **E**, **B**, **H**, the current density **J**, and the charge density ρ, are

$$\nabla \cdot \mathbf{D} = \rho \qquad [\text{Eq. } (6.21)] \tag{6.74a}$$

$$\nabla \cdot \mathbf{B} = 0 \qquad [\text{Eq. } (6.46)] \tag{6.74b}$$

$$\nabla \times \mathbf{E} = -\frac{\partial \mathbf{B}}{\partial t} \qquad [\text{Eq. } (6.50)] \tag{6.74c}$$

$$\nabla \times \mathbf{H} = \mathbf{J} + \frac{\partial \mathbf{D}}{\partial t} \qquad [\text{Eq. } (6.65)] \tag{6.74d}$$

As we shall find in the following chapter, one of the more significant properties of these equations is their implication of wave propagation.

PROBLEMS

6.1. Using the results of Example 6.11, obtain the surface charge density induced on the surface of a grounded conducting sphere of radius a by a point charge q which is a distance d from the origin of the sphere. Express your answer as a function of the polar angle θ with the center of the sphere as the origin; that is, find $\sigma(\theta)$.

6.2. A surface current K flowing in the $y = 0$ plane separates two half spaces each of which carries constant homogeneous **H**-fields. In the right half space $y > 0$,

$$\mathbf{H} = (0, 0, H_0)$$

and in the left half space $y < 0$,

$$\mathbf{H} = (0, H_0, 0)$$

What is the direction and magnitude of **K**?

6.3. A particle of charge q and mass m at time $t = 0$ is located at the position $r = (0, a, 0)$ and has velocity

$$\mathbf{v} = (v_0, 0, 0)$$

The particle moves in a domain of constant homogeneous **B**-field which points in the z-direction and has magnitude B_0.

a. What is the trajectory of the charge, $x = x(t)$, $y = y(t)$, $z = z(t)$?

b. If the particle is an electron with initial energy of 3.1 keV and $B_0 = 10$ gauss, what is the radius a (in Å) of the orbit which the electron executes?

6.4. A torus of sectional radius a and orbital radius R is composed of nonconducting material with high permeability, $\mu \gg \mu_0$. N turns of wire are confined to a small region of the torus, opposite to which there is a narrow air gap. (See Fig. 6.44.)

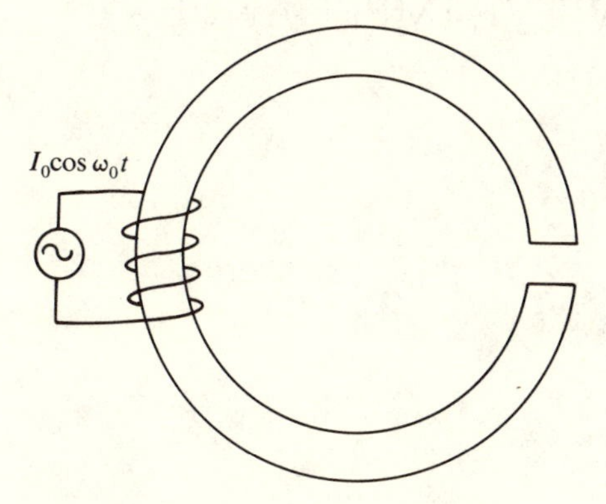

Figure 6.44 Configuration for Problem 6.4.

a. If a current $I = I_0 \cos \omega t$ is driven through the current loop, show that the approximate magnitude of induced electric field in the vicinity of the gap is

$$E \simeq \frac{\omega_0 \mu N I_0 \sin \omega_0 t}{4\pi}$$

with corresponding time-averaged value

$$E_{\text{rms}} = \frac{\omega_0 \mu N I_0}{4\pi\sqrt{2}}$$

b. What is the direction of this **E**-field?

6.5. The four sides of a right square cylinder are maintained at potentials V_0 and 0 as shown in Fig. 6.45.

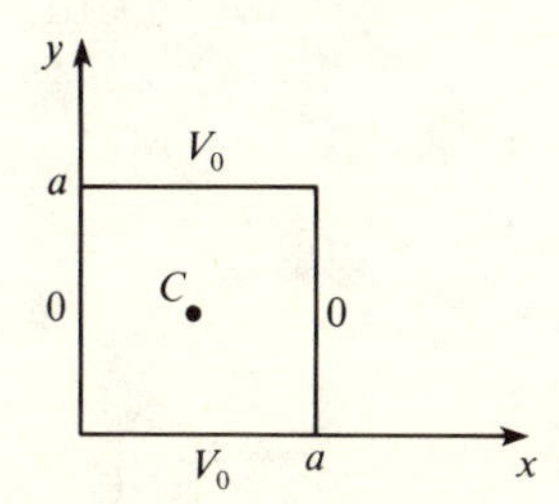

Figure 6.45 Configuration for Problem 6.5.

a. Using the method of separation of variables, obtain $V(x, y)$ everywhere inside the cylinder as an infinite Fourier series.
b. Obtain the potential $V(C)$ at the center of the rectangle from symmetry considerations, and compare this value with that obtained from your answer to part a.
c. If $V_0 = 3$ volts and $a = 1.5$ cm, estimate the components of **E** at $(a, a/2)$, in V/m.

CHAPTER

SEVEN

PLANE WAVES AND WAVE IMPEDANCE

7.1 THE WAVE EQUATION

We now wish to demonstrate that, as indicated at the close of the preceding chapter, Maxwell's equations (6.74) imply wave propagation. In a homogeneous domain free of charge and current with permittivity ϵ and permeability μ, Eqs. (6.74) assume the form

$$\nabla \cdot \mathbf{E} = 0 \tag{7.1a}$$

$$\nabla \cdot \mathbf{H} = 0 \tag{7.1b}$$

$$\nabla \times \mathbf{E} = -\mu \frac{\partial \mathbf{H}}{\partial t} \tag{7.1c}$$

$$\nabla \times \mathbf{H} = \epsilon \frac{\partial \mathbf{E}}{\partial t} \tag{7.1d}$$

Taking the curl of (7.1c) gives

$$\nabla \times (\nabla \times \mathbf{E}) = -\mu \frac{\partial}{\partial t} \nabla \times \mathbf{H}$$

which with (7.1d) gives

$$\nabla \times (\nabla \times \mathbf{E}) = \mu\epsilon \frac{\partial^2}{\partial t^2} \mathbf{E} \tag{7.2}$$

The left-hand side of this equation has the expansion (see Appendix A)

$$\nabla \times (\nabla \times \mathbf{E}) = \nabla(\nabla \cdot \mathbf{E}) - \nabla^2 \mathbf{E}$$

With (7.1a) we obtain

$$\nabla \times (\nabla \times \mathbf{E}) = -\nabla^2 \mathbf{E}$$

and (7.2) becomes

$$\nabla^2 \mathbf{E} - \frac{1}{v^2} \frac{\partial^2 \mathbf{E}}{\partial t^2} = 0 \tag{7.3}$$

$$v^2 = \frac{1}{\mu \epsilon} \tag{7.4}$$

Equation (7.3) represents three independent equations, one for each of the components (E_x, E_y, E_z). The one-dimensional form of (7.3) was encountered previously in (2.7).

In similar manner to the analysis following (2.7), we now find that any vector function of $\omega t - \boldsymbol{\beta} \cdot \mathbf{r}$ is a solution to (7.3). Specifically, a solution appropriate to harmonically varying fields is written

$$E = E_+ \exp j(\omega t - \boldsymbol{\beta} \cdot \mathbf{r}) \tag{7.5}$$

which permits us to identify β as the wavenumber

$$\beta = \frac{2\pi}{\lambda} \tag{7.6}$$

and ω as the angular frequency of the wave. Substituting (7.5) into (7.3) gives

$$\left(\beta^2 - \frac{\omega^2}{v^2}\right)\mathbf{E} = 0$$

We may conclude that (7.5) is a nontrivial solution of (7.3) provided

$$\beta^2 = \omega^2/v^2 \tag{7.7}$$

or equivalently

$$\beta = \pm\omega/v \tag{7.8}$$

At a fixed value of frequency, there are two solutions of the form (7.5), one with $\beta = +\omega/v$ and the other with $\beta = -\omega/v$. Since (7.3) is a linear, homogeneous equation, the superposition of these solutions is also a solution and we may write the more general solution

$$\mathbf{E} = \mathbf{E}_+ \exp[j(\omega t - \boldsymbol{\beta} \cdot \mathbf{r})] + \mathbf{E}_- \exp[-j(\omega t + \boldsymbol{\beta} \cdot \mathbf{r})] \tag{7.9}$$

As we shall find, the $\mathbf{E}_+$ wave propagates in the $\boldsymbol{\beta}$ direction whereas the $\mathbf{E}_-$ wave propagates in the $-\boldsymbol{\beta}$ direction.

Consider for example the E_+ wave (7.5). This solution is called a *plane wave*. The reason for this name is as follows. At any instant t, the electric field given by (7.5) has a constant value wherever the phase

$$\phi = \omega t - \boldsymbol{\beta} \cdot \mathbf{r} \tag{7.10}$$

is constant. So **E** is constant, at any instant, on the surface $\boldsymbol{\beta} \cdot \mathbf{r}$ = constant. This surface is depicted in Fig. 7.1. At later times E will maintain this same value on the surface where ϕ has maintained its starting value. From (7.10), we see that ϕ remains constant as t increases, provided $\boldsymbol{\beta} \cdot \mathbf{r}$ increases. From Fig. 7.1 we see that this condition corresponds to the plane surface of constant phase, moving away from the origin. (See Fig. 7.2.)

If E_+ in (7.5) lies in a constant direction, then the wave is said to be *plane polarized*. Under such circumstances, (7.5) is called a plane-polarized plane wave. The more general case is returned to in Section 7.5.

A similar wave equation to (7.3) follows from (7.1) for the magnetic field **H**:

$$\nabla^2 \mathbf{H} - \frac{1}{v^2} \frac{\partial^2 \mathbf{H}}{\partial t^2} = 0 \tag{7.11}$$

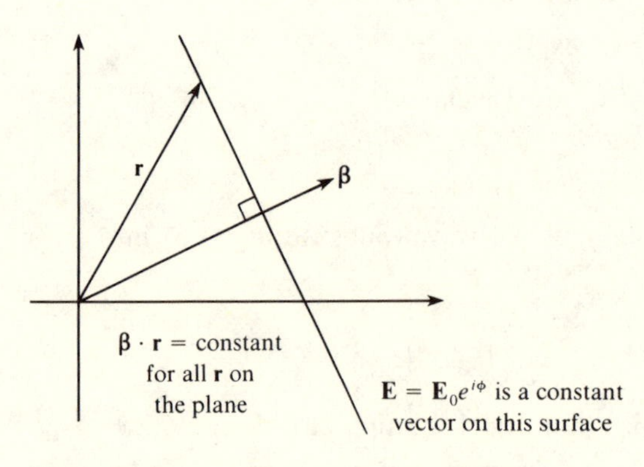

Figure 7.1 Plane-wave parameters: **r** and **β**.

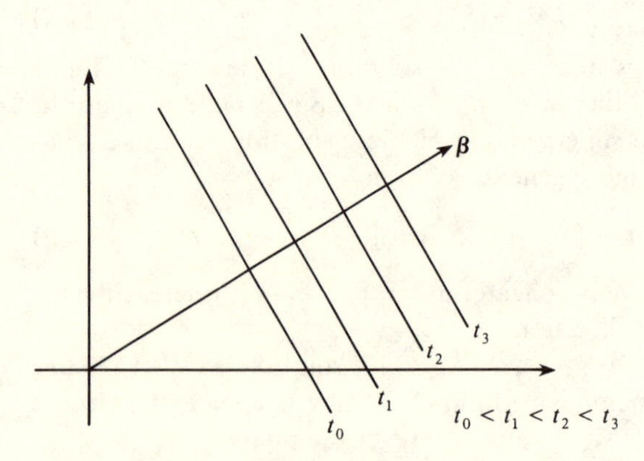

Figure 7.2 Plane-wave surfaces of constant phase.

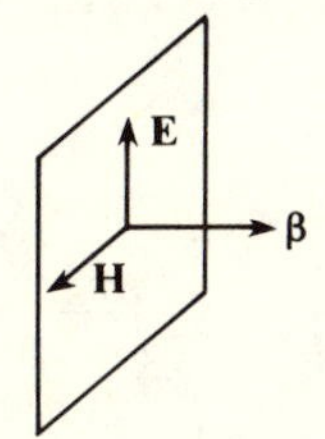

Figure 7.3 The vectors **E**, **H**, and **β** in a transverse wave.

Again we may assume the plane-wave solution

$$\mathbf{H} = \mathbf{H}_+ \exp j(\omega t - \boldsymbol{\beta} \cdot \mathbf{r}) \tag{7.12}$$

This solution, together with **E** given by (7.5) and with (7.1c), gives the relation

$$\boldsymbol{\beta} \times \mathbf{E} = \mu\omega\mathbf{H} \tag{7.13a}$$

The first two equations of (7.1) together with the solutions (7.5) and (7.12) yield

$$\boldsymbol{\beta} \cdot \mathbf{E} = \boldsymbol{\beta} \cdot \mathbf{H} = 0 \tag{7.13b}$$

These latter two equations indicate that the three vectors **β**, **E**, **H** form an orthogonal triad. We may conclude that the fields propagate as a transverse wave. (See Fig. 7.3.)

Wave Speed

Consider again the $\mathbf{E}_+$ wave (7.5). Since **β** is a constant vector, let us orient axes so that $\boldsymbol{\beta} = \mathbf{a}_z\beta$. Then (7.5) assumes the form

$$\begin{aligned} \mathbf{E} &= \mathbf{E}_+ \exp[j(\omega t - \boldsymbol{\beta} \cdot \mathbf{r})] \\ &= \mathbf{E}_+ \exp\left[-j\beta\left(z - \frac{\omega}{\beta}t\right)\right] \end{aligned} \tag{7.14}$$

Comparison with the waveform (2.7) and subsequent description indicates that the speed of the wave (7.14) is

$$v = \omega/\beta \tag{7.15}$$

This is called the *phase speed* of the wave. It is appropriate to a monochromatic wave, that is, a wave with a single, well-defined frequency. With (7.7) and the original identification (7.4), we may write

$$\beta = \frac{2\pi}{\lambda} = \omega\sqrt{\mu\epsilon} \tag{7.16}$$

Taking the magnitudes of both sides of (7.12) and keeping in mind that

β, E, H form an orthogonal triad, we obtain

$$H = \frac{\mu\omega}{\beta}E = \frac{1}{\sqrt{\mu/\epsilon}}E$$
$$= \frac{1}{\eta}E \tag{7.17}$$

Here we have used the parameter

$$\eta \equiv \sqrt{\frac{\mu}{\epsilon}}$$

which is called the *intrinsic wave impedance* of the medium. In free space,

$$\eta_0 = \sqrt{\frac{\mu_0}{\epsilon_0}} = 377\ \Omega$$

This parameter was encountered previously in (2.16).

So we have found that:

1. Maxwell's equations in a homogeneous region of space characterized by the constants μ, ϵ and otherwise free of charge and current have plane-wave solutions for **E** and **H**.
2. The phase velocity of a wave of frequency ω is

$$v = \frac{\omega}{\beta} = \frac{1}{\sqrt{\mu\epsilon}}$$

3. **E**, **H**, and $\boldsymbol{\beta}$ form an orthogonal triad.
4. The propagation vector $\boldsymbol{\beta}$ is normal to planes of constant phase.
5. **E** and **H** lie in these planes and are normal to each other.
6. $\mathbf{E} \times \mathbf{H}$ is in the direction of the propagation vector $\boldsymbol{\beta}$.
7. The magnitudes of the transverse fields are in the ratio $E/H = \eta = \sqrt{\mu/\epsilon}$.

Example 7.1

A plane electromagnetic wave propagates in the z-direction at 60 kHz. The amplitude of the electric field is 10^{-4} V/m.

a. What is the wavelength of this radiation?
b. What is the amplitude of the magnetic field?
c. What are the real electric and magnetic vectors for this wave? The wave is plane polarized with its electric vector in the x-direction.

Ans.

a.

$$\lambda = \frac{c}{f} = \frac{3 \times 10^8}{60 \times 10^3}$$
$$= 0.5 \times 10^4\ \text{m}$$

b.

$$H_0 = \frac{1}{\eta_0} E_0$$
$$= \frac{1}{377} \times E_0$$
$$= 2.65 \times 10^{-7} \text{A/m}$$

c. Since the wave propagates in the z-direction,

$$\boldsymbol{\beta} = \mathbf{a}_z \beta, \qquad \beta = \frac{2\pi}{\lambda} = 1.26 \times 10^{-3}\ \text{m}^{-1}$$

We are told that the electric vector is in the x-direction. With (7.12) there results

$$\mathbf{E} = \mathbf{a}_x E_0 \cos(\omega t - \beta z)$$
$$\mathbf{H} = \mathbf{a}_y \frac{E_0}{\eta_0} \cos(\omega t - \beta z)$$

7.2 POYNTING'S THEOREM

Let us return to the full Maxwell's equations (6.72). Dotting **H** into Faraday's law (6.72c), dotting **E** into the Ampere-Maxwell equation (6.72d), and subtracting the resulting equations gives

$$\mathbf{H} \cdot (\nabla \times \mathbf{E}) - \mathbf{E} \cdot (\nabla \times \mathbf{H}) = -\mathbf{E} \cdot \frac{\partial \mathbf{D}}{\partial t} - \mathbf{H} \cdot \frac{\partial \mathbf{B}}{\partial t} - \mathbf{J} \cdot \mathbf{E} \quad (7.18)$$

With the vector relation from Example 6.6, the left-hand side of this equation may be rewritten $-\nabla \cdot (\mathbf{E} \times \mathbf{H})$. Furthermore, in linear homogeneous media, $\mathbf{B} = \mu \mathbf{H}$ and $\mathbf{D} = \epsilon \mathbf{E}$. With these relations the time derivatives on the right-hand side of (7.18) may be rewritten

$$\mathbf{E} \cdot \frac{\partial \mathbf{D}}{\partial t} + \mathbf{H} \cdot \frac{\partial \mathbf{B}}{\partial t} = \frac{\partial}{\partial t} \left[\tfrac{1}{2} (\mathbf{E} \cdot \mathbf{D} + \mathbf{B} \cdot \mathbf{H}) \right]$$

Substituting these results into (7.18) gives

$$\frac{\partial}{\partial t} \left[\tfrac{1}{2} (\mathbf{E} \cdot \mathbf{D} + \mathbf{B} \cdot \mathbf{H}) \right] + \nabla \cdot (\mathbf{E} \times \mathbf{H}) = -\mathbf{J} \cdot \mathbf{E} \quad (7.19)$$

The vector

$$\mathbf{P} = \mathbf{E} \times \mathbf{H} \quad (7.20)$$

is called the *Poynting vector*, named for John Henry Poynting (1852–1914), and represents radiative power flow per unit area. Its dimensions are W/m^2. The quantity

$$u = \tfrac{1}{2} (\mathbf{E} \cdot \mathbf{D} + \mathbf{B} \cdot \mathbf{H}) \quad (7.21)$$

represents electromagnetic energy density and has dimensions of J/m^3. One may further separate u as

$$u = u_e + u_m \tag{7.22}$$

$$u_e = \tfrac{1}{2}\mathbf{E} \cdot \mathbf{D}, \qquad u_m = \tfrac{1}{2}\mathbf{B} \cdot \mathbf{H}$$

With the relabeling (7.20) and (7.21), we may rewrite (7.19) in the form

$$\frac{\partial}{\partial t}u + \nabla \cdot \mathbf{P} = -\mathbf{J} \cdot \mathbf{E} \tag{7.23}$$

This equation bears comparison with the continuity equation, (6.66). In vacuum the right-hand side of (7.23) is zero and (7.23) has identical form to the continuity equation (6.66). In this case, we may say again that the only way electromagnetic energy contained in a volume can change is by virtue of electromagnetic flux passing through the bounding surface. If currents are present, (7.22) indicates that this electromagnetic energy may also change by virtue of the work done by the fields on charges. This is the essence of the $-\mathbf{J} \cdot \mathbf{E}$ term in (7.22). These notions are made more clear if (7.22) is integrated over a volume V. With Gauss's theorem (6.4), and labeling the field energy

$$U = \int u \, dV \tag{7.24}$$

we obtain

$$\frac{\partial U}{\partial t} + \oiint P \, dS = -\int_V \mathbf{J} \cdot \mathbf{E} \, dV \tag{7.25}$$

The surface integral bounds the volume V. It follows that the electromagnetic energy U contained in V may decrease due to a positive flux of radiation flux out of the volume,

$$\oiint \mathbf{P} \cdot d\mathbf{S} > 0$$

or Joule heating,

$$\int \mathbf{J} \cdot \mathbf{E} \, dV > 0$$

Time-Average Poynting Vector

Consider the plane wave

$$\mathbf{E} = \mathbf{E}_+ e^{i(\omega t - \boldsymbol{\beta} \cdot \mathbf{r})}$$

$$\mathbf{H} = \frac{\hat{\boldsymbol{\beta}} \times \mathbf{E}_+}{\eta} e^{i(\omega t - \boldsymbol{\beta} \cdot \mathbf{r})} \tag{7.26}$$

The real Poynting vector for this field has the value

$$\mathbf{P} = \mathbf{E} \times \mathbf{H} = \frac{\boldsymbol{\beta}}{\mu\omega} E_+^2 \cos^2(\omega t - \boldsymbol{\beta} \cdot \mathbf{r})$$

The time-average power passing through one square meter of flat surface whose surface vector is parallel to $\boldsymbol{\beta}$ is given by

$$\langle \mathbf{P} \rangle = \tfrac{1}{2}\boldsymbol{\beta}\frac{E_+^2}{\mu\omega} = \hat{\boldsymbol{\beta}}\frac{1}{2\eta}E_+^2$$

This expression may also be obtained through the operation

$$\langle \mathbf{P} \rangle = \tfrac{1}{2}\,\mathrm{Re}\,\mathbf{E} \times \mathbf{H}^* \tag{7.27}$$

This form for the time average of the Poynting vector is a valid representation for harmonically varying fields.

In like manner we find for the time average of the stored energy densities (7.22),

$$\langle u_e \rangle = \tfrac{1}{4}\epsilon \mathbf{E} \cdot \mathbf{E}^*$$

$$\langle u_m \rangle = \tfrac{1}{4}\mu \mathbf{H} \cdot \mathbf{H}^* = \frac{1}{4}\frac{\mu}{\eta^2}\mathbf{E} \cdot \mathbf{E}^*$$

$$= \langle u_e \rangle \tag{7.28}$$

The average electric and magnetic energy densities carried in wave field are equal.

The time average of the power-loss term in (7.23) is given by

$$\langle P_{\text{loss}} \rangle = \tfrac{1}{2}\,\mathrm{Re}\,\mathbf{E} \cdot \mathbf{J}^* \tag{7.29}$$

Example 7.2

What are the time averages of **P**, u_e, and u_m for the plane electromagnetic wave described in Example 7.1?

Ans. With (7.23) we obtain

$$\langle \mathbf{P} \rangle = \tfrac{1}{2}\,\mathrm{Re}(\mathbf{E} \times \mathbf{H}^*) = \hat{\mathbf{a}}_z \frac{E_0 H_0}{2}$$

$$= \mathbf{a}_z \frac{10^{-4} \times 2.65 \times 10^{-7}}{2}$$

$$= \mathbf{a}_z 1.33 \times 10^{-11}\ \mathrm{J/m^2}$$

From (7.23) we find

$$\langle u_e \rangle = \langle u_m \rangle = \frac{\epsilon_0}{4}E_0^2 = \frac{8.85 \times 10^{-12}}{4} \times 10^{-8}$$

$$= 2.21 \times 10^{-20}\ \mathrm{J/m^3}$$

7.3 REFLECTION AND TRANSMISSION COEFFICIENTS

We suppose that an electromagnetic plane wave propagating in a dielectric medium with constants (ϵ_1, μ_1) is incident normally on a second dielectric medium with constants (ϵ_2, μ_2). The situation is shown in Fig. 7.4.

Taking $\boldsymbol{\beta}$ to lie in the $+z$ direction, with (7.12) we obtain

$$\mathbf{E}_+ = \mathbf{a}_x E_+ \exp j(\omega t - \beta z)$$
$$\mathbf{H}_+ = \mathbf{a}_y H_+ \exp j(\omega t - \beta z) \tag{7.30}$$

After incidence on the interface separating the two domains, a transmitted wave $(\mathbf{E}_T, \mathbf{H}_T)$ passes into the second dielectric and a reflected wave $(\mathbf{E}_-, \mathbf{H}_-)$ propagates in the $-\boldsymbol{\beta}$ direction. (See Fig. 7.5.) With (7.17) we obtain the relations

$$E_+ = \eta_1 H_+$$
$$E_- = \eta_1 H_-$$
$$E_T = \eta_2 H_T \tag{7.31}$$

The continuity conditions (6.57) for the tangential component of **E**,

$$[E_{\parallel}] = 0$$

and (6.63) for the tangential component of **H**,

$$[H_{\parallel}] = 0$$

together with (7.12), give the additional relations

$$H_+ + H_- = H_T$$
$$E_+ - E_- = E_T \tag{7.32}$$

Combining (7.31) and (7.32) gives the following *steady-state* relations:

$$\frac{E_-}{E_+} = \frac{\eta_2 - \eta_1}{\eta_2 + \eta_1} = \rho \tag{7.33}$$

and

$$\frac{E_T}{E_+} = \frac{2\eta_2}{\eta_2 + \eta_1} = \tau \tag{7.34}$$

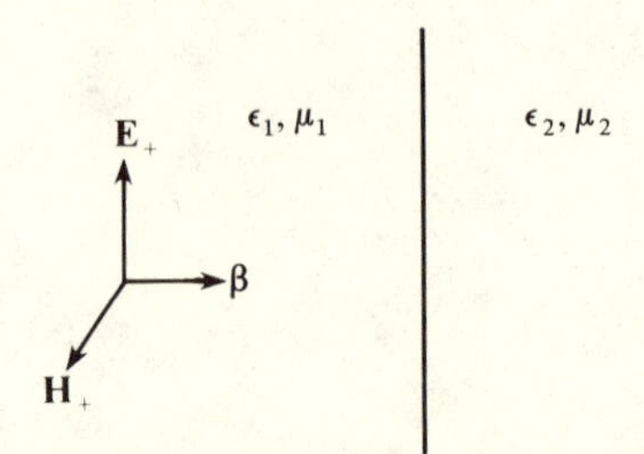

Figure 7.4 Plane wave incident on a half space.

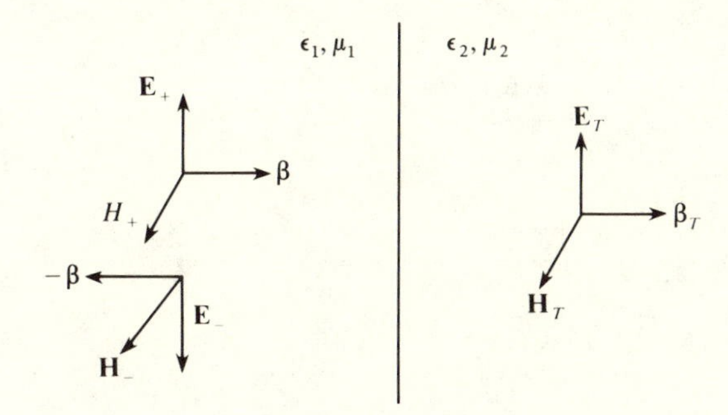

Figure 7.5 Forward, reflected, and transmitted waves.

These latter two relations serve to define the reflection coefficient ρ and transmission coefficient τ relevant to the steady-state, normal reflection and transmission of plane waves from an interface.

From (7.27) we see that the time-average Poynting vector for the forward wave has the value

$$\langle P_+ \rangle = \frac{1}{2\eta_1} |E_+|^2 \tag{7.35}$$

Similarly

$$\langle P_- \rangle = \frac{1}{2\eta_1} |E_-|^2 = \frac{1}{2\eta_1} \left| \frac{E_-}{E_+} \right|^2 |E_+|^2 \tag{7.36}$$

whereas the time-average Poynting vector in the second medium has the value

$$\langle P_T \rangle = \frac{1}{2\eta_2} |E_T|^2 = \frac{1}{2\eta_2} \left| \frac{E_T}{E_+} \right|^2 |E_+|^2 \tag{7.37}$$

From the last four equations we find

$$\langle P_+ \rangle = \langle P_- \rangle + \langle P_T \rangle \tag{7.38}$$

which is the same relation as that found previously (2.33) for transmission lines. With (7.33), the preceding power-flow formulas may be written

$$\langle P_- \rangle = |\rho|^2 \langle P_+ \rangle \tag{7.39a}$$

$$\langle P_T \rangle = (1 - |\rho|^2) \langle P_+ \rangle \tag{7.39b}$$

$$F \equiv \frac{\langle P_T \rangle}{\langle P_+ \rangle} = 1 - |\rho|^2 \tag{7.39c}$$

$$F_- \equiv \frac{\langle P_- \rangle}{\langle P_+ \rangle} = |\rho|^2 \tag{7.39d}$$

The parameter F represents the fractional transmitted power flux, whereas F_- represents the fractional reflected power flux.

Table 7.1

Plane waves	Transmission line
$E_x(z) = E_+ e^{-j\beta z} + E_- e^{j\beta z}$	$V(z) = V_+ e^{-j\beta z} + V_- e^{j\beta z}$
$H_y(z) = \frac{1}{\eta}[E_+ e^{-j\beta z} - E_- e^{j\beta z}]$	$I(z) = \frac{1}{Z_0}[V_+ e^{-j\beta z} - V_- e^{j\beta z}]$
$\beta = \omega\sqrt{\mu\epsilon}$	$\beta = \omega\sqrt{LC}$
$\eta = \sqrt{\frac{\mu}{\epsilon}}$	$Z_0 = \sqrt{\frac{L}{C}}$
$E_- = \rho E_+$	$V_- = \rho V_+$
$\rho = \frac{\eta_2 - \eta_1}{\eta_2 + \eta_1}$	$\rho = \frac{Z_L - Z_0}{Z_L + Z_0}$
$\langle P_- \rangle = \lvert\rho\rvert^2 \langle P_+ \rangle$	$\langle P_- \rangle = \lvert\rho\rvert^2 \langle P_+ \rangle$
$\langle P_+ \rangle = \langle P_- \rangle + \langle P_T \rangle$	$\langle P_+ \rangle = \langle P_- \rangle + \langle P_L \rangle$
$S = \frac{\lvert E_+ \rvert + \lvert E_- \rvert}{\lvert E_+ \rvert - \lvert E_- \rvert}$	$S = \frac{\lvert V_+ \rvert + \lvert V_- \rvert}{\lvert V_+ \rvert - \lvert V_- \rvert}$

Thus we find a precise analogy between the transmission-line concepts developed for the propagating voltage waves V_+, V_-, V_L (Section 2.7) and the fields E_+, E_-, E_T appropriate to a steady-state plane wave incident on an interface separating two dielectric half spaces. Again suppressing the time-dependent factor $e^{i\omega t}$ leads to the analogous relations listed in Table 7.1.

Thus we reach the significant conclusion that the formalism of the Smith chart may be applied to plane waves incident on an interface which separates two dielectric half spaces. The analogy also permits the relation (4.11) to be written for the intrinsic wave impedance,

$$Z(l) = \eta_1 \left(\frac{\eta_2 \cos \beta l + j\eta_1 \sin \beta l}{\eta_1 \cos \beta l + j\eta_2 \sin \beta l} \right) \tag{7.40}$$

This formula gives the ratio $Z = E/H$, l meters in front of the interface.

Dielectric Slabs

An immediate application of the last expression concerns the transmission and reflection of radiation from dielectric slabs. Consider the configuration shown in Fig. 7.6. What is the load impedance Z_L at the first interface? Application of (7.40) gives

$$Z_L = \eta_2 \left(\frac{\eta_3 \cos \beta_2 l + j\eta_2 \sin \beta_2 l}{\eta_2 \cos \beta_2 l + j\eta_3 \sin \beta_2 l} \right) \tag{7.41}$$

Reflection from the first interface will be eliminated if $Z_L = \eta_1$. Following

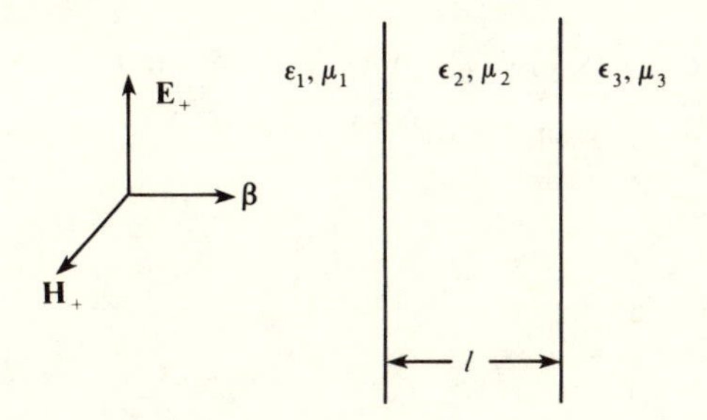

Figure 7.6 Dielectric slabs.

the description of Section 4.2, we set

$$\beta_2 l = \pi/2 \quad \text{and} \quad \eta_2^2 = \eta_1 \eta_3 \tag{7.42}$$

With these values substituted into (7.41) we find that $\eta_L = \eta_1$ and matching is achieved ($\rho = 0$). The width of the slab is $\lambda/4$, and it is called a *quarter-wave coating.*

Example 7.3

a. What is the reflection coefficient for plane-wave radiation incident from air on a medium of intrinsic impedance η, which is coated with N dielectric quarter-wavelength double slabs such as shown in Fig. 7.7?

b. Let the double layers be composed respectively of MgF_2 ($\epsilon_a/\epsilon_0 = 1.96$) and SbS_3 ($\epsilon_b/\epsilon_0 = 9$). What fraction of incident power is transmitted to a slab of glass ($\epsilon = 4\epsilon_0$) coated with two such double layers?

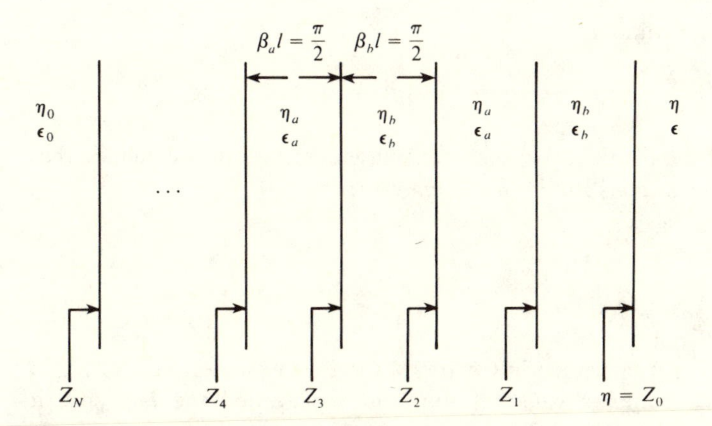

Figure 7.7 Configuration for Example 7.3.

Ans.

a. Since the plates are quarter-wave plates, (7.42) applies and we obtain

$$Z_1 = \frac{\eta_b^2}{Z_0} = \frac{\eta_b^2}{\eta}$$

At the second interface

$$Z_2 = \frac{\eta_a^2}{Z_1} = \frac{\eta\eta_a^2}{\eta_b^2}$$

At the third interface

$$Z_3 = \frac{\eta_b^2}{Z_2} = \frac{\eta_b^4}{\eta_a^2\eta}$$

Continuing in this manner, we obtain for N double layers

$$Z_N = \frac{\eta\eta_a^{2N}}{\eta_b^{2N}}$$

b. For three double layers as described above,

$$Z_3 = \eta\left(\frac{\eta_a}{\eta_b}\right)^6$$

$$= \tfrac{1}{2}\eta_0\left(\frac{\epsilon_b}{\epsilon_a}\right)^3 = \frac{\eta_0}{2}\left(\frac{9}{1.96}\right)^3$$

$$= \eta_0 \times 48.4$$

The reflection coefficient then has the value

$$\rho = \frac{Z_3 - \eta_0}{Z_3 + \eta_0} = \frac{\eta_0(48.4 - 1)}{\eta_0(48.4 + 1)} = 0.96$$

From (7.39) we find

$$\frac{\langle P_T \rangle}{\langle P_+ \rangle} = 1 - |\rho|^2 = 0.08$$

Only 8% of the incident power is transmitted, and we may conclude that the coated glass constitutes an efficient mirror.

7.4 COMPLEX LOAD

In the development presented in Chapter 4 for a transmission line feeding a given load, the complex value of the load represented the fact that it contained resistive and reactive components. We now wish to extend this notion to plane electromagnetic waves incident on a half space as described above. Losses incurred in the target medium are due to either ohmic heating or dielectric loss. If the medium is conducting, we may use Ohm's law (6.71), and the Maxwell-Ampere equation assumes the form (6.73). Inserting the $e^{j\omega t}$ time dependence, we obtain

$$\nabla \times \mathbf{H} = (\sigma + j\omega\epsilon)\mathbf{E}$$
$$= j\omega\epsilon\left(1 - j\frac{\sigma}{\omega\epsilon}\right)\mathbf{E} \tag{7.43}$$

For a lossy dielectric, one sets

$$\epsilon = \epsilon' - j\epsilon'' \tag{7.44}$$

The imaginary component ϵ'' then represents loss due to changes in the polarizability of the dielectric. Substituting (7.44) into the Ampere-Maxwell equation (7.1d), we obtain

$$\nabla \times \mathbf{H} = j\omega\epsilon'\left(1 - \frac{j\epsilon''}{\epsilon'}\right)\mathbf{E} \tag{7.45}$$

Values of ϵ'/ϵ_0 and ϵ''/ϵ_0 for some typical dielectrics are given in Table 7.2. Values for conductivity and the real part of the permittivity for various materials through the microwave range are given in Table 7.3. The frequency

Table 7.2

	ϵ'/ϵ_0			$10^4\ \epsilon''/\epsilon'$		
Material	$f = 10^6$	10^8	10^{10}	$f = 10^6$	10^8	10^{10}
Glass	4.00	4.00	4.00	8	12	21
Fused quartz	3.78	3.78	3.78	2	1	1
Ruby mica	5.4	5.4	—	3	2	—
Titania	100	100	—	3	2.5	—
Neoprene	5.7	3.4	—	950	1600	—

Table 7.3 $f \lesssim 10^5$ **MHz**

Material	σ (mho/m)	ϵ'/ϵ_0	f at which $\sigma = \omega\epsilon'$
Copper	5.80×10^7	—	Optical
Platinum	0.94×10^4	—	Optical
Germanium	2.2	16	2.5 GHz
Seawater	4	81	0.89 GHz
Fresh water	10^{-3}	81	0.22 MHz
Silicon	10^{-3}	12	1.5 MHz
Wet earth	10^{-3}	10	1.8 MHz
Dry earth	10^{-5}	5	3.6 MHz

at which displacement and conduction currents are equal is given in the last column.[1] Equation (7.43) may be placed in the form of (7.45) if we introduce an equivalent conductance,

$$\sigma = \omega\epsilon'' \tag{7.46}$$

With this observation, the form of the right-hand side of (7.45) is appropriate to dielectrics and conductive loss.

It follows that in lossy media, the formalism of Section 7.1 applies, with the modification (7.44). Since ϵ has an imaginary component, β as given by (7.16) is also imaginary. It is conventional in this case to introduce a complex wavenumber γ and reserve β for the imaginary part of γ. Thus, for plane wave propagation in the z direction (7.14) we write

$$\mathbf{E} = \mathbf{E}_+ e^{j\omega t} e^{-\gamma z} \tag{7.47}$$

$$\gamma = \alpha + j\beta$$

The generalization of (7.16) becomes

$$\gamma = j\omega\sqrt{\mu\epsilon'[1 - j(\epsilon''/\epsilon')]} \tag{7.48}$$

Solving for α and β gives

$$\alpha = \omega\sqrt{\frac{\mu\epsilon'}{2}\left[\sqrt{1 + \left(\frac{\epsilon''}{\epsilon'}\right)^2} - 1\right]}$$

$$\beta = \omega\sqrt{\frac{\mu\epsilon'}{2}\left[\sqrt{1 + \left(\frac{\epsilon''}{\epsilon'}\right)^2} + 1\right]} \tag{7.49}$$

The intrinsic impedance also becomes imaginary in lossy material:

$$\eta = \sqrt{\frac{\mu}{\epsilon}} = \sqrt{\frac{\mu}{\epsilon'[1 - j(\epsilon''/\epsilon')]}} \tag{7.50}$$

The complex value of the wave impedance η indicates that **E** and **H** in a plane wave are out of phase in a lossy material.

Two key approximations stem from (7.50). These are $\epsilon''/\epsilon' = \sigma/\omega\epsilon' \ll 1$ and $\epsilon''/\epsilon' \gg 1$.

---

[1] Data for Tables 7.2 and 7.3 were obtained from S. Ramo, J. Whinnery, and T. Van Duzer, "Fields and Waves in Communication Electronics" (Wiley, New York, 1965).

For low-loss materials ($\epsilon''/\epsilon' \ll 1$), the preceding expansions may be Taylor expanded to yield

$$\alpha \simeq \omega\sqrt{\mu\epsilon'}\,\frac{\epsilon''}{2\epsilon'} = \frac{\sigma}{2}\sqrt{\frac{\mu}{\epsilon'}} \tag{7.51a}$$

$$\beta \simeq \omega\sqrt{\mu\epsilon'}\left[1 + \frac{1}{8}\left(\frac{\epsilon''}{\epsilon'}\right)^2\right] \simeq \omega\sqrt{\mu\epsilon'}, \qquad \alpha \simeq \frac{\beta}{2}\,\frac{\sigma}{\omega\epsilon'} \tag{7.51b}$$

$$\eta \simeq \sqrt{\frac{\mu}{\epsilon'}}\left\{\left[1 - \frac{3}{8}\left(\frac{\epsilon''}{\epsilon'}\right)^2\right] + j\frac{\epsilon''}{2\epsilon'}\right\} \tag{7.51c}$$

This limit is appropriate to the case where displacement current far exceeds conduction current.

For conductive loss, the above relations apply with the substitution (7.46). Referring to (7.43), we see that conductivity effects dominate if

$$\frac{\sigma}{\omega\epsilon'} \gg 1 \tag{7.52}$$

which corresponds to the limit $\epsilon''/\epsilon' \gg 1$ in (7.48). There results, for a good conductor,

$$\gamma \simeq j\omega\sqrt{\frac{\mu\sigma}{j\omega}} = (1 + j)\sqrt{\frac{\mu\sigma\omega}{2}}, \qquad \alpha \simeq \beta = \sqrt{\frac{\mu\delta\omega}{2}} \tag{7.53}$$

Similarly (7.51) gives in this same limit

$$\eta = \sqrt{\frac{j\omega\mu}{\sigma}} = (1 + j)\sqrt{\frac{\mu\omega}{2\sigma}} \tag{7.54}$$

Referring to the waveform (7.47), we see that E falls off in z like $e^{-\alpha z}$. It follows that E is reduced by the factor e^{-1} in the distance

$$\delta = \alpha^{-1} = \sqrt{\frac{2}{\mu\sigma\omega}} \tag{7.55}$$

This penetration length in a good conductor is called the *skin depth*.

To sum up, for steady-state wave incidence on a low-loss dielectric, the wave load impedance assumes the form given by (7.52c) and we may write

$$\begin{aligned} Z_L &= \sqrt{\frac{\mu}{\epsilon'}}\left[1 - \frac{3}{8}\left(\frac{\epsilon''}{\epsilon'}\right)^2\right] + j\sqrt{\frac{\mu}{\epsilon'}}\left(\frac{\epsilon''}{2\epsilon'}\right) \\ &\simeq \sqrt{\frac{\mu}{\epsilon'}}\left[1 + \frac{j\sigma}{2\epsilon'\omega}\right] \end{aligned} \tag{7.56}$$

For incidence on a good conductor, the wave load impedance is given by (7.54):

$$Z_L = \sqrt{\frac{\mu\omega}{2\sigma}}\,(1+j) \tag{7.57}$$

These relations complete our analogy between V, I transmission-line relations and E, H plane-wave relations. Note, however, that application of (7.33) and (7.34) assumes no surface charge nor surface current on the separating interface.

Example 7.4

A plane electromagnetic field propagates in a good conductor with conductivity σ at the frequency f in the $+z$ direction. At $z = 0$, the electric field has the value

$$\mathbf{E}_+ = \mathbf{a}_x E_0 e^{j\omega t}$$

a. What is $\mathbf{E}$ for $z > 0$?
b. What is $\mathbf{H}$ for $z > 0$?
c. Show that the power dissipation per unit volume, $\text{Re}\,\frac{1}{2}\mathbf{E}\cdot\mathbf{J}^*$, integrated from $z = 0$ to $z = \infty$, equals the Poynting flux at $z = 0$.
d. What is the phase difference between E and H in the conductor?

Ans.

a. The electric field follows from (7.54):

$$\mathbf{E}(z,t) = \mathbf{a}_x E_0 e^{j\omega t} e^{-(1+j)z/\delta}$$

b. The magnetic field follows from (7.12) and (7.54):

$$\mathbf{H} = \mathbf{a}_y \frac{E_0 e^{j\omega t} e^{-(1+j)z/\delta}}{\eta}$$

$$= \mathbf{a}_y \frac{E_0 e^{j\omega t} e^{-(1+j)z/\delta}}{(1+j)\sqrt{\mu_0\omega/2\sigma}}$$

$$= \mathbf{a}_y \frac{E_0(1-j)}{\omega\mu_0\delta} e^{j\omega t} e^{-(1+j)z/\delta}$$

c. We must calculate the integral

$$P = \int_0^\infty \tfrac{1}{2}\text{Re}\,\mathbf{E}\cdot\mathbf{J}^*\,dz \qquad (\text{W/m}^2)$$

Setting $\mathbf{J} = \sigma \mathbf{E}$ gives

$$P = \int_0^\infty \frac{\sigma E_0^2}{2} e^{-2z/\delta}\, dz$$
$$= \frac{\sigma \delta E_0^2}{4}$$

The time-average incident power flux is given by (7.27). For the case at hand we find

$$|\langle \mathbf{S} \rangle|_{z=0} = \left| \tfrac{1}{2} \mathrm{Re}[\mathbf{E} \times \mathbf{H}^*] \right|_{z=0}$$
$$= \frac{E_0^2}{2\omega\mu_0\delta} \mathrm{Re}\left[(1+j)e^{-2z/\delta}\right]_{z=0}$$
$$= \frac{\sigma\delta E_0^2}{4}$$

which is the desired result.

d. The fact that

$$1 - j = \sqrt{2}\, e^{-j\pi/4}$$

indicates that $\mathbf{H}$ lags $\mathbf{E}$ by $\pi/4$ radians in the conductor.

Example 7.5

A uniform plane wave is normally incident on an expanse of seawater. The electric field is plane polarized in the x-direction and has amplitude 100 V/m. The wave frequency is 3 GHz. If at this frequency, $\epsilon''/\epsilon' \simeq 0.29$, find:

a. The electric and magnetic fields of the incident and reflected waves.
b. The transmitted electric and magnetic fields.
c. The time-average power flux of the three waves.
d. The standing-wave ratio of the incident and reflected waves.
e. The distance over which the power flow beneath the water surface will drop by 20 dB (for ϵ' see Table 7.3).

Ans.

a. Employing the low-loss expression (7.51c), we obtain the load impedance

$$\eta_L \simeq \sqrt{\frac{\mu_0}{\epsilon'}} \left\{ \left[1 - \frac{3}{8} \left(\frac{\epsilon''}{\epsilon'} \right)^2 \right] + j \frac{\epsilon''}{2\epsilon'} \right\}$$
$$= 41.9\,(0.097 + j0.145) + 41.2 \exp j0.15$$

With this value, we may constrict the reflection coefficient

$$\rho = \frac{\eta_L - \eta_0}{\eta_L + \eta_0} = \frac{j6.08 - 336}{j6.08 + 418}$$
$$= -0.80 + j0.026 = -0.80 \exp(-j0.033)$$

For the incident fields we find

$$\mathbf{E}_+ = \mathbf{a}_x E_+ \exp j(\omega t - \beta z)$$

$$\mathbf{H}_+ = \mathbf{a}_y \frac{E_+}{\eta_0} \exp j(\omega t - \beta z)$$

$$\beta = \frac{2\pi}{\lambda} = \frac{2\pi f}{c} = 62.8 \text{ m}^{-1}$$

$$E_+ = 100 \text{ V/m}$$

The amplitudes of the reflected waves are

$$E_- = \rho E_+ = -80 \exp(-j0.033)$$

$$H_- = -\frac{E_-}{\eta_0} = +0.21 \exp(-j0.033)$$

so that

$$\mathbf{E}_- = -\mathbf{a}_x 80 \exp j(\omega t + \beta z - 0.033)$$

$$\mathbf{H}_- = -\mathbf{a}_y 0.21 \exp j(\omega t + \beta z - 0.033)$$

b. With (7.34), the transmission coefficient at the load is

$$\tau = \frac{2\eta_L}{\eta_L + \eta_0} = 1 + \rho$$

The amplitude of the transmitted waves at the water surface have the values

$$E_T = \tau E_+ = 20.2 \exp(j0.13)$$

$$H_T = \frac{E_T}{\eta_L} = 0.48 \exp(-j0.01)$$

The fields in the water are then given by the expressions

$$\mathbf{E}_T = \mathbf{a}_x 20.2 \exp j(\omega t - \beta_T z + 0.13)\, e^{-\alpha z}$$

$$\mathbf{H}_T = \mathbf{a}_y 0.48 \exp j(\omega t - \beta_T z - 0.01)\, e^{-\alpha z}$$

where, from (7.51),

$$\beta_T \simeq \omega\sqrt{\mu\epsilon'} = 565 \text{ m}^{-1}$$

$$\alpha \simeq \frac{\omega\sqrt{\mu\epsilon'}}{2} \frac{\epsilon''}{\epsilon'} = 82 \text{ m}^{-1}$$

c. The time-average power fluxes for the three waves are

$$\langle P_+ \rangle = \tfrac{1}{2}\text{Re}(E_+ H_+^*) = \frac{100 \times 0.27}{2} = 13.5 \text{ W/m}^2$$

$$\langle P_- \rangle = \tfrac{1}{2}\text{Re}(E_- H_-^*) = \frac{80 \times 0.21}{2} = 8.4 \text{ W/m}^2$$

$$\langle P_T \rangle_{z=0} = \tfrac{1}{2}\text{Re}(E_T H_T^*)_{z=0} = \frac{20.2 \times 0.48 \times \cos 0.14}{2} = 4.8 \text{ W/m}^2$$

To within 2% accuracy we find $\langle P_+ \rangle = \langle P_- \rangle + \langle P_T \rangle$.

d. For the standing-wave ratio we find

$$S = \frac{|E_{\max}|}{|E_{\min}|} = \frac{1 + |\rho|}{1 - |\rho|} = \frac{1 + 0.8}{1 - 0.8} = 9$$

e. By definition we have

$$-20 \text{ dB} = 10 \log_{10}\left[\frac{\langle P_T(z) \rangle}{\langle P_T(0) \rangle}\right] = 10 \log_{10} e^{-2\alpha z}$$

$$= -20\alpha z \log_{10} e = -713z$$

Thus, $z = 0.028$ m.

7.5 CIRCULAR AND ELLIPTIC POLARIZATIONS

The fact that the wave equation (7.3) is linear and homogeneous implies the following (as previously discussed concerning Laplace's equation in Sec. 6.3). If E_1 and E_2 are two solutions to (7.3), then the linear superposition $E_1 + bE_2$ is also a solution. Consider for example the real solutions

$$\mathbf{E}_1 = \mathbf{a}_x E_0 \cos(\omega t - \beta z)$$

$$\mathbf{E}_2 = \mathbf{a}_y E_0 \sin(\omega t - \beta z)$$

Then

$$\mathbf{E} = \mathbf{E}_1 + b\mathbf{E}_2$$

$$= \mathbf{a}_x E_0 \cos(\omega t - \beta z) + b\mathbf{a}_y E_0 \sin(\omega t - \beta z)$$

is also a solution to (7.3). The components of this field are

$$E_x = E_0 \cos\phi, \qquad E_y = bE_0 \sin\phi$$

where ϕ is the phase of the wave as given by (7.10). With the above structure of the components of E we obtain

$$E_x^2 + \frac{1}{b^2} E_y^2 = E_0^2$$

This is the equation of an ellipse in the E_x, E_y plane. That is, in the E_x, E_y plane, the tip of the **E**-vector sweeps out an ellipse, and the field is said to be *elliptically polarized*. (See Fig. 7.8.) Since both E_x and E_y vibrate with frequency ω, it follows that **E** rotates with this same frequency. In the sketch above we have assumed that the **E**-vector is aligned with the x-axis at $t = 0$. Furthermore, the plane of the sketch is a fixed value of z, and the phase $\phi = \omega t + \text{const}$.

If $b = 1$, the magnitude $E = E_+$ is constant and the tip of the electric field sweeps out a circle in the E_x, E_y plane. The field is said to be *circularly polarized*. (See Fig. 7.9.)

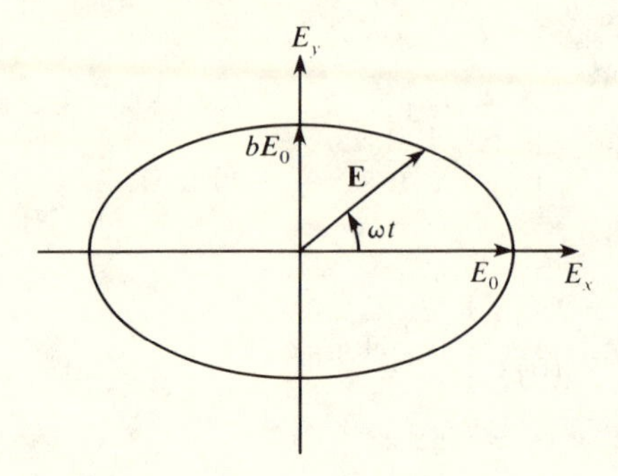

Figure 7.8 Electric vector in a plane of constant z for an elliptically polarized wave.

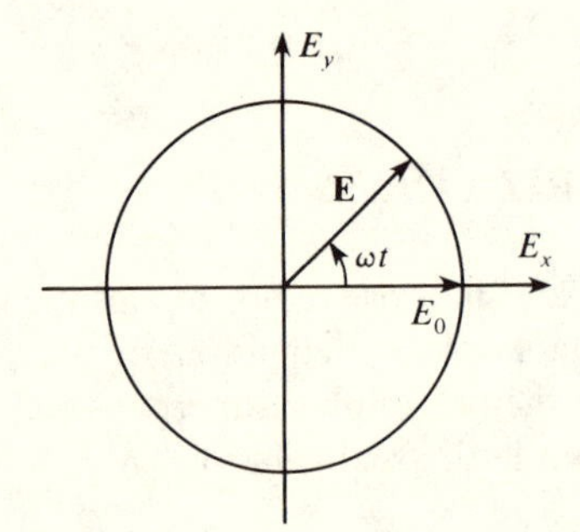

Figure 7.9 Circularly polarized wave.

In general, E_x and E_y may be out of phase by an arbitrary but constant amount and still be solutions to the wave equation (7.3). For example consider the field

$$\begin{aligned} E_x &= E_0 \cos\phi \\ E_y &= bE_0 \cos(\phi + \eta) \end{aligned} \tag{7.58}$$

where

$$\phi = \omega t - \beta z \tag{7.59}$$

In the example immediately above, the phase difference $\eta = \pi/2$. In the present case η is arbitrary. We wish to uncover the polarization of this field. That is, we seek the curve of E in the E_x, E_y plane. Expanding the form of E_y and resubstituting the form for E_x gives

$$\frac{E_y}{b} = E_0[\cos\phi\cos\eta - \sin\phi\sin\eta]$$

$$\frac{E_y}{b} = E_x \cos\eta - \sqrt{E_0^2 - E_x^2}\,\sin\eta$$

Simplifying, we obtain

$$E_x^2 - 2\frac{E_y}{b}E_x \cos\eta + \left(\frac{E_y}{b}\right)^2 = E_0^2 \sin^2\eta$$

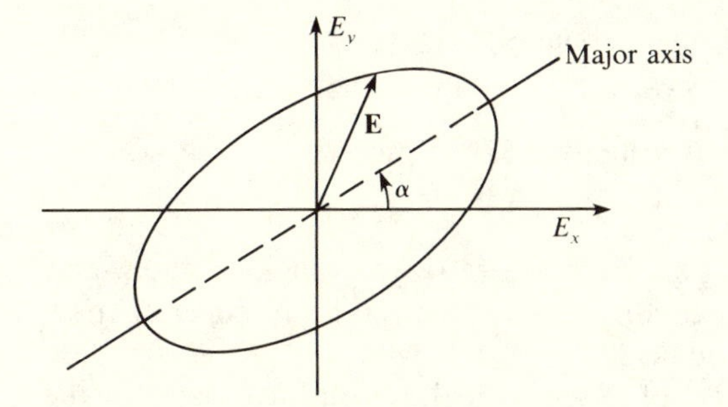

Figure 7.10 Ellipse with major axis rotated through the angle α.

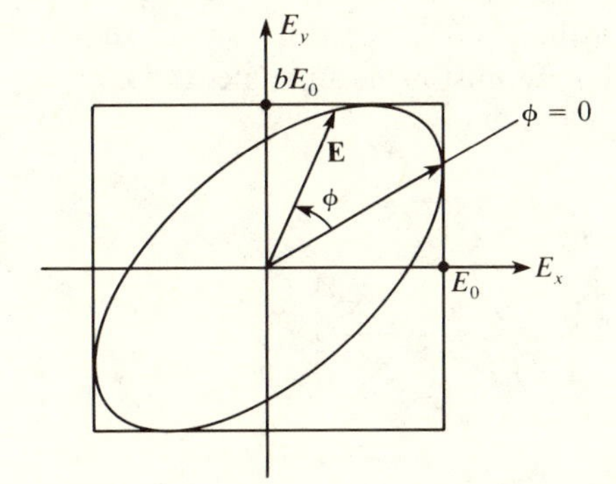

Figure 7.11 Ellipse inscribed in a rectangle.

This is again the equation of an ellipse in the E_x, E_y plane, but whose major axis is rotated through an angle α from the axis given by

$$\tan 2\alpha = \frac{2b\cos\eta}{1 - b^2}$$

(See Fig. 7.10.)

From (7.58) we see that the maximum value of E_x is E_0 which occurs at $\phi = 0$. The maximum value of E_y is bE_0. We may conclude that the ellipse (7.58) is inscribed within a rectangle of edge lengths $2E_0$ and $2bE_0$. (See Fig. 7.11.) When $\eta = 0$, the ellipse collapses to a straight line corresponding to linear polarization.

Example 7.6

A *polarizer* is a sheet of material which contains molecules so aligned that the sheet only passes light with electric vector parallel to a certain characteristic axis of the sheet. The transmitted light is plane polarized.

a. If plane-wave radiation with components

$$E_x = 5\cos\phi \text{ V/m}$$

$$E_y = 3\cos(\phi + 0.524) \text{ V/m}$$

$$\phi = \omega t - \beta z$$

is incident normally on a polarizing sheet, at what angle to the x-axis must the characteristic direction of the polarizer be oriented to pass maximum amplitude electric field?

b. What is the amplitude $\overline{E}_0$ of electric field transmitted through the polarizer at the angle described in part a?

Ans.

a. To effect maximum transmission, the axis of the polarizer must be brought parallel to the major axis of the ellipse of polarization, as shown in Fig. 7.12. Thus, the characteristic axis must be at an angle α from the x axis given, as above by

$$\alpha = \tfrac{1}{2}\tan^{-1}\left[\frac{2b\cos\eta}{1 - b^2}\right]$$

$$= \tfrac{1}{2}\tan^{-1}\left[\frac{2\times(3/5)\times\cos 0.524}{1-(3/5)^2}\right]$$

$$= 0.51 \text{ radian}$$

b. From Fig. 7.10 et seq. we may conclude that the angle θ which $\mathbf{E}$ makes with the x-axis is related to the phase angle ϕ by

$$\psi + \phi = \theta$$

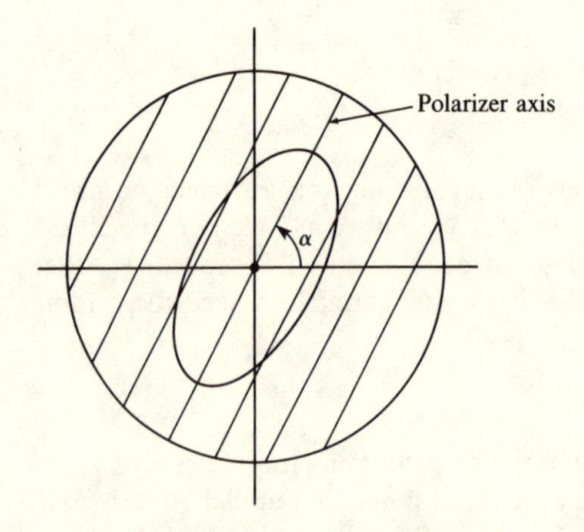

Figure 7.12 Configuration for Example 7.6.

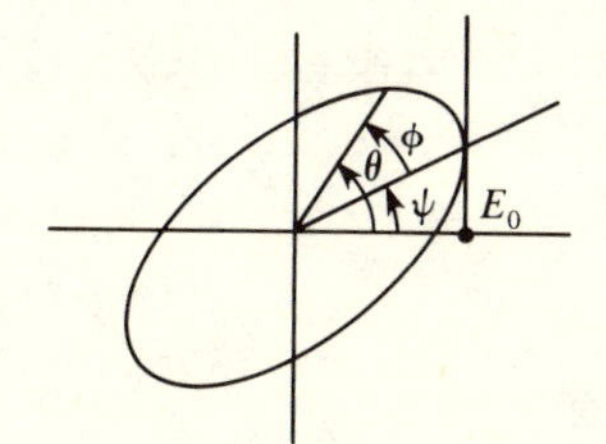

Figure 7.13

(see Fig. 7.13), where

$$\tan\psi = \frac{bE_0\cos\eta}{E_0} = b\cos\eta$$

On the major axis $\theta = \alpha$, so that

$$\begin{aligned}\bar{\phi} &= \alpha - \psi \\ &= 0.51 - \tan^{-1}(b\cos\eta) \\ &= 0.03 \text{ radian}\end{aligned}$$

At this value

$$\bar{E}_x = 5\cos 0.03 = 5.00$$

$$\bar{E}_y = 3\cos 0.554 = 2.55$$

and we find

$$\bar{E}_0 = \sqrt{\bar{E}_x^2 + \bar{E}_y^2} = 5.61 \text{ V/m}$$

Example 7.7

Circularly polarized electromagnetic waves of frequency 15 MHz is normally incident on a low-loss dielectric from air. The amplitude of the electric wave is 100 V/m. The dielectric has permittivity

$$\epsilon = \epsilon_0 + j0.1\epsilon_0$$

and the incident fields are related as

$$\mathbf{E}_+ = \mathbf{a}_x E_+\cos\phi + \mathbf{a}_y E_+\sin\phi$$

$$\mathbf{H}_+ = \mathbf{a}_y H_+\cos\phi - \mathbf{a}_y H_+\sin\phi$$

$$H_+ = \frac{1}{\eta_0}E_+, \qquad \phi = \omega t - \beta z$$

a. What is the fraction of power reflected,

$$F_- = \frac{\langle P_-\rangle}{\langle P_+\rangle}?$$

b. Obtain expressions for the transmitted fields E_T and H_T.

Ans.

a. From (7.39d)

$$F_{-} = |\rho|^2$$

With (7.56) we find

$$\eta_L = \eta_0\left\{\left[1 - \frac{3(0.1)^2}{8}\right] + j\left(\frac{0.1}{2}\right)\right\}$$

$$\simeq \eta_0(1 + j0.05)$$

The incident circularly polarized field is composed of two plane-polarized waves at right angles to each other. We are told that the amplitude of each of these waves is 100 V/m. For each wave

$$\rho = \frac{\eta_L - \eta_0}{\eta_L + \eta_0} = \frac{-j0.05}{2 + j0.05}$$

$$|\rho|^2 = \frac{(0.05)^2}{4 + (0.05)^2} = 6.25 \times 10^{-4}$$

This is the fraction of reflected power for the whole wave.

b. The amplitudes of the transmitted waves are given by (7.34)

$$E_T = \frac{2}{1 + (\eta_0/\eta_L)} E_+$$

$$H_T = \frac{1}{\eta_L} E_T$$

For the problem at hand $\eta_L \simeq \eta_0$ and E_T is very closely in phase with E_+ as is H_T with H_+. Within the dielectric the fields decay as $e^{-\alpha z}$, where from (7.51),

$$\alpha \simeq \frac{\omega}{2c}\frac{\epsilon''}{\epsilon_0} = \frac{2\pi \times 15 \times 10^6 \times 10^{-1}}{2 \times 3 \times 10^{10}}$$

$$= 15.7 \times 10^{-5} \text{ nepers/m}$$

Thus,

$$\mathbf{E}_T \simeq \left[\mathbf{a}_x E_+ \cos\phi + \mathbf{a}_y E_+ \sin\phi\right] e^{-\alpha z}$$

$$\mathbf{H}_T \simeq \frac{1}{\eta_0}\left[\mathbf{a}_y E_+ \cos\phi - \mathbf{a}_y E_+ \sin\phi\right] e^{-\alpha z}$$

PROBLEMS

7.1.

a. Show that the fraction (7.39c) of incident power transmitted to a good conductor when a plane wave is normally incident from a lossless dielectric with characteristic impedance η is

$$F \equiv \frac{\langle P_T \rangle}{\langle P_+ \rangle} \simeq \frac{4}{\eta}\sqrt{\frac{\omega\mu}{2\sigma}}$$

The angular frequency of the incident wave is ω, and the conductivity of the conductor is σ.

b. Evaluate F for copper at frequencies 30 MHz and 36 Hz. Assume $\eta = \eta_0$, $\mu = \mu_0$. For copper, $\sigma = 5.8 \times 10^7$ mho/m.

7.2. Calculate the reflection coefficient ρ and power-flux fraction F of (7.39c) when a plane wave is normally incident on a Plexiglas slab of thickness 1 cm. Assume $\epsilon = 2.8\epsilon_0$ for the slab and $\epsilon = \epsilon_0$ on both sides of the slab. (See Fig. 7.14.) Perform the calculation for incident

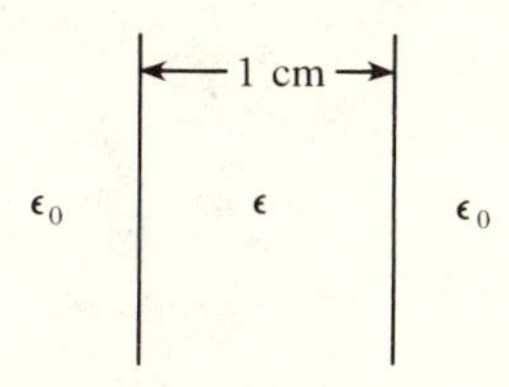

Figure 7.14 Configuration for Problem 7.2.

frequencies 1 and 10 GHz, using first the transform formula (7.40) and second the Smith chart.

7.3.

a. Obtain the reflection and transmission coefficients for a plane wave incident on a perfect conductor ($\sigma = \infty$) from a lossless dielectric medium.

b. What is the standing-wave ratio in front of the conductor?

c. What is the transmission-line configuration analogous to the description of part a?

7.4.

a. Show analytically that the maximum projection of **E** onto the E_x-axis, for the ellipse given by the parametric equation (7.59), is E_0.

b. What is the value of E_y when E_x is maximum?

c. What equation for **H** accompanies (7.59)?

7.5.

a. Write down an explicit form for the Poynting vector **P** for the incident circularly polarized wave described in Example 7.7.

b. What is the value of $\langle P_+ \rangle$ for this wave?

CHAPTER

EIGHT

MODE THEORY OF RECTANGULAR WAVEGUIDES

8.1 WAVE CLASSIFICATIONS

In the preceding chapter an analogy was drawn between plane electromagnetic waves incident on a dielectric or conducting surface and voltage-current waves on a transmission line. This formal comparison permitted the wide and useful techniques of the Smith chart to be extended to problems involving reflection and absorption of plane electromagnetic waves.

In this chapter the theory of rectangular waveguides is developed. Again we find a formal analogy between fields propagating in a guide and voltage-current waves on a transmission line, thereby once more permitting application of results of transmission-line theory.

A waveguide channels electromagnetic waves by confining the waves to a restricted but elongated region of space. The configurations of some typical waveguides are shown in Fig. 8.1.

In the discussion to follow, we shall consider rectangular guides filled with homogeneous dielectric material and perfectly conducting walls. Inside the guide Maxwell's equations give the wave equations (7.3), (7.10):

$$\nabla^2\mathbf{E} - \frac{1}{v^2}\frac{\partial^2\mathbf{E}}{\partial t^2} = 0$$

$$\nabla^2\mathbf{H} - \frac{1}{v^2}\frac{\partial^2\mathbf{H}}{\partial t^2} = 0$$

With the direction of propagation taken along z-axis, we assume the following explicit z-dependence of the fields:

$$\begin{aligned} \mathbf{E}(x, y, z, t) &= \mathbf{E}(x, y)e^{j\omega t - \gamma z} \\ H(x, y, z, t) &= \mathbf{H}(x, y)e^{j\omega t - \gamma z} \end{aligned} \tag{8.1}$$

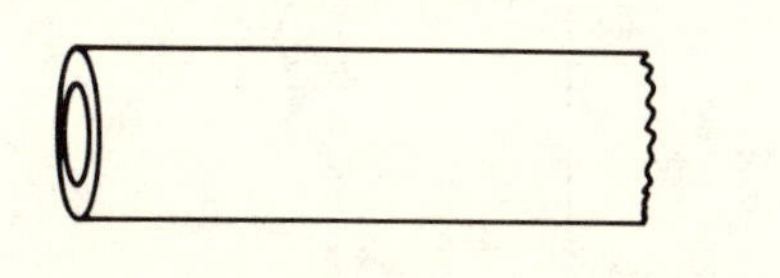

Circular

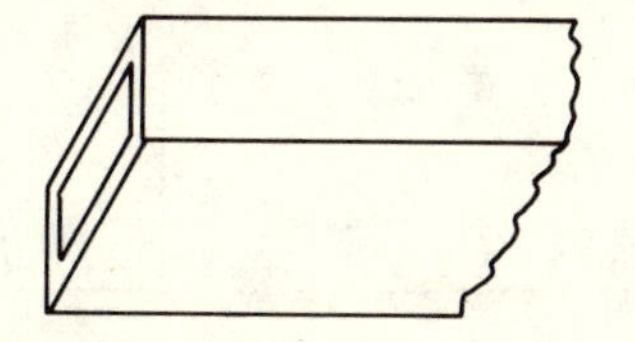

Rectangular

Figure 8.1 Two types of waveguides.

Maxwell's equations (7.1) then yield

$$\frac{\partial H_z}{\partial y} + \gamma H_y = j\omega\epsilon E_x, \qquad \frac{\partial E_z}{\partial y} + \gamma E_y = -j\omega\mu H_x \tag{8.2a}$$

$$\frac{\partial H_z}{\partial x} + \gamma H_x = -j\omega\epsilon E_y, \qquad \frac{\partial E_z}{\partial x} + \gamma E_x = j\omega\mu H_y \tag{8.2b}$$

$$\frac{\partial H_y}{\partial x} - \frac{\partial H_x}{\partial y} = j\omega\epsilon E_z, \qquad \frac{\partial E_y}{\partial x} - \frac{\partial E_x}{\partial y} = -j\omega\mu H_z \tag{8.2c}$$

From (8.2a) and (8.2b) we find

$$\frac{\partial^2 E_z}{\partial x^2} + \frac{\partial^2 E_z}{\partial y^2} + \gamma^2 E_z = -\omega^2\mu\epsilon E_z$$

$$\frac{\partial^2 H_z}{\partial x^2} + \frac{\partial^2 H_z}{\partial y^2} + \gamma^2 H_z = -\omega^2\mu\epsilon H_z \tag{8.3}$$

If (8.3) are solved for E_z and H_z, then the remaining field components are determined through (8.2). With

$$k_c^2 \equiv \gamma^2 + \omega^2\mu\epsilon \tag{8.4}$$

we obtain

$$H_x = -\frac{1}{k_c^2}\left[\gamma\frac{\partial H_z}{\partial x} - j\omega t\frac{\partial E_z}{\partial y}\right]$$

$$H_y = -\frac{1}{k_c^2}\left[\gamma\frac{\partial H_z}{\partial y} + j\omega\epsilon\frac{\partial E_z}{\partial x}\right]$$

$$E_x = -\frac{1}{k_c^2}\left[\gamma\frac{\partial E_z}{\partial x} + j\omega\mu\frac{\partial H_z}{\partial y}\right]$$

$$E_y = -\frac{1}{k_c^2}\left[\gamma\frac{\partial E_z}{\partial y} - j\omega\mu\frac{\partial H_z}{\partial x}\right] \tag{8.5}$$

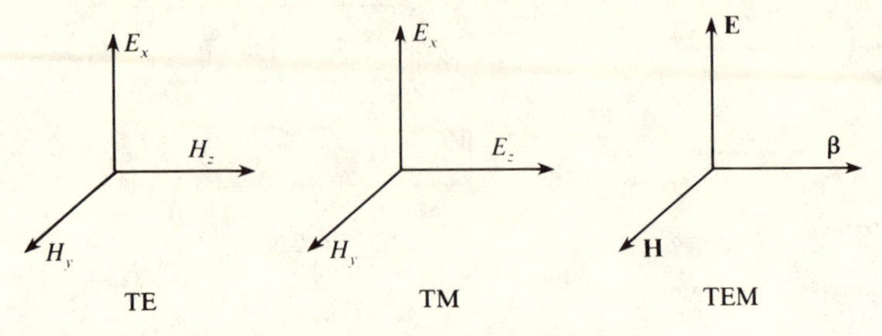

Figure 8.2 Field components in TE, TM, and TEM modes.

TE, TM, and TEM Waves

These last equations indicate an important result for waveguide transmission. Namely, we see that if both E_z and H_z are zero, *all* fields within the guide vanish. This observation suggests a convenient classification of waves in a guide. Namely, waves with $H_z = 0$ are called *transverse magnetic* or *TM modes*, whereas waves with $E_z = 0$ are called *transverse electric* or *TE modes*. Purely transverse electromagnetic waves (TEM) which contain $E_z = H_z = 0$ do not propagate within a hollow homogeneous guide. They do, however, propagate along transmission lines.

This classification is depicted in Fig. 8.2.

Example 8.1

What equations do TEM fields supported by a transmission line satisfy?

Ans. For TEM waves, $E_z = H_z = 0$ Then (7.1) gives

$$(\nabla \times \mathbf{E})_z = 0$$

With Stokes's theorem (6.8), for any closed loop in an x, y plane outside the conductors, we may write

$$\oint_C \mathbf{E} \cdot d\mathbf{l} = 0$$

Thus, in the x, y plane we may set $\mathbf{E} = -\nabla_\perp V$ [see Eq. (6.22)], and with (7.1a) we find

$$\nabla_\perp^2 V_E(x, y) = 0$$

where

$$\nabla_\perp = \mathbf{a}_x \frac{\partial}{\partial x} + \mathbf{a}_y \frac{\partial}{\partial y}$$

$$\nabla_\perp^2 = \frac{\partial^2}{\partial x^2} + \frac{\partial^2}{\partial y^2}$$

In similar manner, (7.1d) yields

$$(\nabla \times \mathbf{H})_z = 0$$

and with $\mathbf{H} = -\nabla_\perp V_H$ we obtain

$$\nabla_\perp^2 V_H(x, y) = 0$$

Thus, the spatial dependences of the **E** and **H** fields of a TEM wave are derivable from *static* potential fields.

8.2 CUTOFF FREQUENCIES

The guide configuration we wish to consider is shown in Fig. 8.3. We return to (8.3) for solution to fields $E_z(x, y)$ and $H_z(x, y)$. Assuming separation of variables (see Sec. 6.3), we write

$$E_z(x, y) = X(x)Y(y) \tag{8.6}$$

and assume $H_z = 0$ appropriate to TM waves. There results

$$\frac{1}{X}\frac{d^2X}{dx^2} + k_c^2 = -\frac{1}{Y}\frac{d^2Y}{dy^2} \equiv k_y^2 \tag{8.7}$$

The solution to these equations has the form

$$X = C_1 \cos k_x x + C_2 \sin k_x x$$
$$Y = C_3 \cos k_y y + C_4 \sin k_y y \tag{8.8}$$

where we have set

$$k_c^2 = k_x^2 + k_y^2 \tag{8.9}$$

The C-constants are determined from boundary conditions. Due to the ideal conductivity of the walls,

$$E_z(0, y) = E_z(a, y) = E_z(x, 0) = E_z(x, b) = 0 \tag{8.10}$$

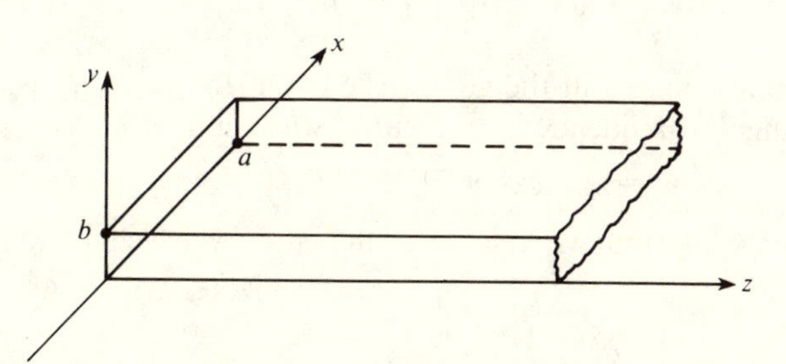

Figure 8.3 The rectangular guide configuration.

Returning to the product form (8.6), we find that boundary conditions at $x = 0$ and $y = 0$ dictate the form

$$E_z(x, y) = C_2C_4 \sin k_x x \sin k_y y \tag{8.11}$$

At $x = a$ (with $C_2C_4 \equiv A$), (8.10) gives

$$E_z = A \sin k_x a \sin k_y y = 0$$

Thus

$$k_x a = m\pi, \qquad m = 1, 2, 3, \ldots \tag{8.12}$$

At $y = b$

$$E_z = A \sin k_x x \sin k_y b$$

so that

$$k_y b = n\pi, \qquad n = 1, 2, 3, \ldots \tag{8.13}$$

Our final expression for $E_z(x, y)$ is

$$E_z(x, y) = A \sin \frac{m\pi x}{a} \sin \frac{n\pi y}{b} \tag{8.14}$$

The remaining field components E_x, E_y, H_x, H_y are obtained from the relations (8.5). Before presenting these solutions, let us return to the form (8.4). Together with the identification (8.9), we find

$$\gamma^2 + \omega^2\mu = k_x^2 + k_y^2 = k_c^2$$

With the preceding expression we obtain

$$\gamma = \sqrt{k_x^2 + k_y^2 - \omega^2\mu\epsilon}$$

With (8.11) and (8.12), there results

$$\gamma = \sqrt{\left(\frac{m\pi}{a}\right)^2 + \left(\frac{n\pi}{b}\right)^2 - \omega^2\mu\epsilon} \tag{8.15}$$

Recalling the general form of the field solutions (8.1), we see that if

$$\omega > \omega_c = \frac{1}{\sqrt{\mu\epsilon}} \sqrt{\left(\frac{m\pi}{a}\right)^2 + \left(\frac{n\pi}{b}\right)^2} \tag{8.16}$$

waves will propagate unattenuated in the guide. The cutoff frequency f_c is simply related to the angular frequency ω_c and cutoff wavenumber k_c as

$$\omega_c = 2\pi f_c = v k_c$$

The cutoff wavenumber k_c is similarly related to the cutoff wavelength λ_c as

$$k_c = \frac{2\pi}{\lambda_c}$$

so we may write

$$f_c\lambda_c = \frac{\omega_c}{k_c} = \frac{1}{\sqrt{\mu\epsilon}} = v \tag{8.17}$$

Corresponding to (8.16) we write

$$\lambda_c = \frac{2}{\sqrt{(m/a)^2 + (n/b)^2}} \tag{8.18}$$

With the identification (7.47)

$$\gamma = \alpha + j\beta$$

we see that for $\omega > \omega_c$, or $k > k_c$, the attenuation constant α is zero.

For frequencies above the cutoff frequency, the phase velocity of the wave propagation in the guide is given by

$$v_\phi = \frac{\omega}{\beta} = \frac{v}{\sqrt{1 - (\omega_c/\omega)^2}} \tag{8.19}$$

$$\lambda_g = \frac{2\pi}{\beta} = \frac{2\pi v/\omega}{\sqrt{1 - (\omega_c/\omega)^2}} = \frac{\lambda}{\sqrt{1 - (\omega_c/\omega)^2}} \tag{8.20}$$

Note that λ_g and ω are the wavelength and angular frequency, respectively, of the electromagnetic wave mode in the waveguide. For an air-filled guide,

$$v = c = \lambda f = \frac{1}{\sqrt{\mu_0\epsilon_0}}$$

The *group velocity* (see Problem 8.2) is given by

$$v_G = \frac{\partial\omega}{\partial\beta} = v\sqrt{1 - \left(\frac{\omega_c}{\omega}\right)^2} \tag{8.21}$$

Comparing these expressions, we see that for $\omega \gg \omega_c$, $v_G \simeq v_\phi \simeq v$.

The cutoff frequency (8.15) corresponds to the (m, n) mode field (8.13). We see that this field vanishes for either $m = 0$ or $n = 0$. Thus the lowest-order TM_{mn} mode which may propagate is the TM_{11} mode.

Explicit Expressions for the Fields

As stated previously, given the solution (8.13), the remaining field components for the TM_{mn} mode are obtained from (8.5). Repeating this construction for the TE_{mn} mode ($E_z = 0$), we obtain the following total solutions:

TM_{mn}	TE_{mn}	
$E_z = A \sin k_x x \sin k_y y$	$H_z = B \cos k_x x \cos k_y y$	(8.22a)
$H_x = j \dfrac{k_y \omega}{k_c \eta \omega_c} A \sin k_x x \cos k_y y$	$E_x = j \dfrac{\eta k_y \omega}{k_c \omega_c} B \cos k_x x \sin k_y y$	(8.22b)
$H_y = -j \dfrac{k_x \omega}{k_c \eta \omega_c} A \cos k_x x \sin k_y y$	$E_y = -j \dfrac{\eta k_x \omega}{k_c \omega_c} B \sin k_x x \cos k_y y$	(8.22c)
$E_x = Z_{TM} H_y$	$H_x = -\dfrac{E_y}{Z_{TE}}$	(8.23a)
$E_y = -Z_{TM} H_x$	$H_y = \dfrac{E_x}{Z_{TE}}$	(8.23b)

In these expressions we have reintroduced the cutoff wavenumber

$$k_c = \sqrt{k_x^2 + k_y^2} = \frac{2\pi}{\lambda_c} = \frac{\omega_c}{v} \tag{8.24}$$

where k_x and k_y are given by (8.12) and (8.13). The field impedances

$$Z_{TM} = \frac{E_x}{H_y} = -\frac{E_y}{H_x} = \eta \left[1 - \left(\frac{\omega_c}{\omega}\right)^2\right]^{1/2},$$

$$Z_{TE} = \frac{E_x}{H_y} = -\frac{E_y}{H_x} = \frac{\eta}{\left[1 - \left(\frac{\omega_c}{\omega}\right)^2\right]^{1/2}} \tag{8.25}$$

where η is the intrinsic impedance as given by (2.15). The preceding expressions are relevant to a forward-propagating wave where field components include the phase factor $\exp(j\omega t - \gamma z)$. For a backward-propagating wave this phase factor becomes $\exp(j\omega t + \gamma z)$, and the signs of Z_{TE} or Z_{TM} are reversed [compare with Eqs. (7.31)].

Attenuation Constant Below Cutoff

Combining (8.15) and (8.16) permits us to write

$$\gamma = \frac{1}{v}\sqrt{\omega_c^2 - \omega^2}$$

It follows that for $\omega < \omega_c$, γ is purely real, so that

$$\gamma = \alpha = \frac{\omega}{v}\sqrt{\left(\frac{\omega_c}{\omega}\right)^2 - 1}$$

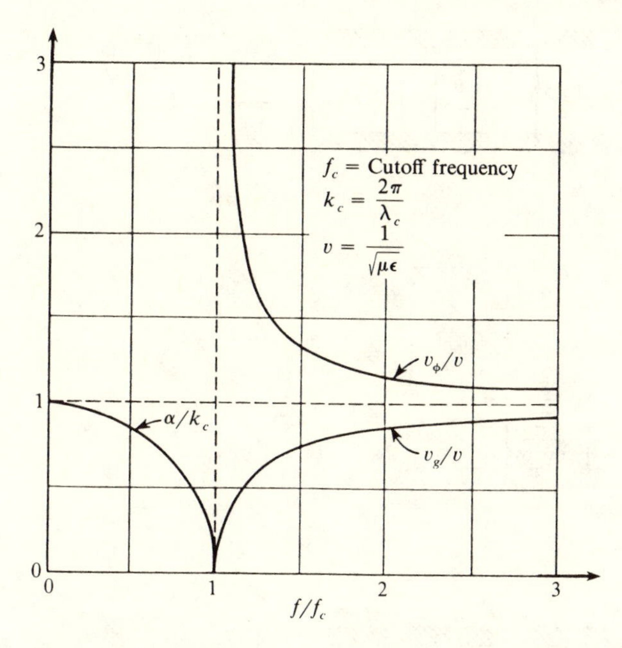

Figure 8.4 The parameters α, v_ϕ, and v_g vs frequency.

For $\omega \ll \omega_c$,

$$\alpha \simeq \frac{\omega_c}{v} = k_c$$

whereas from (8.20), for $\omega \gg \omega_c$,

$$\beta = \frac{\omega}{v}\sqrt{1 - \left(\frac{\omega_c}{\omega}\right)^2} \simeq \frac{\omega}{v}$$

In the high-frequency, short-wavelength limit,

$$\omega = \beta v$$

which is the characteristic of electromagnetic propagation in an unbounded homogeneous medium [see Eq. (7.16)]. The reason for this collapse to free-space propagation is that when the dimensions of the guide a, b grow very large compared to the wavelength $\lambda_g = 2\pi/\beta$ of the propagating mode, fields in the guide become uninfluenced by the walls.

Comparison between these parameters is depicted in Fig. 8.4.

Example 8.2

In an ideal rectangular guide, with $a > b$, make three sketches of the transverse **E** and **H** fields for the TE_{10} mode in the planes $x = a/2$, $y = b/2$, and $z =$ constant.

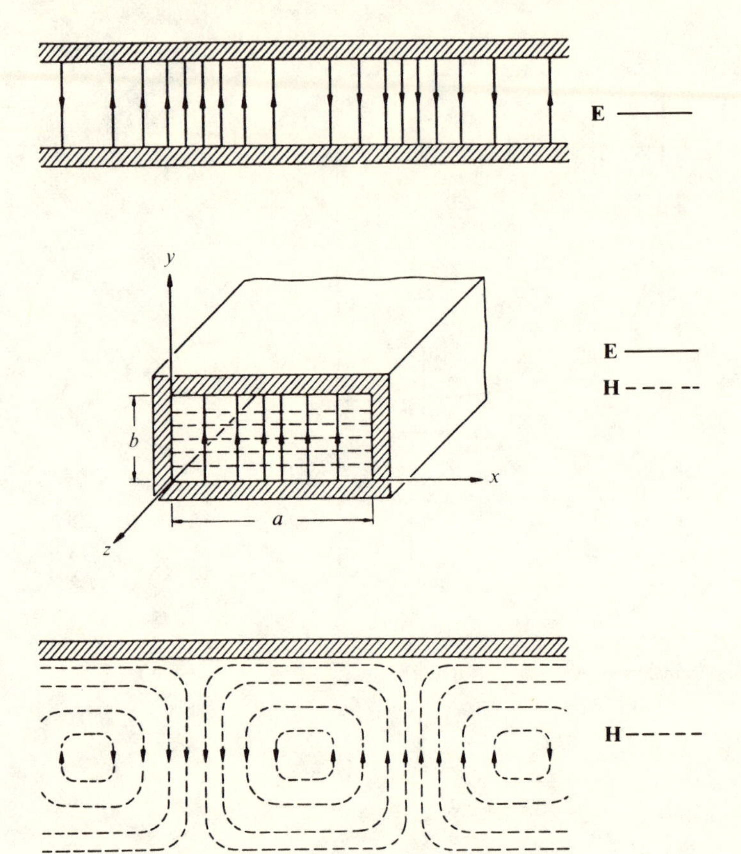

Figure 8.5 **E** and **H** components in the TE_{10} mode.

Ans. For the TE_{10} mode, from (8.22) we find

$$H_z = B\cos\frac{\pi x}{a}, \qquad E_y = -\frac{j\eta\omega}{\omega_c}B\sin\frac{\pi x}{a}, \qquad H_x = -\frac{E_y}{Z_{TE}}$$

$$E_x = H_y = 0$$

Thus, the surviving field components, apart from the unwritten factor $e^{-j\beta z}$, depend only on x. The factor j in the expressions above indicates a $\pi/2$ phase difference between field components. We see that in the TE_{10} mode, electric field lines are straight lines parallel to the y-axis. The magnetic field lines form closed loops.[1] (See Fig. 8.5.)

[1] For additional field-line configurations, see Appendix D.

Example 8.3

An air-filled rectangular guide with dimensions 2.29 cm × 1.02 cm is employed for propagation in the X-band (8.0–12.0 GHz).

a. Determine the cutoff frequencies of the five lowest modes which can propagate in the guide.
b. What is the guide wavelength for the TE_{10} mode at 8 GHz?
c. What is the field impedance of the TE_{10} mode at 8 GHz?
d. What is the attenuation constant α at 6 GHz?

Ans.

a. For an air-filled guide, the cutoff frequencies are given by the expression [see Eq. (8.16)]

$$f_c = \frac{c}{2\pi}\left[\left(\frac{m\pi}{a}\right)^2 + \left(\frac{n\pi}{b}\right)^2\right]^{1/2}$$

where $a = 2.29 \times 10^{-2}$ m and $b = 1.02 \times 10^{-2}$ m. There results:

Mode	m	n	f_c (GHz)
TE_{10}	1	0	6.55
TE_{20}	2	0	13.10
TE_{01}	0	1	14.71
TE_{11}, TM_{11}	1	1	16.10

These are the five modes of lowest frequency.

b.

$$\lambda_g = \frac{\lambda}{\sqrt{1-(f_c/f)^2}} = \frac{3 \times 10^8/8 \times 10^9}{\sqrt{1-\left(\frac{6.55}{8}\right)^2}} = 6.53 \text{ cm}$$

c.

$$Z_{TE} = \frac{\eta_0}{\sqrt{1-(f_c/f)^2}} = 656.5 \text{ ohms}$$

d.

$$\alpha = \frac{2\pi f}{c}\sqrt{\left(\frac{f_c}{f}\right)^2 - 1} = 55 \text{ nepers/m}$$

The term "X-band" was introduced in the preceding example. A list of names for waveguide frequency bands from 3 MHz to 300 GHz is presented in Table 8.1.

Table 8.1 Waveguide frequency bands

Band	Frequency range
HF	3–30 MHz
VHF	30–300 MHz
UHF	300–1000 MHz
L-band	1.0–2.0 GHz
S-band	2.0–4.0 GHz
C-band	4.0–8.0 GHz
X-band	8.0–12.0 GHz
Ku-band	12.0–18.0 GHz
K-band	18.0–27.0 GHz
Ka-band	27.0–40.0 GHz
Millimeter	40–300 GHz

8.3 TRANSMISSION-LINE ANALOGY

In this section we wish to discover the equivalent distributed circuit for the transmission line which corresponds to the field equations (8.2). The general form of a distributed network is shown in Fig. 8.6. The voltage-current equations for this network are [compare Eq. (2.3)]

$$\frac{\partial I}{\partial z} = -(Y_1 + Y_2)V, \qquad \frac{\partial V}{\partial z} = -(Z_1 + Z_2)I \tag{8.26}$$

We must cast the field equations (8.2) in this form to find the equivalent Z, Y and I, V transmission-line parameters.

TE Waves

For TE waves, $E_z = 0$ and (8.2c) gives

$$(\nabla \times \mathbf{H})_z = 0 \tag{8.27}$$

It follows [see Eqs. (6.22), (6.23)] that in the transverse x, y plane one may

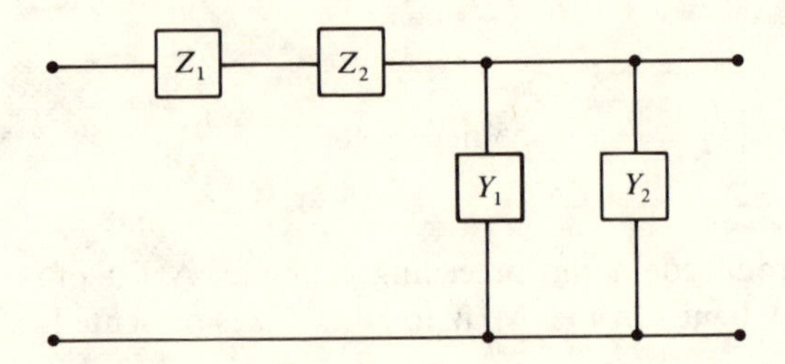

Figure 8.6 Distributed network.

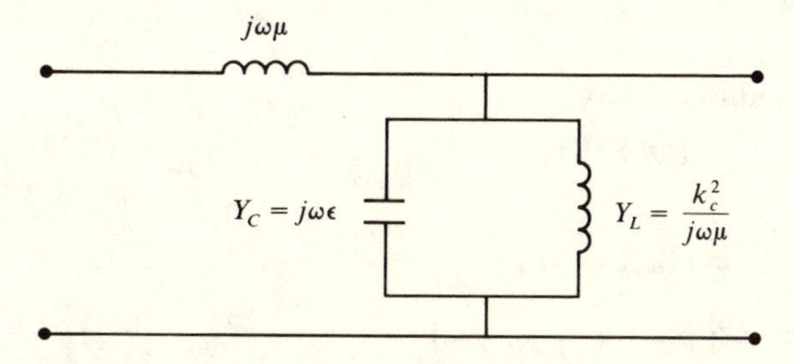

Figure 8.7 Distributed network for TE waves.

define a potential function U such that

$$\mathbf{H}_\perp = -\nabla_\perp U(x, y) \tag{8.28}$$

Equivalently,

$$H_x = -\frac{\partial U}{\partial x}, \qquad H_y = -\frac{\partial U}{\partial y} \tag{8.29}$$

Substituting these relations into Eqs. (8.2), after some manipulation of terms, we obtain

$$\frac{\partial}{\partial x}\left(\frac{j\omega\mu}{k_c^2} H_z\right) = -j\omega\mu U \tag{8.30a}$$

$$\frac{\partial}{\partial z} U = -\left(\frac{k^2}{j\omega\mu} + j\omega\epsilon\right)\left(\frac{j\omega\mu}{k_c^2} H_z\right) \tag{8.30b}$$

Comparing these equations with (8.26) permits the analogy

TE:
$$V \leftrightarrow j\frac{\omega\mu H_z}{k_c^2}, \qquad I \leftrightarrow U \tag{8.31}$$

$$Z = j\omega\mu, \qquad Y = j\omega\epsilon + \frac{k^2}{j\omega\mu}$$

Thus we are able to draw the equivalent distributed network, as shown in Fig. 8.7. The characteristic impedance of this line is

$$Z_{\mathrm{TE}} = \sqrt{\frac{Z}{Y}} = \sqrt{\frac{j\omega\mu}{j\omega\epsilon + \left(k_c^2/j\omega\mu\right)}}$$

$$Z_{\mathrm{TE}} = \frac{\eta}{\sqrt{1 - (\omega_c/\omega)^2}} \tag{8.32}$$

in agreement with (8.25).

TM Waves

For TM waves $H_z = 0$ and (8.2c) gives

$$(\nabla \times \mathbf{E})_z = 0 \tag{8.33}$$

and we may set

$$\mathbf{E}_\perp = -\nabla_\perp V(x, y) \tag{8.34}$$

Again, with this relation and Eqs. (8.2), we find

$$\frac{\partial}{\partial z}\left(\frac{j\omega}{k_c^2}E_z\right) = -j\omega\epsilon V \tag{8.35a}$$

$$\frac{\partial}{\partial z}V = -\left(j\omega\mu + \frac{k^2}{j\omega\epsilon}\right)\left(\frac{j\omega\epsilon}{k_c^2}E_z\right) \tag{8.35b}$$

Comparing these equations with (8.26) permits the analogy

TM: $\qquad V \leftrightarrow V, \qquad I \leftrightarrow \dfrac{j\omega\epsilon}{k_c^2}E_z$

Equation (8.35) permit us to draw the equivalent distributed network shown in Fig. 8.8. The characteristic impedance of this line is

$$Z_{\text{TM}} = \sqrt{\frac{Z}{Y}} = \sqrt{\frac{j\omega\mu + (k^2/j\omega)}{j\omega}}$$

$$Z_{\text{TM}} = \eta\sqrt{1 - \left(\frac{\omega_c}{\omega}\right)^2} \tag{8.37}$$

in agreement with (8.25).

These analogous relations are summarized in Table 8.2.

Note that only two scalar variables enter the transmission-line analysis: I and V. The waveguide analysis, on the other hand, involves five variables. For TM modes, for example, these are $E_z, \mathbf{E}_\perp, \mathbf{H}_\perp$. However, as noted previously, with Eqs. (8.5), knowledge of E_z determines $\mathbf{E}_\perp$ and $\mathbf{H}_\perp$. So

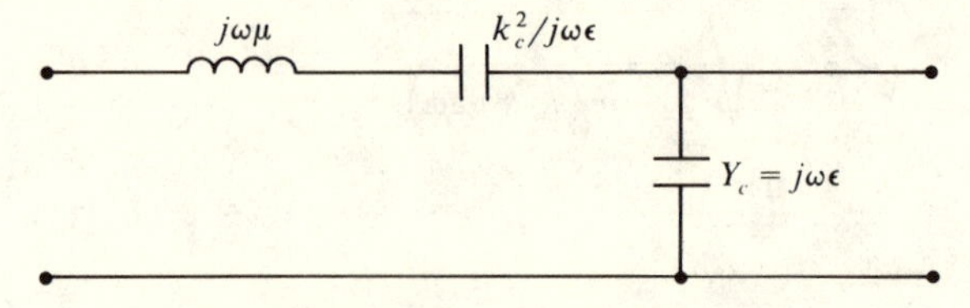

Figure 8.8 Distributed network for TM waves.

Table 8.2

Transmission line	Waveguide
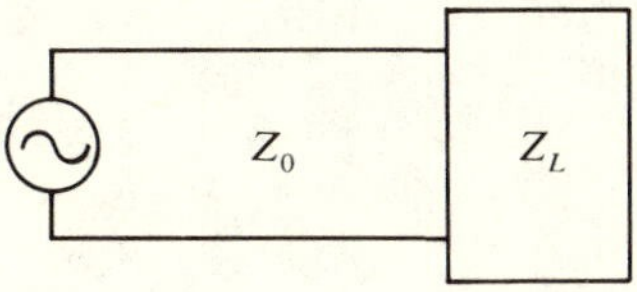	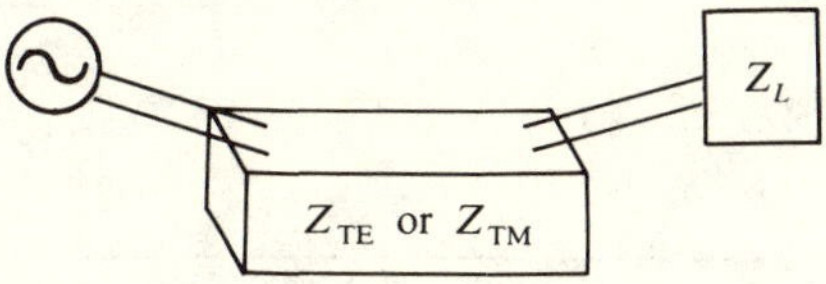
	TE
V	$\dfrac{j\omega\mu H_z}{k_c^2}$
I	$U \quad (\mathbf{H}_\perp = -\nabla_\perp U)$
Z_0	$Z_{\text{TE}} = \dfrac{\eta}{\sqrt{1-(\omega_c/\omega)^2}}$
	TM
V	$V \quad (\mathbf{E}_\perp = -\nabla_\perp V)$
I	$j\dfrac{\omega\epsilon}{k_c^2}E_z$
Z_0	$Z_{\text{TM}} = \eta\sqrt{1-(\omega_c/\omega)^2}$
	TE *or* **TM**
$f_c = 0$	$f_c = \dfrac{v}{2\pi}\left[\left(\dfrac{m\pi}{a}\right)^2 + \left(\dfrac{n\pi}{b}\right)^2\right]^{1/2}$ $\omega_c = vk_c$

knowledge of the current I in the transmission-line analysis, with (8.36), determines E_z and therefore $\mathbf{E}_\perp$ and $\mathbf{H}_\perp$. Thus the transmission-line analysis gives all desired field information.

This analogy is very relevant to practical waveguide problems. Thus, for example, the waveguide configuration involving a sudden change in geometry as shown in Fig. 8.9 has the equivalent network[2] shown in Fig. 8.10. The *iris partitions* with edges (a) perpendicular to **E** and (b) parallel to **E** as shown in Fig. 8.11 have the equivalent networks shown in Fig. 8.12.

With the above analogy presented, all the formalism attendant to transmission lines (e.g., standing-wave ratio, reflection coefficient) becomes

[2] For a more extensive list of equivalent discontinuity networks, see, J. R. Whinnery and H. W. Jamieson, "Equivalent Circuits for Discontinuities in Transmission Lines," Proc. IRE, vol. 32, pp. 98–114, 1944. Various types of "waveguide couplers" are described in Appendix D.

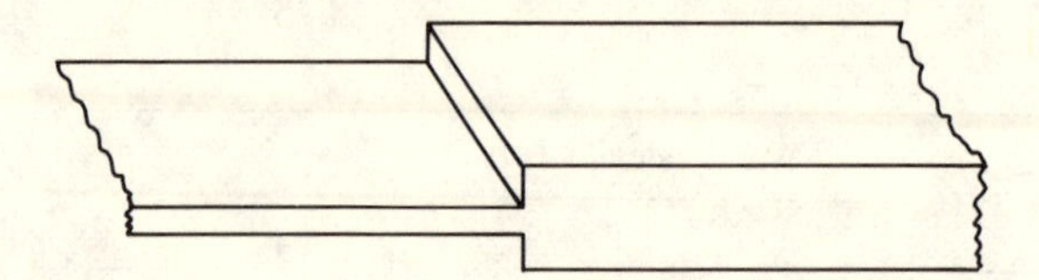

Figure 8.9

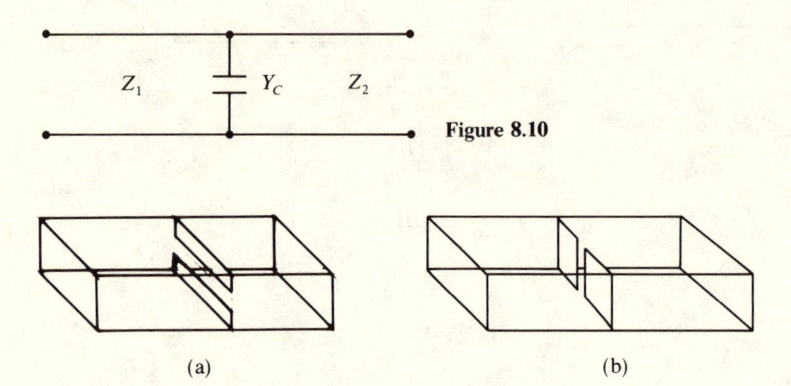

Figure 8.10

Figure 8.11 The iris partition.

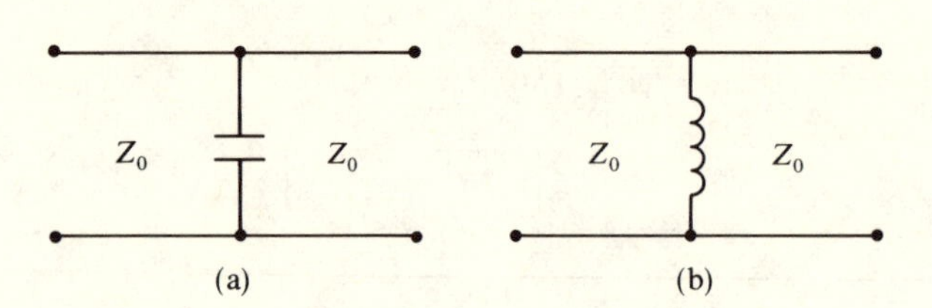

Figure 8.12 Equivalent networks for the iris partition.

relevant to waveguides. A simple but revealing example is offered in the following problem.

Example 8.4

A waveguide is operated in the TE_{10} mode at the frequency 13 GHz. The guide has cross section 1.25 cm × 0.6 cm.

a. What is the guide cutoff frequency?
b. What is the wave impedance at the operating frequency?
c. The guide is terminated in a load of impedance $2\eta_0$. What is the standing-wave ratio S in the guide?

d. What is the reflection coefficient at the load?
e. Does the standing-wave pattern have a maximum or a minimum at the load?
f. What is the fractional time-average power delivered to the load?

Ans.

a.

$$f_c = \frac{c}{2a} = \frac{3 \times 10^8}{2.5 \times 10^{-2}} = 12 \text{ GHz}$$

b.

$$Z_{\text{TE}} = \frac{\eta_0}{1 - (f_c/f)^2} = \frac{377}{\left[1 - \left(\frac{12}{13}\right)^2\right]^{1/2}} = \frac{13\eta_0}{5} = 980 \text{ ohms}$$

c. Recalling (4.34), we write

$$S = \frac{Z_{\text{TE}}}{R_L} = \frac{13\eta/5}{2\eta_0} = \frac{13}{10} = 1.3$$

d. With (4.29) we write

$$\rho = \frac{S-1}{S+1} = \frac{1.3-1}{1.3+1} = 0.13$$

e. Since the impedance of the load is real and less than Z_{TE}, Table 4.1 applies and the standing-wave pattern has a minimum at the load. $|V|_{\min} = |V|_{\max}/S$ (see Problem 8.1).
f. From (2.32),

$$\langle P_L \rangle = \langle P_+ \rangle (1 - |\rho_L|^2)$$

$$F = \frac{\langle P_L \rangle}{\langle P_+ \rangle} = (1 - |\rho_L|^2)$$

$$= 1 - (0.13)^2 = 0.98$$

8.4 POWER LOSS. LOW-LOSS APPROXIMATION

As stated previously, time-average radiative power flow (7.27) per unit area is given by

$$\langle \mathbf{P} \rangle = \tfrac{1}{2} \operatorname{Re} \mathbf{E} \times \mathbf{H}^*$$

It follows that in a waveguide such as described above, the average power flow in the axial z-direction per unit area is given by

$$\mathbf{a}_z\langle P_z\rangle = \tfrac{1}{2}\mathrm{Re}\left(\mathbf{E}_\perp \times \mathbf{H}_\perp^*\right)$$

$$\langle P_z\rangle = \tfrac{1}{2}\mathrm{Re}\left(E_x H_y^* - E_y H_x^*\right)$$

$$= \tfrac{1}{2}\mathrm{Re}\left(\frac{E_x E_x^*}{Z} + \frac{E_y E_y^*}{Z}\right)$$

$$= \frac{1}{2Z}\mathbf{E}_\perp \cdot \mathbf{E}_\perp^* \tag{8.38}$$

In this expression Z represents either Z_{TM} or Z_{TE} and is assumed to be real. Equivalently, we assume that γ is pure imaginary, so that $\mathbf{E}_\perp$ and $\mathbf{H}_\perp$ are in phase. Thus this analysis may be termed a *low-loss approximation*. The total radiative power flow through the guide in the z-direction is given by the surface integral

$$W_T = \iint \langle \mathbf{P}\rangle \cdot d\mathbf{S}_\perp = \iint \langle P_z\rangle\, dx\, dy$$

$$= \frac{1}{2Z}\iint |\mathbf{E}_\perp|^2 dx\, dy = \frac{Z}{2}\iint |\mathbf{H}_\perp|^2 dx\, dy \tag{8.39}$$

With the fields (8.22) we find (see Example 8.6)

$$W_T^{(\mathrm{TM})} = \Lambda |E_{0z}|^2 \tag{8.39a}$$

$$W_T^{(\mathrm{TE})} = \eta^2 \Lambda |H_{0z}|^2 \qquad (m, n) = (1,0), (0,1) \tag{8.39b}$$

$$W_T^{(\mathrm{TE})} = 2\eta^2 \Lambda |H_{0z}|^2 \qquad (m, n) = (1,0), (0,1) \tag{8.39c}$$

where

$$\Lambda \equiv \frac{ab}{8}\frac{1}{\eta}\left(\frac{\omega}{\omega_c}\right)^2 \sqrt{1 - \left(\frac{\omega_c}{\omega}\right)^2}$$

The time-average power loss per unit area is given by (7.29):

$$\langle P_{\mathrm{loss}}\rangle = \tfrac{1}{2}\mathrm{Re}\, \mathbf{E} \cdot \mathbf{J}^* \tag{8.40}$$

In a guide filled with a lossless homogeneous dielectric, loss of energy in the wave is due to finite resistivity of the walls. Writing

$$E_z = R_s J_z \tag{8.41}$$

for current flow in the walls of the guide, (8.40) becomes

$$\langle P_{\mathrm{loss}}\rangle = \tfrac{1}{2}\mathrm{Re}\, R_s\left(J_z J_z^* + J_\perp J_\perp^*\right)$$

$$= \tfrac{1}{2}R_s\left[|J_z|^2 + |J_\perp|^2\right] \tag{8.42}$$

From Ampere's law (6.45) we see that the axial component of current in the guide wall stems from the transverse component of $\mathbf{H}$ at the wall, whereas the transverse component of current stems from the axial component of **H** at the wall. With (6.63) we write

$$|J_\perp| = |H_z|$$
$$|J_z| = |H_\perp| \tag{8.43}$$

where J (with dimensions A/m) is assumed to be a surface current. The preceding expression (8.42) becomes

$$\langle P_{\text{loss}} \rangle = \tfrac{1}{2} R_s \left[|H_\perp|^2 + |H_z|^2 \right]$$

The power loss per unit length of guide is then given by

$$W_{\text{loss}} = \frac{R_s}{2} \oint \left[|H_\perp|^2 + |H_z|^2 \right] dl \tag{8.44}$$

where the integration is taken about the inner surface perimeter of the guide in a plane of constant z. The variable l denotes the arc length along this perimeter. The integration over a unit z-interval in going from (8.39) to (8.44) gives unit because again we assume, within the low-loss approximation, that γ is pure imaginary so that field amplitudes are independent of z.

Relation to Attenuation Coefficient α

In steady state, the time average of Poynting's equation (7.23) yields

$$\frac{\partial}{\partial z} \langle P_z \rangle = \langle \mathbf{J} \cdot \mathbf{E} \rangle \tag{8.45}$$

Here we have neglected $\nabla_\perp \cdot \langle \mathbf{P}_\perp \rangle$, consistent with the low-loss approximation. Writing

$$\langle P_z \rangle = \frac{1}{2Z} |E_\perp|^2 e^{-2\alpha z} \tag{8.46}$$

the preceding equation gives

$$2\alpha \langle P_z \rangle = \langle \mathbf{J} \cdot \mathbf{E} \rangle$$

Integrating this result over a volume of the guide equal to the cross sectional area times a unit length, and assuming that $\langle P_z \rangle$ varies slowly with z, we obtain the result

$$\alpha = \frac{W_{\text{loss}}}{2W_T} \tag{8.47}$$

[Note that W_{loss} is the power loss per unit length. Thus in (8.47), α has dimensions of m^{-1}, consistent with (8.46). In the companion equation for circuits, (1.38), α has dimensions of sec^{-1}, consistent with (1.37).] Together with the preceding expression for W_{loss} and W_T, we find

$$\alpha = \frac{\dfrac{R_s}{2} \oint \left[|H_\perp|^2 + |H_z|^2 \right] dl}{\dfrac{Z}{2} \int\int |H_\perp|^2 dx\, dy}$$

For frequencies greater than cutoff, with the fields (8.22), we have the following results (see Example 8.5):

1. TE waves:

$$\alpha_{m0} = \frac{\tilde{R}}{2}\left[1 + \frac{2b}{a}\left(\frac{\omega_c}{\omega}\right)^2\right] \tag{8.49}$$

$$\alpha_{mn} = \tilde{R}\left\{\left(1 + \frac{b}{a}\right)\left(\frac{\omega_c}{\omega}\right)^2 + \left[1 - \left(\frac{\omega_c}{\omega}\right)^2\right]\frac{\frac{b}{a}\left(\frac{b}{a}m^2 + n^2\right)}{\frac{b^2m^2}{a^2} + n^2}\right\} \tag{8.50}$$

2. TM waves:

$$\alpha_{mn} = \tilde{R}\frac{m^2(b/a)^3 + n^2}{m^2(b/a)^2 + n^2} \tag{8.51}$$

In these expressions we have set

$$\tilde{R} \equiv \frac{2R_s}{b\eta\sqrt{1 - (\omega_c/\omega)^2}} \tag{8.52}$$

and consistent with (7.57), we may set

$$R_s = \sqrt{\frac{\mu\omega}{2\sigma}}$$

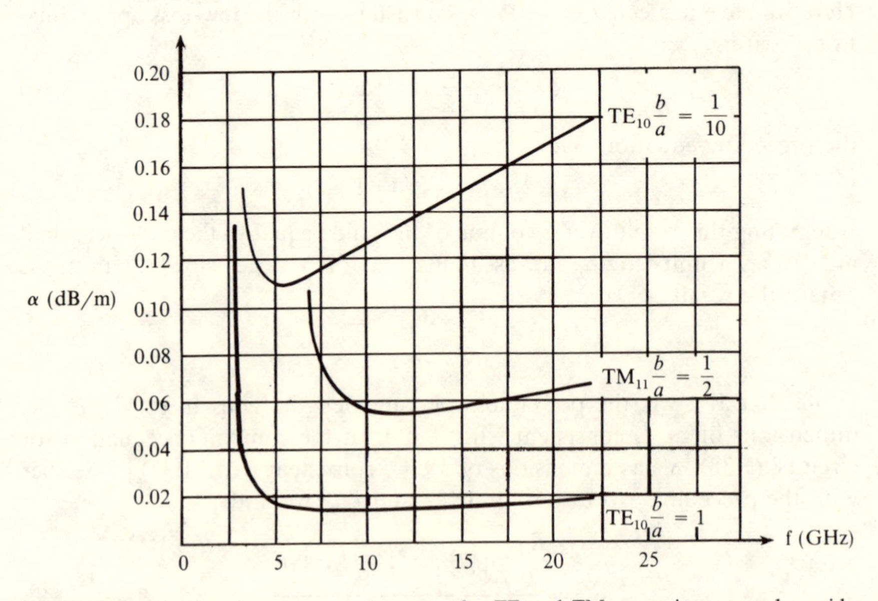

Figure 8.13 Attenuation constant vs frequency for TE and TM waves in rectangular guides with several edge-length ratios.

For the dominant TE_{10} mode in rectangular guides, α reduces to

$$\alpha = \frac{R_s}{b\eta\sqrt{1-(\omega_c/\omega)^2}}\left[1+\frac{2b}{a}\left(\frac{\omega_c}{\omega}\right)^2\right] \text{ nepers/m}$$

$$\alpha(\text{dB/m}) = 8.7\alpha(\text{nepers/m}) \tag{8.53}$$

The loss in dB/m was defined in Example 7.5.

Attenuation vs frequency for a rectangular guide with copper walls and $a = 2$ in. is shown in Fig. 8.13 for a few typical dimensions.

Example 8.5

Establish the formula (8.49) for the attenuation coefficient α_{m0} relevant to TE_{m0} waves.

Ans. We must evaluate the integrals in (8.48).

In the integration for W_{loss} in (8.44), we first note the differential relations for the perimeter parameter l, shown in Fig. 8.14. In the TE_{m0} mode, with $k_c = k_x$, we obtain

$$H_\perp = H_x = j\eta\frac{\omega}{\omega_c}\frac{B\sin k_x x}{Z}, \qquad H_z = B\cos k_x x$$

so that

$$W_{\text{loss}} = \frac{R_s}{2}\left[2\int_0^a dx\left(\left|j\eta\frac{\omega}{\omega_c}\frac{B}{Z}\sin k_x x\right|^2 + |B\cos k_x x|^2\right)\right] + \frac{R_s}{2}2bB^2$$

$$W_T = \frac{Z}{2}\int_0^a\int_0^b dx\,dy\left|j\eta\frac{\omega}{\omega_c}\frac{B}{Z}\sin k_x x\right|^2$$

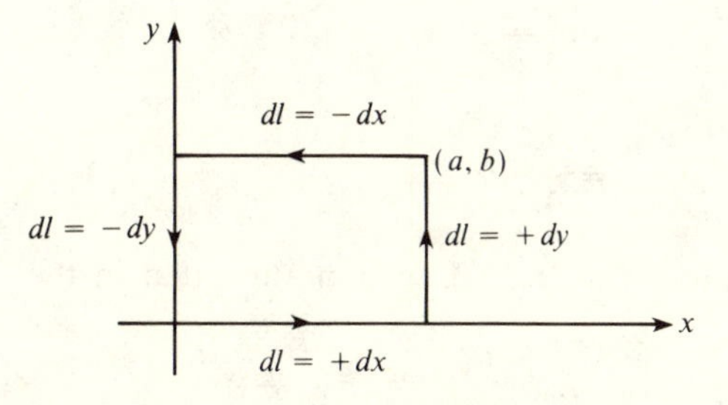

Figure 8.14 Integration parameters for Example 8.5.

It follows that

$$\alpha = \frac{W_{\text{loss}}}{2W_T} = \frac{R_s Z}{b\eta^2}\left[\left(\frac{y}{z}\right)^2 + \left(\frac{\omega_c}{\omega}\right)^2 + 2ab\left(\frac{\omega_c}{\omega}\right)^2\right]$$

But for a TE mode,

$$\left(\frac{\eta}{Z}\right)^2 + \left(\frac{\omega_c}{\omega}\right)^2 = 1$$

whence

$$\alpha = \frac{R_s}{b\eta}\left[1 + \frac{2b}{a}\left(\frac{\omega_c}{\omega}\right)^2\right]$$

in agreement with (8.49).

Example 8.6

Show that the time-average power flow in the z-direction through a rectangular guide operating in the TE_{10} mode is given by

$$W_T = \frac{E_0^2}{4Z}ab$$

where E_0 is the amplitude of $\mathbf{E}_\perp$. Show further that this result agrees with (8.39c).

Ans. We must evaluate the integral (8.39):

$$W_T = \frac{1}{2}\int_0^a dx \int_0^b dy\, \text{Re}[\mathbf{E} \times \mathbf{H}^*]$$

$$= \frac{1}{2Z}\int\int dx\, dy\, |\mathbf{E}_\perp|^2$$

In the TE_{10} mode, we write

$$\mathbf{E}_\perp = \mathbf{a}_y E_0 \sin\frac{\pi x}{a}.$$

$$W_T = \frac{E_0^2 b}{2Z}\int_0^a \sin^2\frac{\pi x}{a}\, dx$$

$$= \frac{E_0^2 b}{2Z}\frac{a}{2} = \frac{E_0^2}{4Z}ab$$

To show the equivalence of this result to (8.39c), we note first that in the TE_{10} mode,

$$H_{0z} = \frac{\omega_c}{\omega}\frac{E_0}{\eta}$$

Substituting this form together with the relation

$$Z_{\mathrm{TE}} = \frac{\eta}{\sqrt{1-(\omega_c/\omega)^2}}$$

into the above answer for W_T, we obtain (8.39c).

Lossy Dielectric

If the waveguide is filled with a lossy dielectric, then with (7.44) we write

$$\epsilon = \epsilon' - j\epsilon''$$

Substituting this form into (8.15),

$$\gamma = j\sqrt{\omega^2 - k_c^2}$$

gives

$$\gamma = j\sqrt{\omega^2\mu' - k_c^2 - j\omega^2\mu''} \tag{8.54}$$

In the low-loss approximation, with $\epsilon'' = \sigma/\omega$, Taylor series expansion of the radical, as was performed previously in obtaining (7.51), gives

$$\alpha \simeq \frac{\alpha\eta}{2}\left[1 - \left(\frac{\omega_c}{\omega}\right)^2\right]^{-1/2} \tag{8.55a}$$

$$\beta \simeq \frac{\omega}{v}\left[1 - \left(\frac{\omega_c}{\omega}\right)^2\right]^{1/2} = \frac{2\pi}{\lambda_0}\sqrt{1 - \left(\frac{\omega_c}{\omega}\right)^2} \tag{8.55b}$$

Note that the mode wavelength, $2\pi/\beta$, maintains the value (8.20) that it had in the lossless case.

Example 8.7

Obtain expressions for the phase difference between $\mathbf{E}_\perp$ and $\mathbf{H}_\perp$ in both the TE and TM modes, for a lossy waveguide.

Ans. We first rewrite the wave impedance in terms of γ:

$$Z_{\mathrm{TE}} = \frac{\eta\omega}{\sqrt{\omega^2 - \omega_c^2}} = \frac{j\eta\omega}{v\gamma} = \frac{j\mu\omega}{\gamma}$$

$$= \frac{-\mu\omega}{j\alpha - \beta} = \frac{\mu\omega}{\alpha^2 + \beta^2}(\beta + j\alpha)$$

It follows that

$$Z_{\mathrm{TE}} = \frac{\mu\omega}{\alpha^2 + \beta^2}|\gamma|e^{j\theta}, \qquad \tan\theta = \frac{\alpha}{\beta}$$

Table 8.3 Formulas for a rectangular waveguide[a]

Dispersion relation	$\gamma^2 = \frac{1}{v^2}(\omega_c^2 - \omega^2),\ v^2 = 1/\mu\epsilon$
Cutoff frequency,	$f_c = \frac{\omega_c}{2\pi} = \frac{v}{\lambda_c} = \frac{vk_c}{2\pi} = \frac{v}{2}\sqrt{\left(\frac{m}{a}\right)^2 + \left(\frac{n}{b}\right)^2}$
wavenumber, etc.	$f_c\lambda_c = f\lambda = v$
Guide wavelength[b]	$\lambda_g = \frac{2\pi}{\beta} = \frac{\lambda}{\sqrt{1-(\omega_c/\omega)^2}} \geq \lambda$
	$\frac{1}{\lambda^2} = \frac{1}{\lambda_g^2} + \frac{1}{\lambda_c^2}$
Attenuation constant:	
$\omega < \omega_c$	$\alpha = \frac{\omega}{v}\sqrt{\left(\frac{\omega_c}{\omega}\right)^2 - 1}$
$\omega > \omega_c$:	
TE_{m0} mode (resistive loss)	$\alpha = \frac{2R_s}{b\eta\sqrt{1-(\omega_c/\omega)^2}}\left[1 + \frac{2b}{a}\left(\frac{\omega_c}{\omega}\right)^2\right]$
Low-loss dielectric	$\alpha \simeq \frac{\sigma\eta}{2}\frac{1}{\sqrt{1-(\omega_c/\omega)^2}}$
Wave phase velocity	$v_\phi = \frac{v}{\sqrt{1-(\omega_c/\omega)^2}}$
Wave group velocity	$v_G = v\sqrt{1-(\omega_c/\omega)^2}$
	$v_\phi v_G = v^2$
Wave impedance	$Z_{TE} = \frac{\eta}{\sqrt{1-(\omega_c/\omega)^2}}$
	$Z_{TM} = \eta\sqrt{1-(\omega_c/\omega)^2}$
	$Z_{TE}Z_{TM} = \eta^2$

[a] All expressions except the last three apply to both TE and TM modes.
[b] Wavelength of mode in the waveguide.

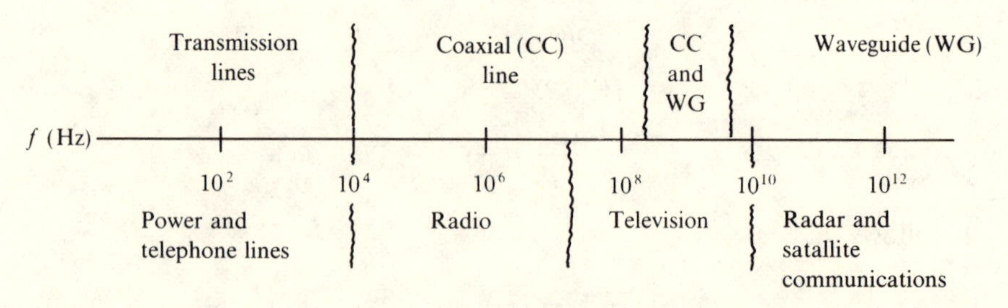

Figure 8.15 Relevant transmission systems for various operating frequency bands.

Thus, in a TE mode, the phase difference between $\mathbf{E}_\perp$ and $\mathbf{H}_\perp$ is proportional to $\tan^{-1}(\alpha/\beta)$, which in the lossless case ($\alpha = 0$) vanishes.

In a TM mode,

$$Z_{\text{TM}} = \frac{\eta}{\omega}\sqrt{\omega^2 - \omega_c^2} = \frac{\eta}{j\omega}v\gamma = \frac{\gamma}{j\omega}$$

$$Z_{\text{TM}} = \frac{-j\alpha + \beta}{\omega(\epsilon' - j\epsilon'')}$$

$$Z_{\text{TM}} = \frac{|\gamma\epsilon| e^{j\theta} e^{j\psi}}{\omega(\epsilon'^2 + \epsilon''^2)}$$

$$\tan\psi = \frac{\epsilon''}{\epsilon'}, \qquad \tan\theta = -\frac{\alpha}{\beta}$$

For a low-loss dielectric, α is given by (8.55), and the total phase difference between $\mathbf{E}_\perp$ and $\mathbf{H}_\perp$ goes as $\psi + \theta$. If losses are due solely to wall resistance, then $\epsilon'' = 0$, and α is given by (8.51).

We conclude this chapter with a brief recapitulation of important formulas for a rectangular waveguide, listed in Table 8.3.

Having reached this point in our discourse, the reader may well ask, "What are the efficient working frequency ranges for the various transmission devices discussed?" In all, we have described transmission lines, coaxial cables, and waveguides. The chart in Fig. 8.15 delineates their efficient frequency ranges and corresponding types of electromagnetic transmission systems. (See also Table 8.1 and the graph of α vs f in Fig. 8.13.)

PROBLEMS

8.1. In Example 8.4e it was concluded that there was a voltage minimum in the equivalent standing-wave pattern for the given waveguide configuration. Employing the correspondence equations (8.31), discuss the maxima and minima of the components of the field variables **E** and **H** at this location.

8.2. Show that the velocity of propagation of wave energy in the TE_{10} mode is equal to the group velocity of the wave. Hint: This velocity is given by the following expression:

$$v_E = \frac{\iint \langle P_z \rangle \, dx\, dy}{\iint [\langle u_e \rangle + \langle u_m \rangle] \, dx\, dy}$$

Both integrals are evaluated on a plane of constant z.

8.3. A wave propagating in the TE_{10} mode of a rectangular waveguide is terminated in a load such that the standing-wave ratio is found to have the value 2. Successive equivalent voltage minima in the standing-wave pattern are located at distances 0.5, 2.5, and 4.5 cm from the load.

If the guide dimensions are 2.29 cm × 1.02 cm, find:

a. The guide wavelength.
b. The wave frequency. What band does this frequency fall in?
c. The wave impedance.
d. The load impedance (use the Smith chart here).

8.4. A rectangular waveguide is operated in the TE_{10} mode at a frequency of 10 GHz. The guide is terminated in a load of $150\pi(3 + 4j)$ ohms. The cutoff frequency for the TE_{10} mode is 6 GHz.

a. Determine the width a of the guide.
b. What is the reflection coefficient at the load?
c. What fraction of the incident power is absorbed in the load?
d. How far must one move from the load to obtain a purely real input impedance? What is the wave impedance at that point?
e. What is the standing-wave ratio S?
f. If the narrow side of the guide (b) has a width of 1 cm and the amplitude of the electric field in the guide is 10^{-3} V/m, calculate the time-average power flow through the guide, $\langle P_z \rangle$.

8.5. A 0.5-cm × 1.0-cm air-filled waveguide is coupled to a 1.0-cm × 1.5-cm air-filled waveguide through a junction which has an equivalent shunt capacitance of 1.5 pF. The equivalent transmission-line configuration is shown in Fig. 8.16. The system transmits in the TE_{11} mode of guide A at 5 GHz.

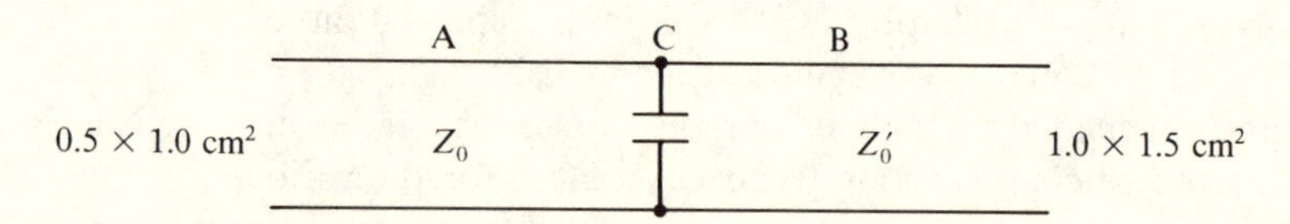

Figure 8.16 Configuration for Problem 8.5.

a. What are the equivalent characteristic admittances Y_0 and Y_0' of the guides?
b. What is the input admittance Y_L at C?
c. What is the reflection coefficient at C?
d. What is the fraction F of incident power reflected at C?
e. At what minimum distance from C, toward the generator, can a shunt inductance L be placed to match the guides? What is the value of L? (Recall Sec. 5.5.)

8.6. Show that TEM waves cannot propagate in a hollow guide.

8.7. An air-filled waveguide has dimensions $a \times a$. What is the smallest value of a for propagation in the TE_{10} mode at 10 MHz (i.e., in the HF band)?

CHAPTER

NINE

MICROSTRIP, STRIPLINES, AND OTHER PRACTICAL TRANSMISSION LINES

In Chapter 2 the elementary concepts of transmission lines were discussed, and the coaxial and twin-lead lines were given as typical structures for transmitting electrical energy from a generator to a load. Other examples of commonly used transmission lines are discussed in this chapter. These include the microstrip line, the stripline, the slotline, the coplanar line, and some variations of the microstrip line. Several of these structures have been found to be very useful as microwave integrated-circuit elements, the microstrip and coplanar being the most popular. Some properties of these transmission lines are described in detail because of their practical importance.

9.1 MICROSTRIP LINE

The microstrip transmission line shown in Fig. 9.1 consists of a conductor strip separated from a ground plane by a dielectric substrate. If a high-dielectric substrate is used, the electrical field lines concentrate in the dielectric material and the result is a very compact circuit. Since some of the field lines are actually outside the substrate, as shown in the figure, *the effective dielectric constant* (ϵ_{re}) is lower than that of the substrate (ϵ_r). Typically, substrates include glass-impregnated Teflon, alumina, special plastics, polyimide semiinsulating gallium arsenide, and in some cases, semiinsulating silicon. The first four are useful as microwave and high-speed

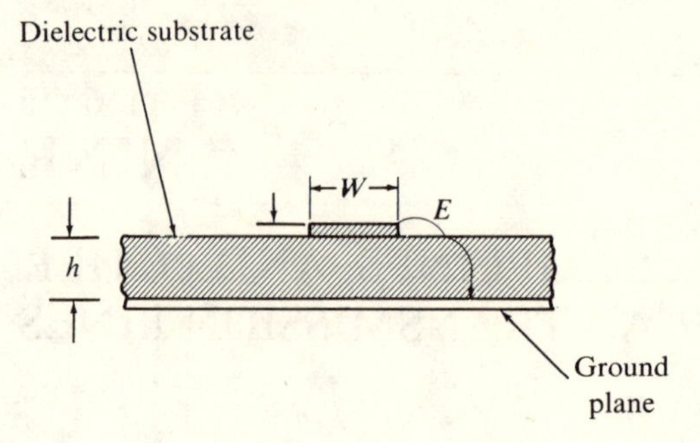

Figure 9.1 Cross section of microstrip line.

logic circuit elements. Gallium arsenide and silicon substrates are used in monolithic microwave circuits in which the microstrip circuits are integrated along with transistors, diodes, and other active elements on the semiconductor chip.

An approximate formula for the characteristic impedance of the microstrip line, assuming that the conducting-strip thickness t is small, is

$$Z_0 = \frac{60}{\sqrt{\epsilon_{re}}} \ln\left(\frac{8h}{w} + \frac{w}{4h}\right) \qquad \text{for} \quad 0.2 < w/h \le 3 \tag{9.1}$$

$$Z_0 = 120 \frac{\pi}{\sqrt{\epsilon_{re}}} \left[\frac{w}{h} + 1.39 + \tfrac{2}{3}\ln\left(\frac{w}{h} + 144\right)\right]^{-1} \qquad \text{for} \quad 0.2 \le w/h \le 3 \tag{9.2}$$

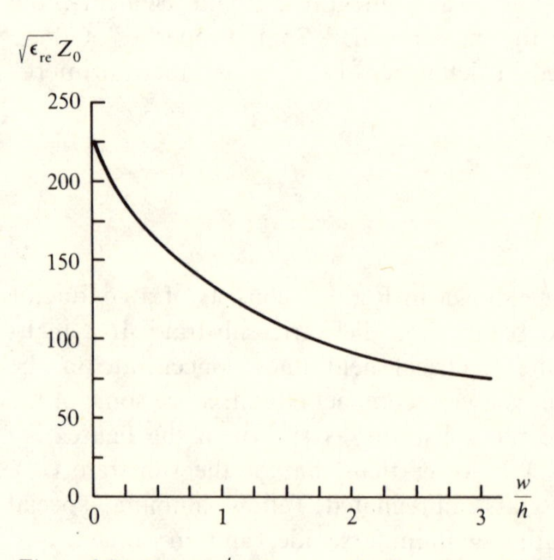

Figure 9.2a Plot of $\sqrt{\epsilon_{re}}\, Z_0$ vs. w/h (Eqs. 9-1 and 9-2).

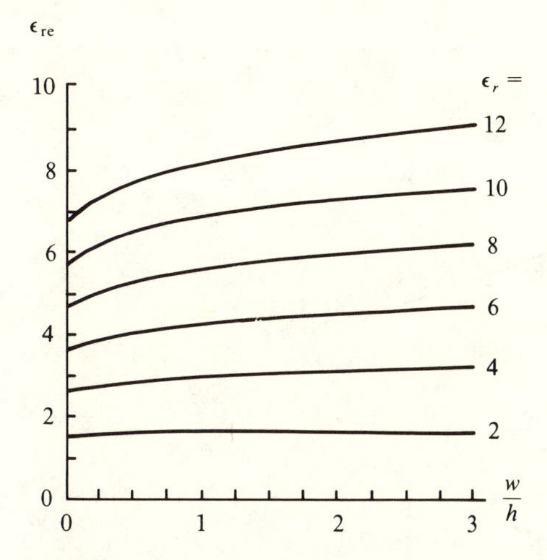

Figure 9.2b Plot of the effective dielectric constant vs. w/h with the relative dielectric constant as a parameter (Eq. 9-3).

where the *effective relative constant* ϵ_{re} is given by

$$\epsilon_{re} = \frac{\epsilon_r + 1}{2} + \frac{\epsilon_r - 1}{2}\left(1 + \frac{10h}{w}\right)^{-1/2} \tag{9.3}$$

and

$$h = \text{dielectric-substrate thickness}$$
$$w = \text{conducting-strip width}$$

These equations, shown graphically in Fig. 9.2a and b, are valid within a few percent for microstrips operated at or below 2 GHz.

Example 9.1

A 50-ohm microstrip transmission line is to be connected to a load of 30 ohms. Design a quarter-wavelength microstrip section which, when connected in cascade between the 50-ohm line and the 30-ohm line, will eliminate reflections from the load at a frequency of 1000 MHz. The relative dielectric constant of the microstrip substrate to be used is 10.2, and the thickness of the substrate is 0.025 inches (0.635 mm).

Ans. We want to determine the required values of the strip conductor width w and the length of the section, l. To find w we make use of Eqs. (9.1), (9.2), and (9.3). In these equations $\epsilon_r = 10.2$, $h = 0.025$ in. $= 0.635$ mm, and $Z_0 = (Z_0' Z_L)^{1/2}$, where in this case, $Z_0' = 50$ ohms and $Z_L = 30$ ohms. Thus $Z_0 = 38.73$ ohms. Since we do not have an equation that gives w directly as a function of Z_0, the simplest approach is to graph Eqs. (9.1)

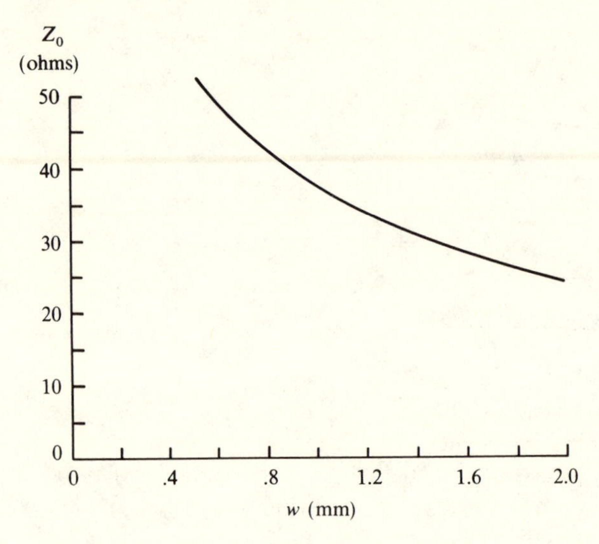

Figure 9.3 Graph of Z_0 vs. width for $\epsilon_r = 10.2$ and $h = 0.635$ mm (Example 9.1).

and (9.2) for a range of values of w. From the graph, shown in Fig. 9.3, we find that, for $Z_0 = 38.73$, w is approximately 0.95 mm. Using Eqs. (9.2) and (9.3), we find $w = 0.945$ by an interpolation process.

To find the required length l of the matching section we make use of the fact that

$$\beta l = \frac{2\pi f}{v} l = \frac{2\pi f}{c} \epsilon_{re}^{1/2} l = \frac{2\pi}{4}$$

or

$$l = \frac{c}{4f} \epsilon_{re}^{-1/2}$$

where $c = 3 \times 10^{10}$ cm/sec, $f = 1000$ MHz, and ϵ_{re} is given by Eq. (9.3). Since

$$\frac{w}{h} = \frac{0.945}{0.635} = 1.488$$

then

$$\epsilon_{re} = 7.26$$

and

$$l = \frac{3 \times 10^{10}}{4 \times 10^3 \times 10^6} (7.26)^{-1/2} = 2.784 \text{ cm}$$

Summarizing, the matching-section parameters are

$$w = 0.945 \text{ mm}$$
$$h = 0.635 \text{ mm}$$
$$l = 27.8 \text{ mm}$$
$$\epsilon_r = 10.2$$
$$\epsilon_{re} = 7.26$$

9.2 STRIPLINE

A stripline transmission line consists, as shown in Fig. 9.4, of a thin, narrow strip of highly conducting metal which is imbedded in a dielectric medium between two grounded conductors. All of the electric and magnetic fields are confined between the two conductors, and operation is in the TEM.

Stripline is very useful for broadband circuits, since it operates in the TEM. Its major disadvantage is that once constructed, it is difficult to make circuit modifications, since the center strip is completely encased between the two grounded conductors.

An approximate expression for the characteristic impedance of stripline, assuming that the thickness of the central conducting strip is small, is

$$Z_0 = \frac{30\pi^2}{\sqrt{\epsilon_r}}\left[\ln\left(2\frac{1+\sqrt{k}}{1-\sqrt{k}}\right)\right]^{-1} \text{ ohms} \qquad (0.2 \le k < 3) \tag{9.4}$$

where

$$k = \tanh\left(\frac{\pi w}{2b}\right) \tag{9.5}$$

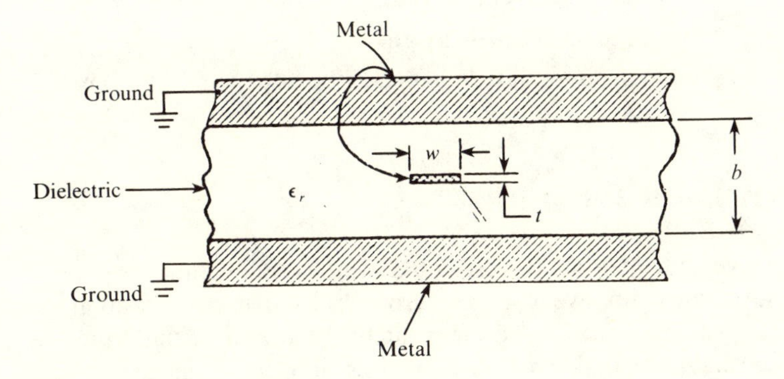

Figure 9.4 Strip line cross section.

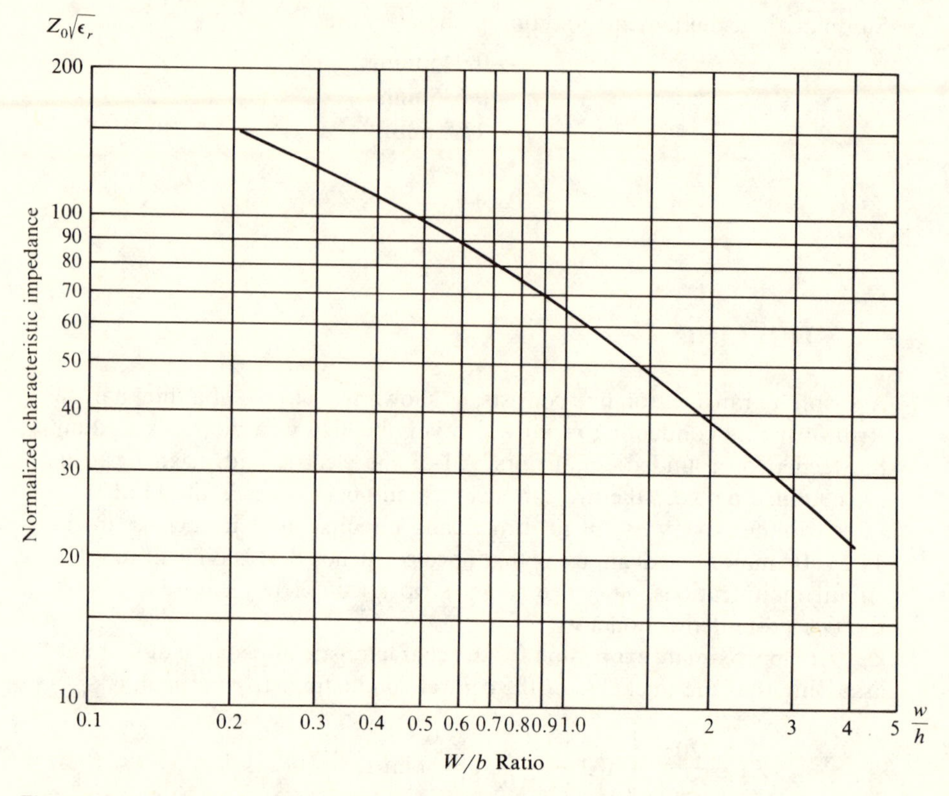

Figure 9.5 Normalized characteristic impedance vs. W/b ratio for strip line ($t/b \ll 1$).

and

$$\epsilon_r = \text{relative dielectric constant of the dielectric}$$
$$w = \text{width of the central conducting strip}$$
$$b = \text{thickness of the dielectric}$$

Figure 9.5 is a plot of the normalized characteristic impedance as a function of w/b.

9.3 COPLANAR WAVEGUIDE

Coplanar waveguide (CPW) consists of a center conducting strip on a dielectric substrate with two coplanar grounds located parallel to it, as shown in Fig. 9.6. The electric and magnetic field pattern for this structure, shown in the figure, contains a longitudinal component of magentic field. The propagation is not of the TEM type at high frequencies but is

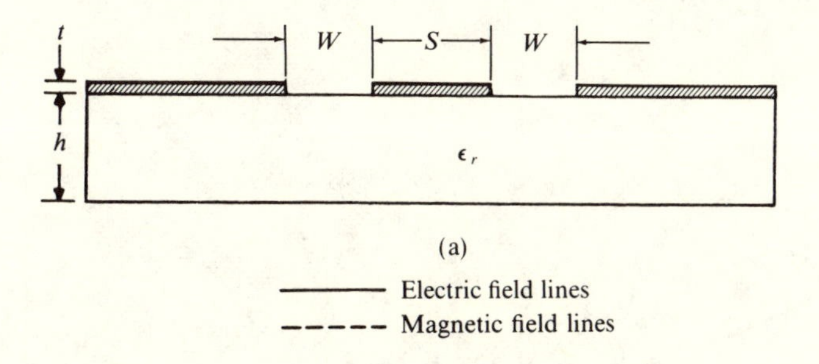

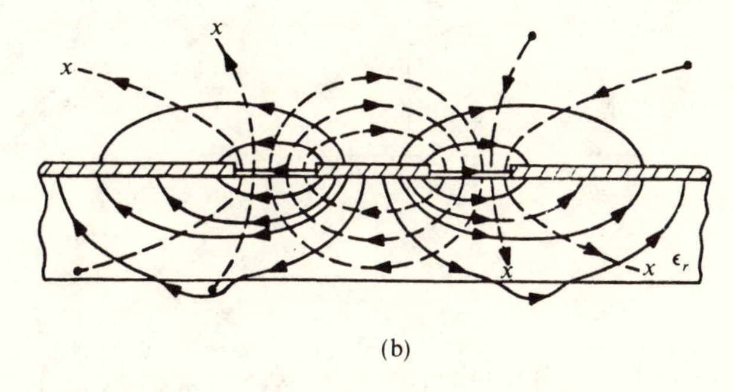

Figure 9.6 Coplanar waveguide: (a) cross section; (b) field lines.

quasi-TEM at lower frequencies. This line has the special advantage that active or passive circuit components can be readily incorporated in either a series or shunt connection. Since microwave integrated circuits are basically coplanar in structure, coplanar waveguide lines are used widely as circuit elements and as interconnecting lines.

An approximate formula for the characteristic impedance of coplanar waveguide, assuming t is small, $0 < k < 1$, and $h \gg w$, is

$$Z_0 = \frac{30\pi^2}{\sqrt{(\epsilon_r + 1)/2}} \left[\ln\left(2 \frac{1 + \sqrt{k}}{1 - \sqrt{k}} \right) \right]^{-1} \text{ ohms} \qquad (9.6)$$

where

$$k = \frac{s}{s + 2w} \qquad (9.7)$$

and

s = center-strip width

w = slot width

ϵ_r = relative dielectric constant of the dielectric substrate

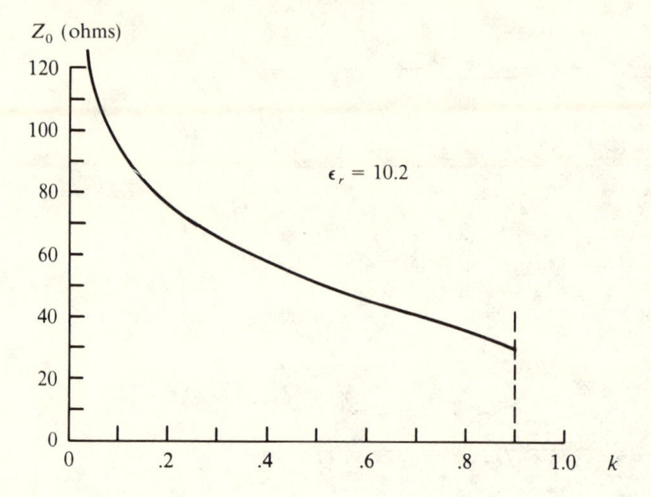

Figure 9.7 Plot of Z_0 vs. k for $\epsilon_r = 10.2$.

Figure 9.7 is a plot of Eq. (9.7) for the case $\epsilon_r = 10.2$

An empirical equation for the effective relative dielectric constant ϵ_{re} of Eq. (9.6) is

$$\epsilon_{re} = \frac{\epsilon_r + 1}{2}\left[\tanh\left(1.785 \log\frac{h}{w} + 1.75\right) + \frac{kw}{h}\left(0.04 - 0.7k + (1 - 0.1\epsilon_r)\frac{(0.25 + k)}{100}\right)\right] \quad (9.8)$$

9.4 COPLANAR-STRIP TRANSMISSION LINE

The structure of coplanar-strip (CPS) transmission line, shown schematically in Fig. 9.8, is similar to that of coplanar waveguide. As shown in the figure, two parallel conductors lie on the top of a dielectric substrate. This structure is roughly similar to that of the twin lead transmission line.

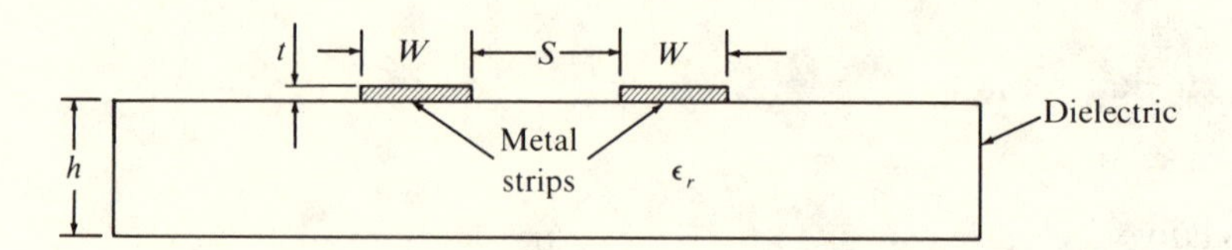

Figure 9.8 Cross section of coplanar strip transmission line.

9.5 SLOTLINE

Slotline, shown schematically in Fig. 9.9, consists of a dielectric substrate metallized on one side, with a narrow slot etched in the metallization. The electric field component of a slotline is principally directed from one of the two conductors to the other through both the substrate and the air. The electromagnetic wave propagates along the slot in a non-TEM mode, but there is no cutoff frequency because of the two-conductor structure.

The slotline has been analyzed carefully, and design curves are available in the literature. No simple formula for its impedance has been derived, however.

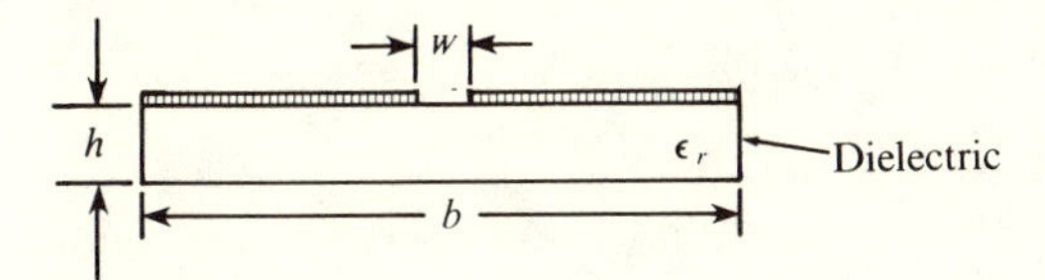

Figure 9.9 Cross section of a slotline.

9.6 FINLINE

The finline transmission-line structure is shown schematically in Fig. 9.10. It consists of a metallic slot pattern on a dielectric substrate centrally located inside a shielding waveguide. The usual waveguide modes do not exist in this case, because the two metallic strips are soldered to the upper and lower waveguide walls. The waves that propagate in this structure are not the TEM type.

The finline is an interesting structure because it has low internal losses and because operation at frequencies up to 100 GHz is practical.

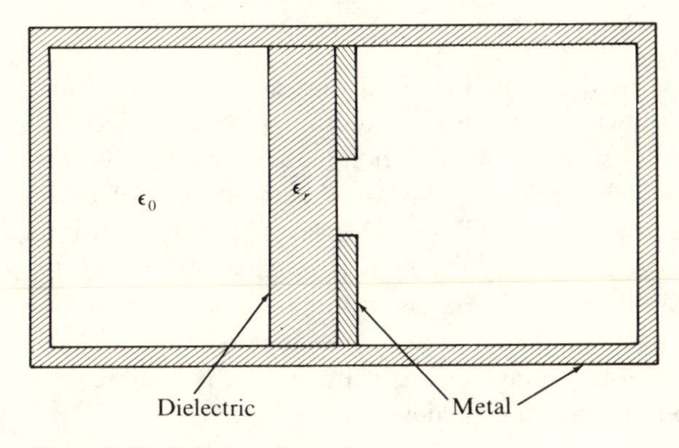

Figure 9.10 Finline configuration.

9.7 INVERTED MICROSTRIP TRANSMISSION LINES

The inverted microstrip differs from the microstrip transmission line in that the usual ground plane is removed from the insulating substrate and the structure is inverted and located opposite a separate ground plane, as shown in Fig. 9.11. In this case air is substituted for the usual substrate, the primary role of the latter simply being to provide a mechanical support for the strip. Most of the fields lie in the air between the strip and the ground plane. Since the effective dielectric constant is close to that of air, the guided wavelength is longer than that for microstrip. Also, a wider stripline length is required for a given value of characteristic impedance. Because of the low losses in this structure, operation up to 100 GHz is practical.

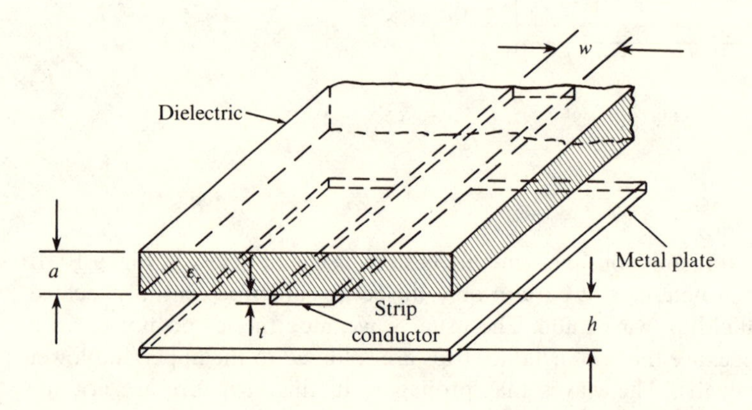

Figure 9.11 Inverted microstrip.

9.8 IMAGELINE

An imageline, shown schematically in Fig. 9.12, has a very simple waveguide-type structure. As shown, it consists of a continuous strip of dielectric, usually of rectangular cross section, which is bonded to a metallic ground plane. Essentially it is a dielectric waveguide which, on account of reflections at the air-dielectric interface, the TE and TM waves are trapped within it. This "trapping" can occur only at very high frequencies for which the wavelengths are comparable with the cross-sectional dimensions of the waveguide. Typically, experimental operation is in hundreds of GHz. While the losses are very low, a disadvantage of imageline is that it is difficult to construct circuits incorporating solid-state devices.

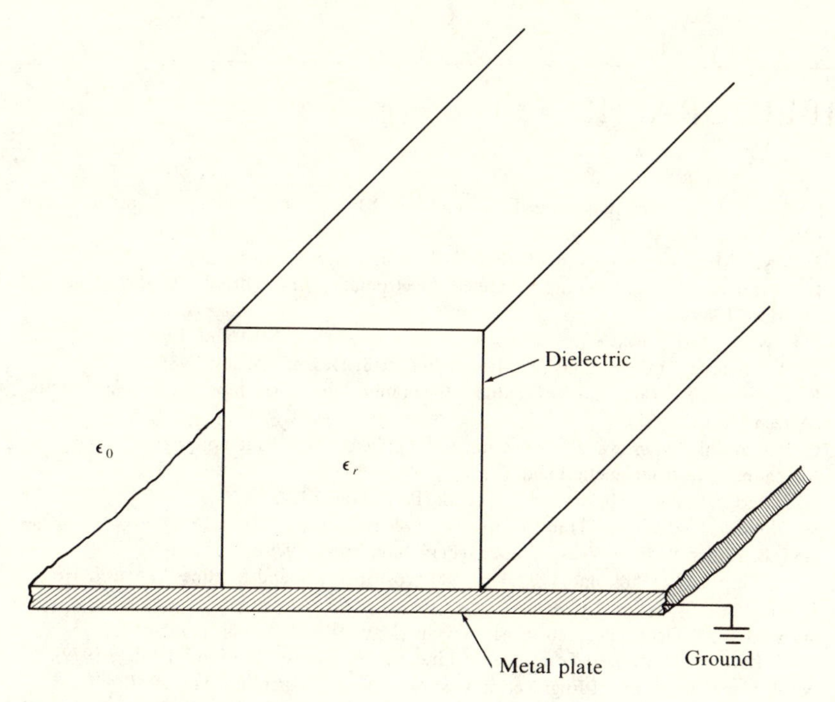

Figure 9.12 Image line configuration.

PROBLEMS

9.1. You wish to fabricate a quarter-wave microstrip resonant structure for operation at 2 GHz. You have two copper-clad dielectric substrates available, both having a 0.635-mm substrate thickness, but one having a dielectric constant of 2.3 and the other 10.2. Calculate the physical lengths of the quarter-wave section required for each substrate.

9.2. A 10-ohm transmission line is to be fabricated on an $\epsilon_r = 10.2$ dielectric substrate of thickness 0.635 mm. Which configuration would be easier to make: a stripline or a coplanar waveguide if both have the same center strip width? Explain your answer.

9.3.

a. Measurements of the capacitance of a 3-cm length of microstrip yield a value of 0.64 pF. The relative dielectric constant of the microstrip substrate is 10.2. Calculate the characteristic impedance of the microstrip.
b. Find the standing wave ratio S on the microstrip if a load impedance $Z_L = 100 + j65$ terminates the line.
c. Calculate the maximum and minimum values of the voltage across the microstrip if the average power delivered to the load impedance is 10 mW.

BIBLIOGRAPHY

1. S. Ramo, J. R. Whinnery, and T. Van Duzer, "Fields and Waves in Communication Electronics," 2nd ed. (Wiley, 1984).
2. J. Frey, "Microwave Integrated Circuits" (Artech House, 1985).
3. L. C. Shen and J. A. Kong, "Applied Electromagnetism" (Brooks/Cole Engineering Division, 1983).
4. J. F. White, "Microwave Semiconductor Engineering" (Van Nostrand, 1982).
5. T. C. Edwards, "Foundations for Microstrip Circuit Design" (Wiley, 1981).
6. K. C. Gupta, R. Garg, and R. Chadha, "Computer-Aided Design of Microwave Circuits" (Artech House, 1981).
7. O. P. Gandhi, "Microwave Engineering and Applications" (Pergamon, 1981).
8. M. Zahn, "Electromagnetic Field Theory" (Wiley, 1979).
9. A. J. Baden Fuller, "Microwaves," 2nd ed. (Pergamon, 1979).
10. W. Sinnema, "Electronic Transmission Technology" (Prentice-Hall, 1979).
11. A. J. B. Fuller, "Microwaves," 2nd ed. (Pergamon Press, 1979).
12. K. C. Gupta, R. Garg, and I. J. Bahl, "Microstrip Lines and Slotlines" (Artech House, 1979).
13. W. Sinnema, "Electronic Transmission Technology" (Prentice-Hall, 1979).
14. L. N. Dworsky, "Modern Transmission Line Theory and Applications" (Wiley, 1979).
15. N. N. Rao, "Elements of Engineering Electromagnetics" (Prentice-Hall, 1977).
16. E. C. Jordan, K. G. Balmain, "Electromagnetic Waves and Radiating Systems," 2nd ed. (Prentice-Hall, 1968).
17. J. D. Jackson, "Classical Electrodynamics" (Wiley, 1975).
18. C. T. A. Johnk, "Engineering Electromagnetic Fields and Waves" (Wiley, 1975).
19. H. Howe, Jr., "Stripline Circuit Design" (Artech House, 1974).
20. R. G. Brown, R. A. Sharpe, W. L. Hughes, and R. E. Past, "Lines, Waves and Antennas," 2nd ed. (Wiley, 1973).
21. S. R. Seshardi, "Fundamentals of Transmission Lines and Electromagnetic Fields" (Addison-Wesley, 1971).
22. K. I. Thomassen, "Introduction to Microwave Fields and Circuits" (Prentice-Hall, 1971).
23. D. T. Paris and F. K. Hurd, "Basic Electromagnetic Theory" (McGraw-Hill, 1969).
24. R. E. Collins, "Foundations of Microwave Engineering" (McGraw-Hill, 1966).
25. J. L. Altman, "Microwave Circuits" (Van Nostrand, 1964).
26. W. L. Weeks, "Electromagnetic Theory for Engineering Applications" (Wiley, 1964).
27. P. Moon and D. E. Spencer, "Field Theory for Engineers" (Van Nostrand, 1961).
28. R. N. Ghose, "Microwave Circuit Theory and Analysis" (McGraw-Hill, 1963).
29. R. Plansey and R. E. Collins, "Principles and Applications of Electromagnetic Fields" (McGraw-Hill, 1961).
30. R. B. Adler, L. J. Chu, and R. M. Fano, "Electromagnetic Energy Transmission and Radiation" (Wiley, 1960).
31. J. B. Walsh, "Electromagnetic Theory and Engineering Applications" (Ronald Press, 1960).
32. R. E. Collins, "Field Theory of Guided Waves" (McGraw-Hill, 1960).
33. W. Jackson, "High Frequency Transmission Lines" (Methuen, 1951).

APPENDIX

A

FORMULAS OF VECTOR ANALYSIS AND VECTOR CALCULUS

$$\nabla \cdot \mathbf{A} \times \mathbf{B} = \mathbf{B} \cdot \nabla \times \mathbf{A} - \mathbf{A} \cdot \nabla \times \mathbf{B} \tag{A1}$$

$$\nabla(\mathbf{A} \cdot \mathbf{B}) = (\mathbf{A} \cdot \nabla)\mathbf{B} + (\mathbf{B} \cdot \nabla)\mathbf{A} + \mathbf{A} \times (\nabla \times \mathbf{B}) + \mathbf{B} \times (\nabla \times \mathbf{A}) \tag{A2}$$

$$\nabla \cdot (\nabla \times \mathbf{A}) = 0 \tag{A3}$$

$$\nabla \times (\nabla \times \mathbf{A}) = \nabla(\nabla \cdot \mathbf{A}) - (\nabla \cdot \nabla)\mathbf{A} \tag{A4}$$

$$\nabla \times (\mathbf{A} \times \mathbf{B}) = (\mathbf{B} \cdot \nabla)\mathbf{A} + \mathbf{A}(\nabla \cdot \mathbf{B}) - (\mathbf{A} \cdot \nabla)\mathbf{B} - \mathbf{B}(\nabla \cdot \mathbf{A}) \tag{A5}$$

$$\mathbf{A} \times (\mathbf{B} \times \mathbf{C}) = \mathbf{B}(\mathbf{A} \cdot \mathbf{C}) - \mathbf{C}(\mathbf{A} \cdot \mathbf{B}) \tag{A6a}$$

$$\mathbf{A} \times (\mathbf{B} \times \mathbf{C}) + \mathbf{B} \times (\mathbf{C} \times \mathbf{A}) + \mathbf{C} \times (\mathbf{A} \times \mathbf{B}) = 0 \tag{A6b}$$

$$\mathbf{A} \cdot \mathbf{B} \times \mathbf{C} = \begin{vmatrix} A_x & A_y & A_z \\ B_x & B_y & B_z \\ C_x & C_y & C_z \end{vmatrix} \tag{A7a}$$

$$\mathbf{A} \cdot \mathbf{B} \times \mathbf{C} = \mathbf{B} \cdot \mathbf{C} \times \mathbf{A} = \mathbf{C} \cdot \mathbf{A} \times \mathbf{B} \tag{A7b}$$

$$(\mathbf{A} \times \mathbf{B}) \cdot (\mathbf{C} \times \mathbf{D}) = (\mathbf{A} \cdot \mathbf{C})(\mathbf{B} \cdot \mathbf{D}) - (\mathbf{A} \cdot \mathbf{D})(\mathbf{B} \cdot \mathbf{C}) \tag{A8}$$

$$(\mathbf{A} \times \mathbf{B}) \times (\mathbf{C} \times \mathbf{D}) = [\mathbf{D} \cdot \mathbf{A} \times \mathbf{B}]\mathbf{C} - [\mathbf{C} \cdot \mathbf{A} \times \mathbf{B}]\mathbf{D} \tag{A9}$$

$$\nabla(\phi\psi) = \phi\nabla\psi + \psi\nabla\phi \tag{A10}$$

$$\nabla \cdot (\phi\mathbf{A}) = \mathbf{A} \cdot \nabla\phi + \phi\nabla \cdot \mathbf{A} \tag{A11}$$

$$\nabla \times (\phi\mathbf{A}) = \phi\nabla \times \mathbf{A} - \mathbf{A} \times \nabla\phi \tag{A12}$$

If $\mathbf{r}$ is the radius vector from the origin and $\mathbf{A}$ is a constant vector, then

$$\nabla \cdot \mathbf{r} = 3 \tag{A13}$$

$$\nabla \times \mathbf{r} = 0 \tag{A14}$$

$$\nabla r = \mathbf{r}/r \tag{A15}$$

$$\nabla r^{-1} = -\mathbf{r}/r^3 \tag{A16}$$

$$\nabla \cdot (\mathbf{r}r^{-3}) = -\nabla^2 r^{-1} = 4\pi\delta(\mathbf{r}) \tag{A17}^*$$

$$\nabla \cdot (\mathbf{A}r^{-1}) = \mathbf{A} \cdot (\nabla r^{-1}) = -(\mathbf{A} \cdot \mathbf{r})r^{-3} \tag{A18}$$

$$\nabla \times [\mathbf{A} \times (\mathbf{r}/r^3)] = -\nabla(\mathbf{A} \cdot \mathbf{r}/r^3) \qquad \text{for} \quad r \neq 0 \tag{A19}$$

$$\nabla^2 \mathbf{A} r^{-1} = \mathbf{A}\nabla^2 r^{-1} = 0 \qquad \text{for} \quad r \neq 0 \tag{A20}$$

$$(\nabla^2 + k^2)\frac{e^{ikr}}{r} = (\nabla^2 + k^2)\frac{\cos kr}{r} = -4\pi\delta(\mathbf{r}) \tag{A21}^*$$

PROPERTIES OF INTEGRALS OVER VECTOR FUNCTIONS

The Line Integral

$$\int_{\mathbf{r}_1}^{\mathbf{r}_2} \mathbf{A} \cdot d\mathbf{l} = -\int_{\mathbf{r}_2}^{\mathbf{r}_1} \mathbf{A} \cdot d\mathbf{l} \tag{A22}$$

If $\mathbf{A} = \nabla\phi$, or $\nabla \times \mathbf{A} = 0$, then the line integral

$$\int_{r_1}^{r_2} \mathbf{A} \cdot d\mathbf{l} = \phi(\mathbf{r}_2) - \phi(\mathbf{r}_1) \tag{A23}$$

is independent of the path of integration from $\mathbf{r}_1$ to $\mathbf{r}_2$. If the path of integration is a closed curve, then

$$\oint \mathbf{A} \cdot d\mathbf{l} = \oint \nabla\phi \cdot d\mathbf{l} = 0 \tag{A24}$$

Gauss's Theorem

Let S be a surface bounding a region of volume V. Then,

$$\int_V \nabla \cdot \mathbf{A}\, dV = \int_S \mathbf{A} \cdot d\mathbf{S} \tag{A25}$$

Stokes's Theorem

Let S be an open surface bounded by the closed, nonintersecting curve C. Then

$$\int_C \mathbf{A} \cdot d\mathbf{l} = \int_S (\nabla \times \mathbf{A}) \cdot d\mathbf{S} \tag{A26}$$

*Properties of the Dirac delta function in one dimension were previously described in Chapter 4. See, for example, (4.90).

Green's First Identity

$$\int_V (\phi\nabla^2\psi + \nabla\phi \cdot \nabla\psi)\, dV = \int_S (\phi\nabla\psi) \cdot d\mathbf{S} \tag{A27}$$

Green's Second Identity

$$\int_V (\phi\nabla^2\psi - \psi\nabla^2\phi)\, dV = \int_S (\phi\nabla\psi - \psi\nabla\phi) \cdot d\mathbf{S} \tag{A28}$$

Other Identities

$$\int_V \nabla \times \mathbf{A}\, dV = \int_S d\mathbf{S} \times \mathbf{A} \tag{A29}$$

$$\int_C \phi\, d\mathbf{l} = \int_S d\mathbf{S} \times \nabla\phi \tag{A30}$$

APPENDIX

B

DIFFERENTIAL VECTOR RELATIONS IN ORTHOGONAL COORDINATE FRAMES

The line element in orthogonal coordinates is

$$ds^2 = (h_1\,dx_1)^2 + (h_2\,dx_2)^2 + (h_3\,dx_3)^2 \tag{B1}$$

The ∇ operations are

$$\nabla\psi = \frac{1}{h_1}\frac{\partial\psi}{\partial x_1}\mathbf{a}_1 + \frac{1}{h_2}\frac{\partial\psi}{\partial x_2}\mathbf{a}_2 + \frac{1}{h_3}\frac{\partial\psi}{\partial x_3}\mathbf{a}_3 \tag{B2}$$

$$\nabla\cdot\mathbf{A} = \frac{1}{h_1h_2h_3}\left[\frac{\partial}{\partial x_1}(h_2h_3A_1) + \frac{\partial}{\partial x_2}(h_1h_3A_2) + \frac{\partial}{\partial x_3}(h_1h_2A_3)\right] \tag{B3}$$

$$\begin{aligned}\nabla\times\mathbf{A} = {} & \frac{1}{h_2h_3}\left[\frac{\partial}{\partial x_2}(h_3A_3) - \frac{\partial}{\partial x_3}(h_2A_2)\right]\mathbf{a}_1 \\ & + \frac{1}{h_1h_3}\left[\frac{\partial}{\partial x_3}(h_1A_1) - \frac{\partial}{\partial x_1}(h_3A_3)\right]\mathbf{a}_2 \\ & + \frac{1}{h_1h_2}\left[\frac{\partial}{\partial x_1}(h_2A_2) - \frac{\partial}{\partial x_2}(h_1A_1)\right]\mathbf{a}_3\end{aligned} \tag{B4}$$

$$\nabla^2\psi = \frac{1}{h_1h_2h_3}\left[\frac{\partial}{\partial x_1}\left(\frac{h_3h_2}{h_1}\frac{\partial\psi}{\partial x_1}\right) + \frac{\partial}{\partial x_2}\left(\frac{h_3h_1}{h_2}\frac{\partial\psi}{\partial x_2}\right) + \frac{\partial}{\partial x_3}\left(\frac{h_1h_2}{h_3}\frac{\partial\psi}{\partial x_3}\right)\right] \tag{B5}$$

Cartesian coordinates:

$$h_1 = 1, \quad h_2 = 1, \quad h_3 = 1, \qquad x_1 = x, \quad x_2 = y, \quad x_3 = z$$

Cylindrical coordinates:

$$h_1 = 1, \quad h_2 = \rho, \quad h_3 = 1, \qquad x_1 = \rho, \quad x_2 = \phi, \quad x_3 = z$$

Spherical coordinates:

$$h_1 = 1, \quad h_2 = r\sin\theta, \quad h_3 = r, \qquad x_1 = r, \quad x_2 = \phi, \quad x_3 = \theta$$

Differential operator relations in three coordinate frames

Cartesian coordinates	Cylindrical coordinates	Spherical coordinates
dx, dy, dz	$d\rho, \rho\, d\phi, dz$	$dr, r\, d\theta, r \sin\theta\, d\phi$
	Unit vector and elementary cross relations	
$\mathbf{a}_x, \mathbf{a}_y, \mathbf{a}_z$	$\mathbf{a}_\rho, \mathbf{a}_\phi, \mathbf{a}_z$ $x = \rho \cos\phi$ $y = \rho \sin\phi$ $z = z$ $\rho^2 = x^2 + y^2$ $\tan\phi = y/x$ $z = z$ $\mathbf{a}_\rho = \cos\phi\, \mathbf{a}_x + \sin\phi\, \mathbf{a}_y$ $\mathbf{a}_\phi = -\sin\phi\, \mathbf{a}_x + \cos\phi\, \mathbf{a}_y$ $\mathbf{a}_z = \mathbf{a}_z$ $\mathbf{a}_x = \cos\phi\, \mathbf{a}_\rho - \sin\phi\, \mathbf{a}_\phi$ $\mathbf{a}_y = \sin\phi\, \mathbf{a}_\rho + \cos\phi\, \mathbf{a}_\phi$ $\mathbf{a}_z = \mathbf{a}_z$	$\mathbf{a}_r, \mathbf{a}_\theta, \mathbf{a}_\phi$ $x = r \sin\theta \cos\phi$ $y = r \sin\theta \sin\phi$ $z = r \cos\theta$ $r^2 = x^2 + y^2 + z^2$ $\cos\theta = z/\sqrt{x^2 + y^2 + z^2}$ $\phi = \tan^{-1}(y/x)$ $\mathbf{a}_r = \sin\theta \cos\phi\, \mathbf{a}_x + \sin\theta \sin\phi\, \mathbf{a}_y + \cos\theta\, \mathbf{a}_z$ $\mathbf{a}_\theta = \cos\theta \cos\phi\, \mathbf{a}_x + \cos\theta \sin\phi\, \mathbf{a}_y - \sin\theta\, \mathbf{a}_z$ $\mathbf{a}_\phi = -\sin\phi\, \mathbf{a}_x + \cos\phi\, \mathbf{a}_y$ $\mathbf{a}_x = \sin\theta \cos\phi\, \mathbf{a}_r + \cos\theta \cos\phi\, \mathbf{a}_\theta - \sin\phi\, \mathbf{a}_\phi$ $\mathbf{a}_y = \sin\theta \sin\phi\, \mathbf{a}_r + \cos\theta \sin\phi\, \mathbf{a}_\theta + \cos\phi\, \mathbf{a}_\phi$ $\mathbf{a}_z = \cos\theta\, \mathbf{a}_r - \sin\theta\, \mathbf{a}_\theta$
	The gradient of ψ, $\nabla\psi$	
$(\nabla\psi)_x = \dfrac{\partial\psi}{\partial x}$ $(\nabla\psi)_y = \dfrac{\partial\psi}{\partial y}$ $(\nabla\psi)_z = \dfrac{\partial\psi}{\partial z}$	$(\nabla\psi)_\rho = \dfrac{\partial\psi}{\partial\rho}$ $(\nabla\psi)_\phi = \dfrac{1}{\rho}\dfrac{\partial\psi}{\partial\phi}$ $(\nabla\psi)_z = \dfrac{\partial\psi}{\partial z}$	$(\nabla\psi)_r = \dfrac{\partial\psi}{\partial r}$ $(\nabla\psi)_\theta = \dfrac{1}{r}\dfrac{\partial\psi}{\partial\theta}$ $(\nabla\psi)_\phi = \dfrac{1}{r\sin\theta}\dfrac{\partial\psi}{\partial\phi}$
	The divergence of $\mathbf{A}$, $\nabla \cdot \mathbf{A}$	
$\dfrac{\partial A_x}{\partial x} + \dfrac{\partial A_y}{\partial y} + \dfrac{\partial A_z}{\partial z}$	$\dfrac{1}{\rho}\dfrac{\partial(\rho A_r)}{\partial\rho} + \dfrac{1}{\rho}\dfrac{\partial A_\phi}{\partial\phi} + \dfrac{\partial A_z}{\partial z}$	$\dfrac{1}{r^2}\dfrac{\partial(r^2 A_r)}{\partial r} + \dfrac{1}{r\sin\theta}\dfrac{\partial(\sin\theta A_\theta)}{\partial\theta} + \dfrac{1}{r\sin\theta}\dfrac{\partial A_\phi}{\partial\phi}$
	The Laplacian of ψ, $\nabla^2\psi$	
$\dfrac{\partial^2\psi}{\partial x^2} + \dfrac{\partial^2\psi}{\partial y^2} + \dfrac{\partial^2\psi}{\partial z^2}$	$\dfrac{1}{\rho}\dfrac{\partial}{\partial\rho}\left(\rho\dfrac{\partial\psi}{\partial\rho}\right) + \dfrac{1}{\rho^2}\dfrac{\partial^2\psi}{\partial\phi^2} + \dfrac{\partial^2\psi}{\partial z^2}$	$\dfrac{1}{r^2}\dfrac{\partial}{\partial r}\left(r^2\dfrac{\partial\psi}{\partial r}\right) + \dfrac{1}{r^2\sin\theta}\dfrac{\partial}{\partial\theta}\left(\sin\theta\dfrac{\partial\psi}{\partial\theta}\right) + \dfrac{1}{r^2\sin^2\Theta}\dfrac{\partial^2\psi}{\partial\phi^2}$
	The curl of $\mathbf{A}$, $\nabla \times \mathbf{A}$	
$(\nabla \times \mathbf{A})_x = \left(\dfrac{\partial A_z}{\partial y} - \dfrac{\partial A_y}{\partial z}\right)$ $(\nabla \times \mathbf{A})_y = \left(\dfrac{\partial A_x}{\partial z} - \dfrac{\partial A_z}{\partial x}\right)$ $(\nabla \times \mathbf{A})_z = \left(\dfrac{\partial A_y}{\partial x} - \dfrac{\partial A_x}{\partial y}\right)$	$(\nabla \times \mathbf{A})_\rho = \left(\dfrac{1}{\rho}\dfrac{\partial A_z}{\partial\phi} - \dfrac{\partial A_\phi}{\partial z}\right)$ $(\nabla \times \mathbf{A})_\phi = \left(\dfrac{\partial A_\rho}{\partial z} - \dfrac{\partial A_z}{\partial\rho}\right)$ $(\nabla \times \mathbf{A})_z = \dfrac{1}{\rho}\left(\dfrac{\partial(\rho A_\phi)}{\partial\rho} - \dfrac{\partial A_\rho}{\partial\phi}\right)$	$(\nabla \times \mathbf{A})_r = \dfrac{1}{r\sin\theta}\left(\dfrac{\partial(\sin\theta A_\phi)}{\partial\theta} - \dfrac{\partial A_\theta}{\partial\phi}\right)$ $(\nabla \times \mathbf{A})_\theta = \dfrac{1}{r\sin\theta}\dfrac{\partial A_r}{\partial\phi} - \dfrac{1}{r}\dfrac{\partial(rA_\phi)}{\partial r}$ $(\nabla \times \mathbf{A})_\phi = \dfrac{1}{r}\left(\dfrac{\partial(rA_\theta)}{\partial r} - \dfrac{\partial A_r}{\partial\theta}\right)$

APPENDIX

C

SIMPLE PROPERTIES OF COMPLEX FUNCTIONS

In this appendix we review some elementary properties of complex functions (i.e., functions of a complex variable).

We recall that if x and y are real variables, then

$$z = x + jy \tag{C1}$$

is a complex variable, with $j = \sqrt{-1}$. The function $f(z)$ is called a complex function. It has a real ($\text{Re} f$) and an imaginary ($\text{Im} f$) part, and can always be written as

$$f(z) = \text{Re} f + j \,\text{Im} f \tag{C2}$$

For instance

$$f(z) = z^2 = (x + jy)^2 = x^2 - y^2 + j^2xy \tag{C3}$$

In the complex (x, y) plane, $z = x + jy$ may be represented as a vector emanating from the origin with components (x, y). The length of the vector

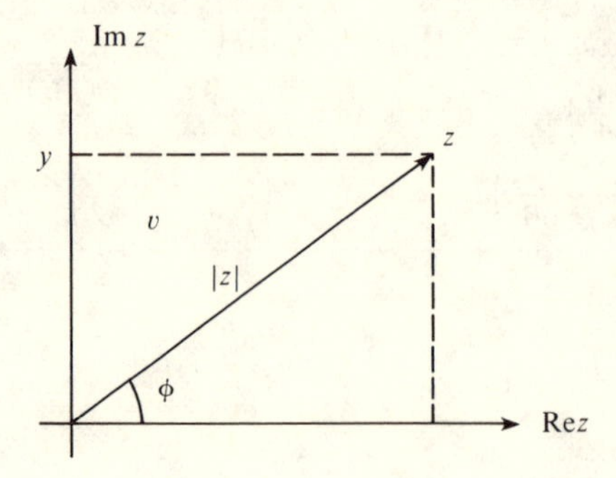

Figure C.1

is written $|z|$. If z makes an angle θ with the real (x) axis, then (Fig. C.1)

$$z = |z|\cos\theta + j|z|\sin\theta$$
$$= |z|(\cos\theta + j\sin\theta) = |z|e^{j\theta}$$
$$|z| = \sqrt{x^2 + y^2} \tag{C4}$$

Similarly, for $f(z)$ (Fig. C.2),

$$f(z) = |f(z)|e^{j\alpha} \tag{C5}$$

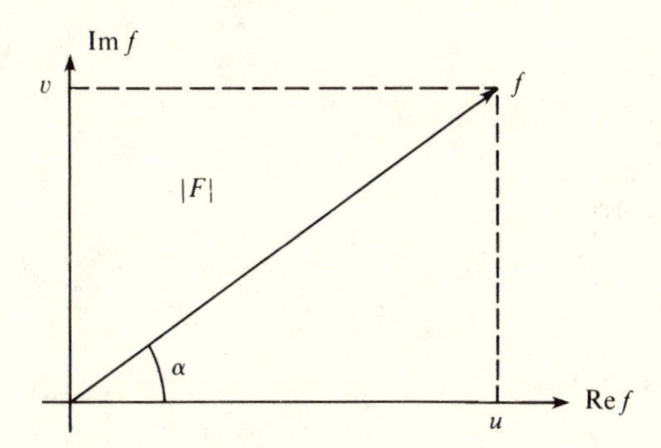

Figure C.2

where

$$|f|\cos\alpha = \operatorname{Re} f$$
$$|f|\sin\alpha = \operatorname{Im} f$$
$$|f|^2 = (\operatorname{Re} f)^2 + (\operatorname{Im} f)^2$$

The function $f(z)$ when written as

$$f(z) = u + jv \tag{C6}$$

is a vector in complex (u, v) space.

The complex conjugate of z is written z^*:

$$z^* = x - jy$$
$$f^* = u - jv \tag{C7}$$

We note the following properties:

$$\operatorname{Re} f = \frac{f + f^*}{2} \tag{C8a}$$

$$\operatorname{Im} f = \frac{f - f^*}{2j} \tag{C8b}$$

A complex function f is real if and only if

$$f = f^*$$

Two complex functions f and g are equal if and only if

$$\operatorname{Re} f = \operatorname{Re} g$$
$$\operatorname{Im} f = \operatorname{Im} g \tag{C9}$$

If α is any real angle, then

$$|e^{j\alpha}| = 1 \tag{C10}$$

Let us show that

$$|f|^2 = |f^2| \tag{C11}$$

Writing

$$f = |f|e^{j\alpha}$$

gives

$$f^2 = |f|^2 e^{j2\alpha}$$
$$|f^2| = |f|^2 |e^{j2\alpha}| = |f|^2$$

Note also that

$$(f^*)^* = f \tag{C12}$$
$$|f|^2 = f^* f \tag{C13}$$

If f and g are two complex functions, then

$$(fg)^* = f^* g^* \tag{C14}$$

and

$$|fg| = |f|\,|g| \tag{C15}$$

If

$$f = |f| \exp j\alpha_1$$
$$g = |g| \exp j\alpha_2$$

then

$$|f + g|^2 = |f|^2 + |g|^2 + 2|fg|\cos(\alpha_1 - \alpha_2). \tag{C16}$$

APPENDIX

D

MICROWAVE COMPONENTS: STRUCTURES AND ELECTRICAL PROPERTIES

This appendix is a brief compilation of some of the more important microwave components in present-day use. It is intended as a reference for those planning further studies of the field of microwaves and as an introduction to the practical aspects of microwave engineering. The topics included are: transmission lines and waveguides, resonant cavities and junctions, directional couplers, and other components. Because of its wide use, special emphasis has been given to the directional coupler. Discussion of microstrips, slot lines, etc. have been omitted in this appendix, since Chapter 9 covers the properties of these lines in detail.

D.1 TRANSMISSION LINES AND WAVEGUIDES

As discussed previously (Fig. 8.15), transmission lines and waveguides are widely used as means to transfer electrical signals from one point to another. For example, the output of a TV transmitter can be connected efficiently to a remotely located transmitting antenna by means of a transmission line or waveguide. Also, signals from a receiving antenna can be connected by a coaxial-cable system to many separate receivers.

Figure D.1 illustrates four practical transmission-line structures: the coaxial line, the two-wire line, the balanced shielded line, and the parallel-plate line. The formulas given correspond to operation in the TEM mode.

Figure D.2 illustrates the rectangular waveguide, which is categorized with transmission lines, since its properties, like those of the transmission

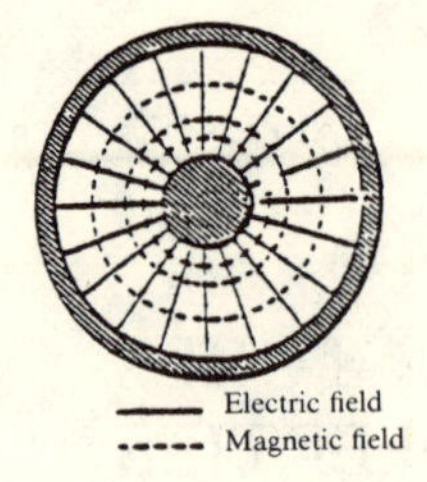

$$Z_0 = \frac{60}{\sqrt{\epsilon_r}} \ln \frac{D}{d} \qquad \text{(ohms)}$$

(a) Coaxial line

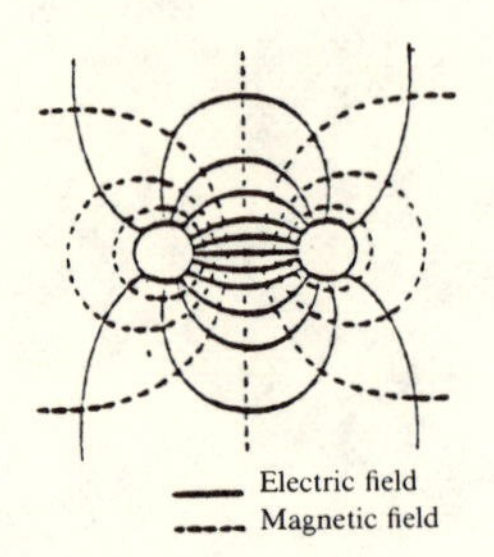

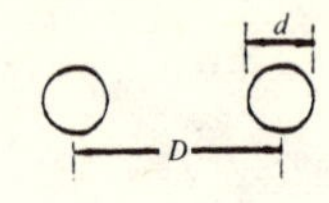

$$Z_0 = \frac{120}{\sqrt{\epsilon_r}} \ln \frac{2D}{d} \qquad \text{(ohms)}$$

$$d/D \ll 1$$

(b) Two wire line

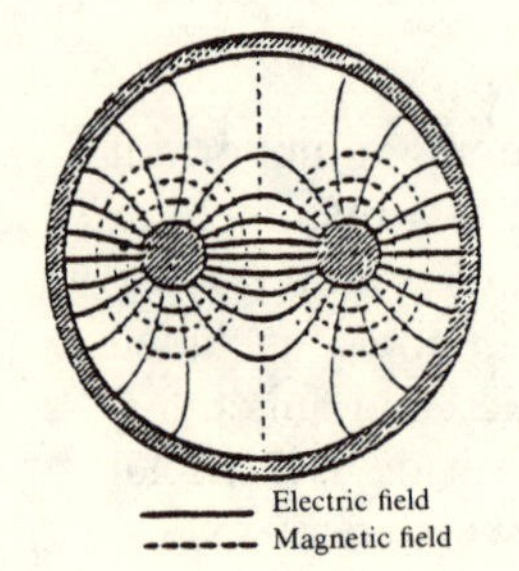

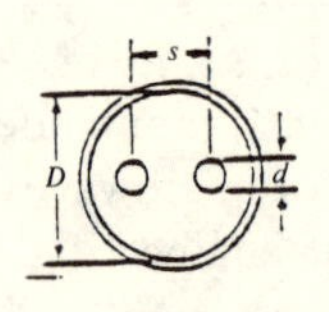

$$Z_0 = \frac{120}{\sqrt{\epsilon_r}} \left\{ \ln\left[2b\left(\frac{1-a^2}{1+a^2}\right) - \frac{1+4b^2}{16b^2}(1-4a^2)\right\} \qquad \text{(ohms)}$$

$$\begin{cases} a = \dfrac{s}{D} \\ b = \dfrac{s}{d} \end{cases}$$

(c) Balanced shielded line

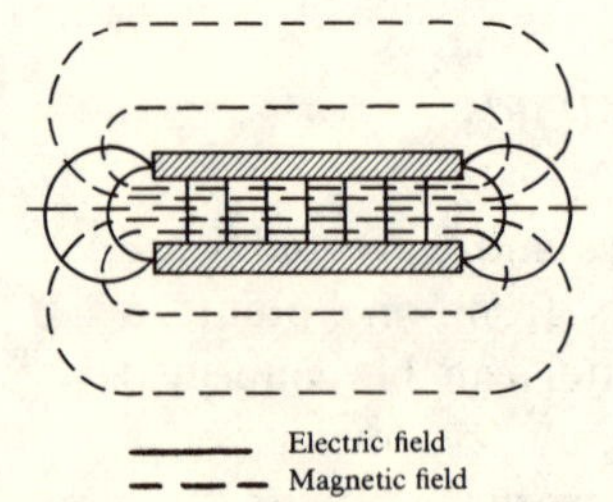

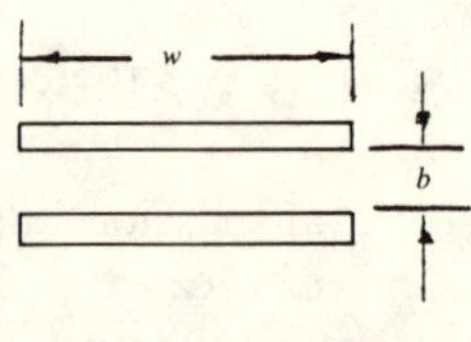

$$Z_0 = \frac{377}{\sqrt{\epsilon_r}} \frac{b}{w} \qquad \text{(ohms)}$$

(d) Parallel plate line

Figure D.1 Transmission lines.

TE_{10} wave (dominant mode)

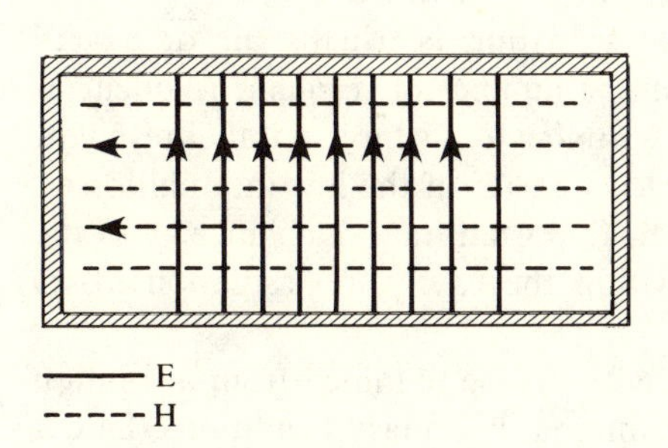

Coordinates

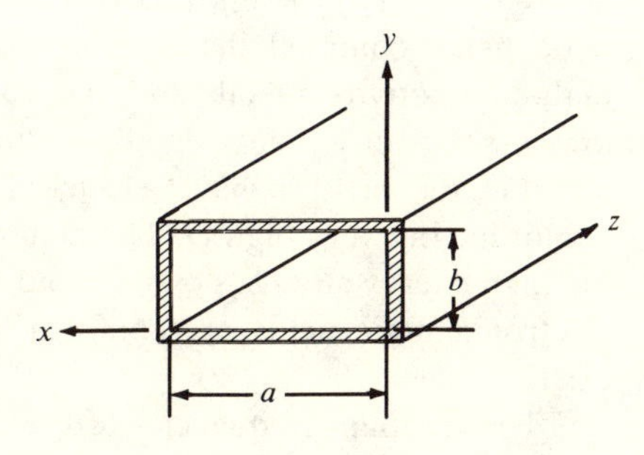

Field equations

$$E_z = E_x = H_y = 0$$

$$E_y = -\frac{j\eta\omega}{\omega_c} B \sin\frac{\pi x}{a}$$

$$H_x = \frac{j\eta\omega}{\omega z_{TE}} B \sin\frac{\pi x}{a}$$

$$H_z = B \cos\frac{\pi x}{a}$$

Cut-off frequency

$$f_c = \frac{1}{2a\sqrt{\mu\epsilon}}$$

where

$$z_{TE} = \frac{\eta}{\left[1 - \left(\frac{\omega_c}{\omega}\right)^2\right]^{1/2}}$$

Figure D.2 Rectangular waveguide (TE_{10} mode).

lines, are determined from traveling-wave theory. Two types of waves are possible in waveguides: TE waves and TM waves. Unlike the coaxial and open-wire lines, TEM waves cannot be propagated in waveguide. The TE waves are characterized by the fact that the electric vector is always perpendicular to the propagation direction, whereas for TM waves, the magnetic field is perpendicular to the propagation direction. There are actually an infinite set of modes possible for the TE waves and TM waves. In this appendix it is assumed, unless otherwise indicated, that the waves are in the *dominant* mode, i.e., the mode of operation which has the lowest cutoff frequency. It is also assumed that the waveguides contain lossless dielectrics and have metallic walls of infinite conductivity.

D.2 RESONANT CAVITIES

A resonant cavity is essentially a hollow conducting box with electromagnetic fields confined inside. The walls of the cavity are usually of very high-conductivity metal, and the volume within is usually air or a very low-loss dielectric. Near one of the infinite number of resonant frequencies possible, the cavity behaves electrically similar to a simple parallel resonant circuit having very high Q. The high Q is a result of the low internal losses, the high energy storage possible, and the important fact that, since all of the electromagnetic fields are confined within the cavity, no radiation losses exist.

The resonant frequencies of the cavity can be obtained from a solution of Maxwell's equations subject to the internal boundary conditions. Essentially, standing-wave patterns exist within the cavity which arise from reflections of the traveling waves from the cavity walls. One or more small openings in the cavity walls coupled to external coaxial lines or waveguides are provided so that the cavity can be excited by an external source of power.

Resonant cavities have many important applications, such as oscillator and amplifier resonators, filter-circuit elements, frequency meters, stable-frequency reference resonators, etc. Figure D.3a, b, and c are examples of rectangular, cylindrical, and coaxial resonators.

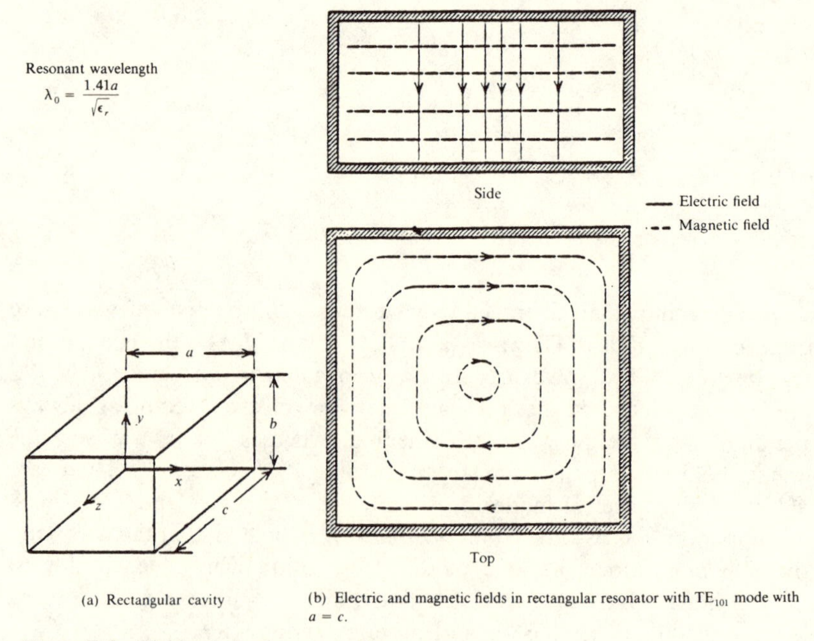

(a) Rectangular cavity

(b) Electric and magnetic fields in rectangular resonator with TE_{101} mode with $a = c$.

Figure D.3a Fields of simple rectangular resonator.

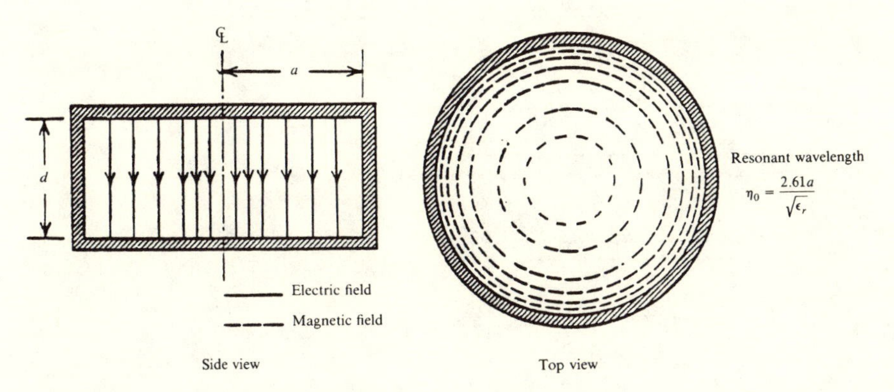

Figure D.3b Circular cylindrical resonator.

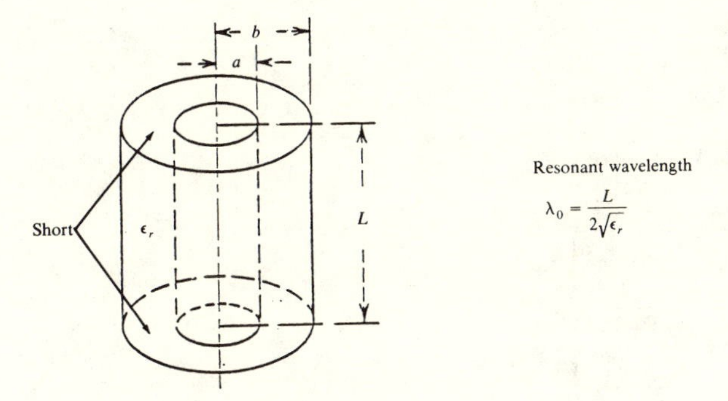

Figure D.3c Coaxial line resonator.

D.3 JUNCTIONS AND OTHER COMPONENTS

There are a wide variety of devices, consisting of modified sections of transmission lines and waveguides, that are useful as junctions connecting different components. An important example is the junction between a coaxial line or a waveguide, to a resonant cavity as illustrated in Fig. D.4a, b, and c.

Figure D.4a illustrates electric coupling between a coaxial line and a cavity by means of a probe. The center conductor of the coaxial line is shown penetrating the central portion of a $\lambda/2$ resonant coaxial cavity where the electric field is a maximum. Part of the electric field within the cavity terminates on the center conductor, and excites electromagnetic waves in the coaxial line. The deeper the penetration of the center conductor, the stronger the coupling between the resonator and the coaxial line.

Figure D.4b is an example of magnetic coupling between a coaxial line and a cavity by means of a coupling loop. As shown, the magnetic field,

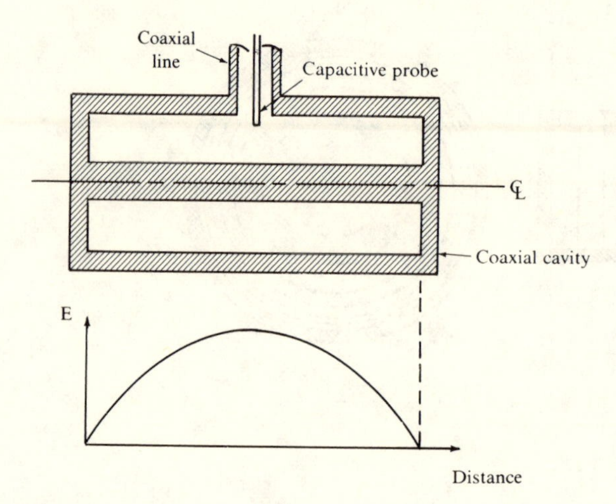

Figure D.4a Capacitive coupling.

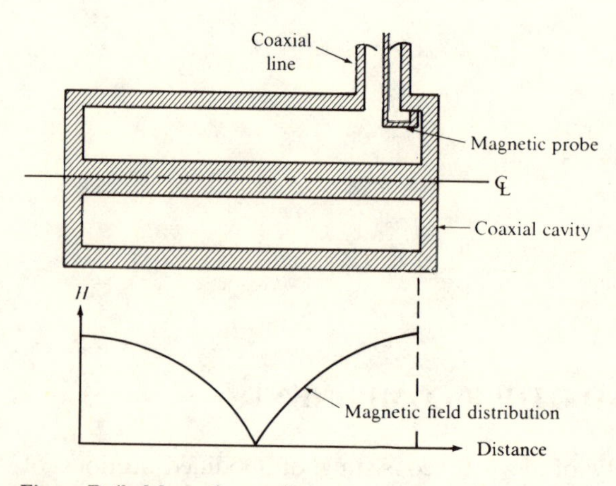

Figure D.4b Magnetic coupling.

which is a maximum at either end of the cavity, links the loop and induces a voltage which excites waves in the coaxial line. For this case, the larger the area of the loop, the larger the induced loop voltage and coupling.

Figure D.4c shows a TE_{10} waveguide magnetically coupled to a cylindrical resonator. In this case, some of the magnetic field within the cavity leaks though an iris cut into the sides of the waveguide and resonator walls, exciting electromagnetic waves in the waveguide. The larger the iris size, the stronger the degree of coupling.

It should be noted that the coupling systems shown in Fig. D.4a, b, and c are reciprocal, so that coupling can occur in either direction, i.e., electro-

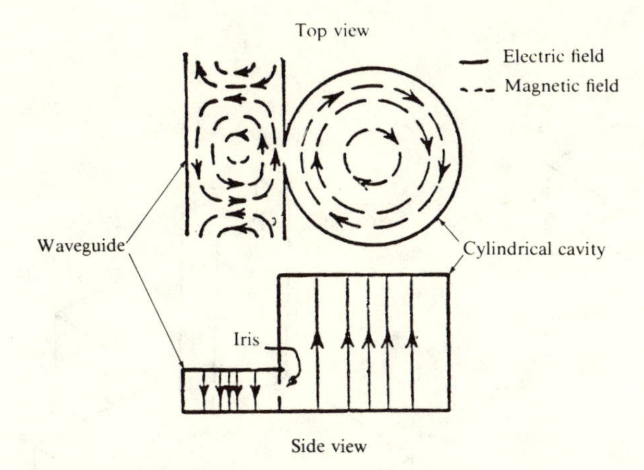

Figure D.4c Iris coupling (magnetic).

magnetic waves in the cavities can excite waves in the coaxial lines or waveguide, as discussed, or waves in the coaxial lines or waveguides can excite fields in the cavity resonators.

Figure D.5 illustrates waveguide hybrid structures. Shown in Fig. D.5a and b are schematic drawings of an *E*-plane tee and an *H*-plane tee, respectively. If a signal is applied to port 1 of the *E*-plane tee, the signals appearing at ports 2 and 3 will be equal in magnitude but 180° out of phase with respect to each other. In the case of the *H*-plane tee, the signals at ports 2 and 3 will also be equal in magnitude, but in this case they will be in phase with each other. The *magic tee*, shown in Fig. D.5c, is essentially a combination of Fig. D.5a and b. A special feature of the magic tee is that the junctions of the four arms are carefully designed so that, with ports 2, 3, and 4 match terminated, microwave power transmitted through port 1 will split evenly into ports 2 and 3, with practically no power into port 4. As with the *E*-plane tee, the signal appearing at port 3 will be 180° out of phase with the signal at port 2. If the signal is applied to port 4, however, no power will appear at port 1, and the input power will appear equally divided and in phase at ports 2 and 3.

The magic tee and the *E*- and *H*-plane tees can be used for power division and the connection of auxiliary lines or components. The magic tee is especially useful in impedance bridges, balanced mixers, phase detectors, discriminators, etc.

Figure D.6 shows a coaxial line to a TE_{01} waveguide adaptor. In this case an electric field from the extended center of the coaxial line fringes into the waveguide and launches the TE_{01} mode. The distance *a* is made close to $\frac{1}{4}$ wavelength. An excellent, very low-VSWR transition over a 15 to 20% bandwidth can be achieved with this structure. It should be noted that this

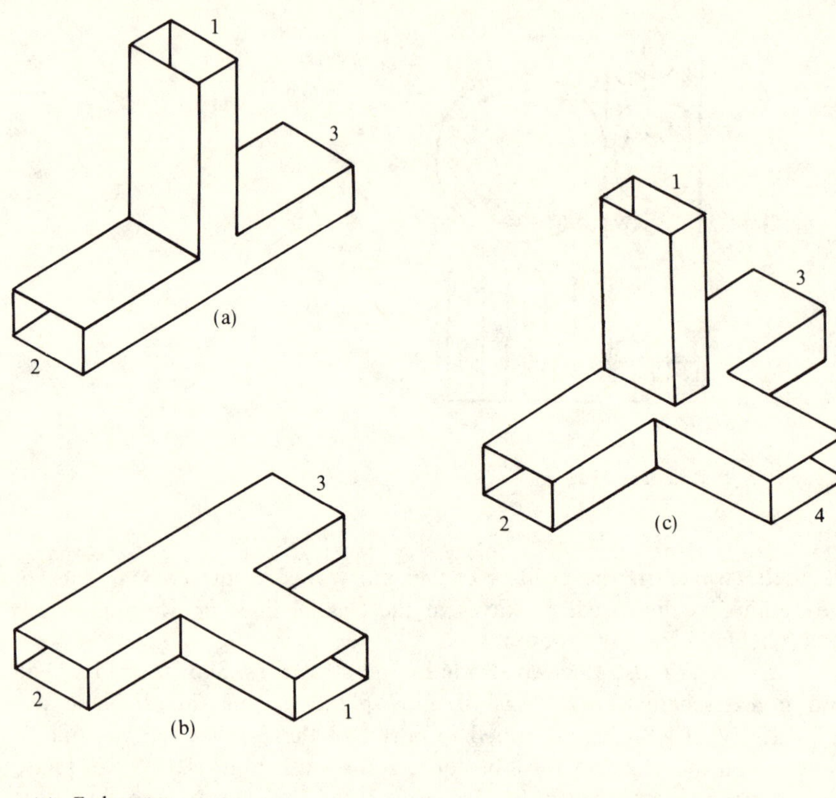

(a) *E*-plane tee
(b) *H*-plane tee
(c) Magic tee

Figure D.5 Waveguide tees.

structure is also reciprocal, so that a signal entering the waveguide will couple out into the coaxial line.

Figure D.7a, b, c, d, and e show typical waveguide corners, bends, and a twist. These structures permit directional changes and offsets in waveguide runs. Bent sections can introduce undesirable discontinuities in the waveguide. Consequently slight modifications in the design are introduced to minimize reflections. For example, the lengths L of the double-mitered bends shown in parts a and b are adjusted to minimize reflections. When enough physical space is available, gradual circular bends such as those shown in Fig. D.6c and d are generally used. Figure D.7e shows a continuous waveguide twist which rotates the polarization of the waves by 90°. If L is relatively large, as shown, very little reflection occurs as a result of the twist.

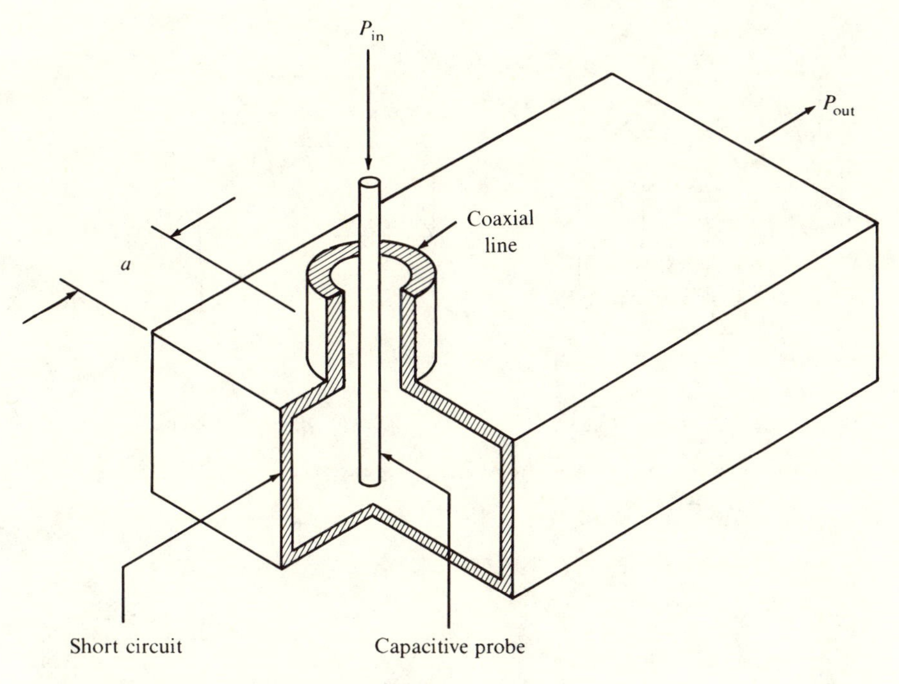

Figure D.6 Coaxial line to waveguide adaptor.

D.4 DIRECTIONAL COUPLERS

An interesting microwave circuit element having wide application is the directional coupler shown schematically in Fig. D.8a, b, and c. Because of its great importance in the microwave field, a detailed review of the basic operation of the directional coupler, a discussion of its limitations, and a brief review of its application are presented in this appendix.

Basically, a microwave directional coupler is a four-port component which, when connected to a waveguide or transmission line, can be used to sample either the incident wave or the reflected wave or both. An ideal directional coupler, shown schematically in Fig. D.8a, is defined as a junction of four transmission lines such that when all ports are terminated in their characteristic impedances, there is no coupling between ports (1) and (3) and between (2) and (4). Thus, a signal incident on port (1) will appear only at ports (2) and (4). Similarly if the signal were incident on port (2), output signals would appear at ports (1) and (3) only.

Usually port (3) is terminated in its characteristic impedance internally, and port (4) is used to sample the wave entering port (1). The same directional coupler can also be used to sample the reflected wave if it is connected to the transmission line in the reverse direction. In some cases,

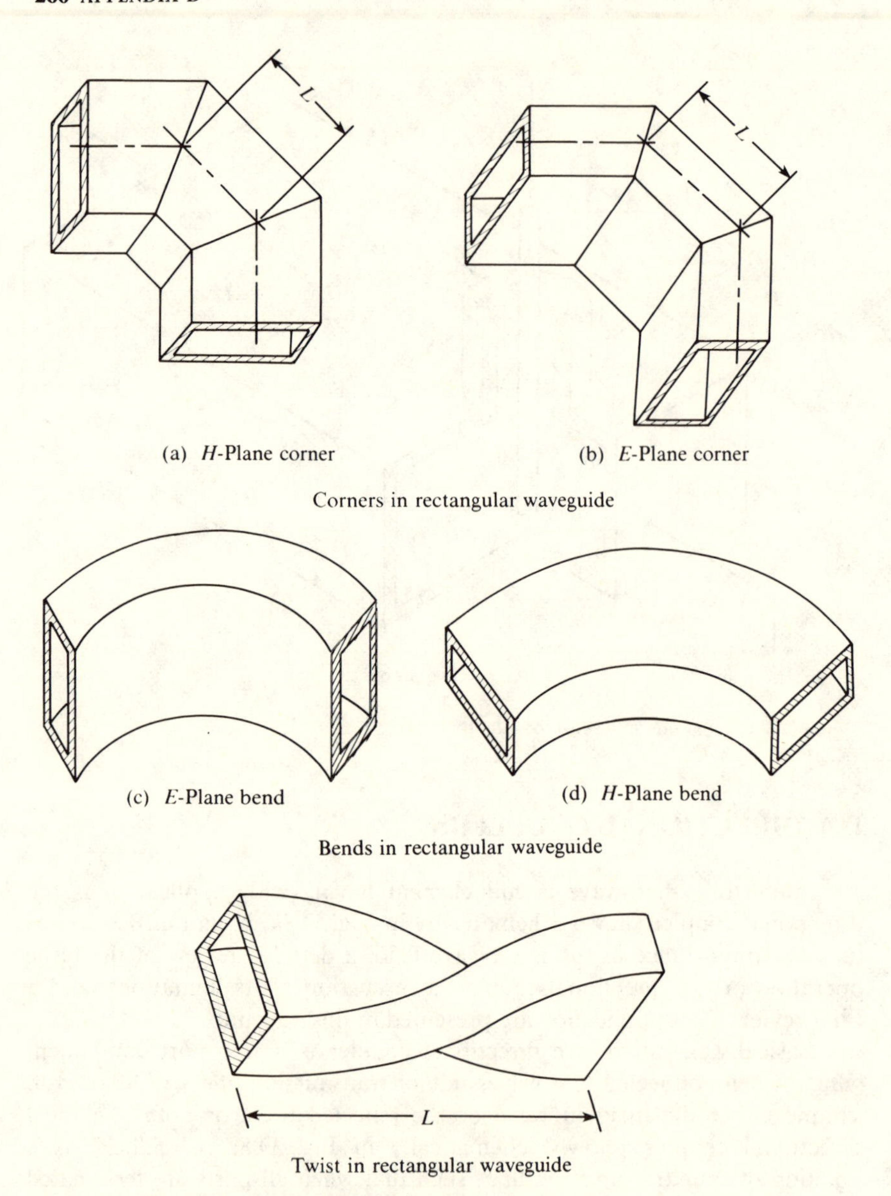

Figure D.7 Waveguide corners, bends, and twist.

however, it can be operated as a dual-directional coupler in which port (4) is used to sample the incident wave, as discussed, and any reflected wave impinging on port (2) will be sampled at port (3). In this case, special care must be taken in the design of the coupler to be sure that the circuits used to sample the signals at ports (3) and (4) are properly terminated. To avoid this problem, practical dual directional couplers often consist of two single-ended couplers connected back to back.

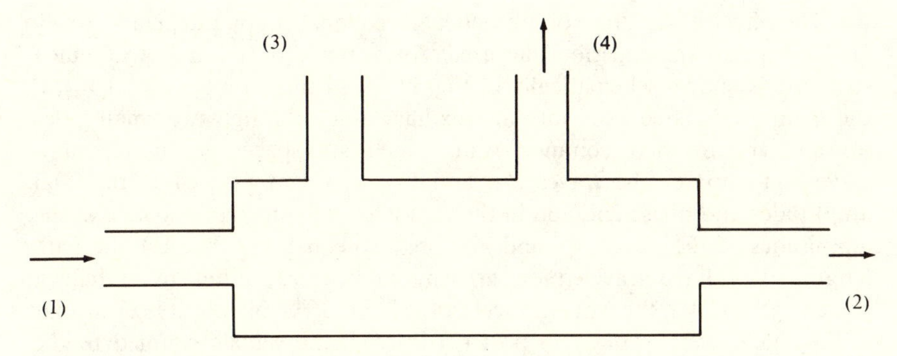

Figure D.8a Ideal directional coupler shown with a signal incident on port (1).

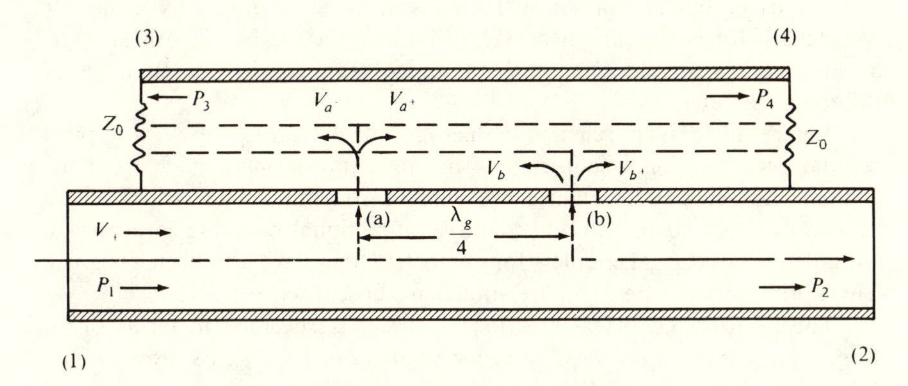

Figure D.8b Two-hole waveguide directional coupler.

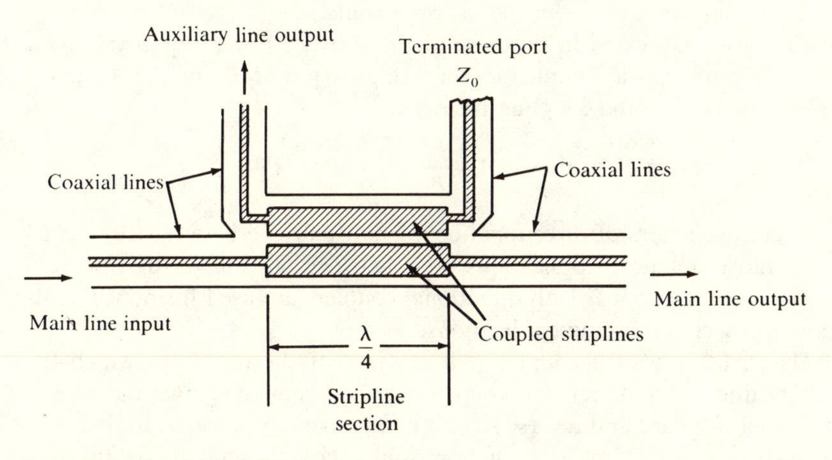

Figure D.8c Coaxial line directional coupler.

The physical structure of a practical directional coupler depends largely on the type of transmission line used. An example of a simple waveguide structure is shown schematically in Fig. D.8b. In this case the signal V_+ in the main guide is coupled into an auxiliary guide through two small holes (a) and (b) in their common wall, which are spaced a quarter of a wavelength apart. The waves V_{a+} and V_{b+} arrive at port (4) with equal amplitudes and phase and add in the characteristic-impedance load Z_0. The amplitudes of the waves V_{a-} and V_{b-} are also equal, but, because the path lengths they have traveled on arriving at port (3) differ by a half a wavelength (180°), the waves cancel in the load Z_0. Similarly, it is clear that if the signal were applied to port (2) instead, the signals coupled to the auxiliary guide would add at port (3) and cancel at port (4).

The two-hole waveguide directional coupler, while very useful, has a rather narrow bandwidth, since the holes must be separated by a quarter wavelength for optimum performance. To increase the bandwidth of operation, couplers are designed with multiple-offset, equally spaced holes of different diameters.

Figure D.8c is a schematic diagram of the structure of a typical coaxial-line directional coupler. As shown, a transition is made from the coaxial-line sections to quarter-wavelength striplines. The striplines are physically close to one another, and unidirectional coupling is achieved. Broadband directional couplers for microstrip lines have also been designed. Their principles of operation are similar to those discussed above.

The performance of a directional coupler is specified in terms of its standing-wave ratio, its *coupling factor* γ_c, its *directivity* γ_d, its *insertion loss* γ_L, and sometimes its *isolation* γ_i. The standing-wave ratio of a directional coupler is defined as the value measured at its input port, with all other ports terminated with reflectionless loads (Z_0). This characteristic is important in applications, such as impedance measurements, where multiple reflections can cause errors in the measurements.

The ratio, expressed in decibels, of the forward power P_1 in the main arm of the directional coupler to the sampled power P_4 in the auxiliary output is defined as the coupling factor:

$$\gamma_c = 10 \log \frac{P_1}{P_4} \qquad \text{(dB)}$$

Coupling factors of directional couplers usually lie between 10 and 20 dB for most practical applications, and between 30 and 40 dB for very high-power applications. 3-dB directional couplers are used for power-splitting and power-combining applications.

The directivity of a coupler is a measure of the degree to which the auxiliary line of the directional coupler can discriminate against the reverse signal when forward and reverse signals of equal amplitude exist in the main line. Basically it is a measure of how well a coupler can isolate the two signals and sets the limits on how accurately it can perform in a specific measurement system.

In a practical coupler, such as shown in Fig. D.86, the power at port (3) will not be zero, because V_{a-} and V_{b-} waves may not cancel completely. This is due to slight differences in the coupling hole sizes, improper hole spacing, as well as small reflections from port (4). It is evident that these effects can produce a measurement error, since power from a reflected signal incident on port (2) can couple to port (4) and add to that produced by the incident signal. Defined mathematically, the directivity γ_d, is

$$\gamma_d = 10 \log \frac{P_4}{P_4'} \qquad \text{(dB)}$$

where

P_4 = power at port (4) for an input power of P_1 at port (1)

P_4' = power at port (4) for an input power of P_1 at port (2)

In the measurement of γ_d it is necessary to adjust the relative phase of the reflected signal for the worst case, i.e., the value required to make the measured value of P_4/P_4' a maximum. Values of 30 dB or greater are typical of high quality multihole directional couplers.

The insertion loss of a directional coupler is equal to 10 times the logarithm of the ratio of the power out of the main line to the power fed into the main line. Referring to Figure D.8b,

$$\gamma_L = 10 \log \frac{P_2}{P_1} \qquad \text{(dB)}$$

This loss is due in part to small electrical losses in the main line and to the loss of the power coupled to the auxiliary line. This loss is usually considered unimportant at low frequencies (i.e., less than 20 GHz), but at high frequencies, where the signal power is usually quite low, it can be critical.

The isolation γ_i is another quantity which indicates how much of the incident power leaks into the isolated port. In terms of the input power P_1 and the power to the isolated port, P_3,

$$\gamma_i = 10 \log \frac{P_1}{P_3} \qquad \text{(dB)}$$

It should be noted that $\gamma_i \sim \gamma_c + \gamma_d$.

There are many practical applications of directional couplers. Dual directional couplers—two separate couplers, one connected to sample the incident power of a system and the other to sample the reflected power—can be used to measure the impedance of the system. Basically, the reflection coefficient can be found by taking the complex ratio of the two signals in the auxiliary arms; knowing the reflection coefficient, the impedance can also be found. Several "network analyzer" systems based upon this approach have been developed commercially. Directional couplers are also useful in injecting power into a transmission line for power flow in one direction. They are also useful to extract small samples within a microwave system so that its spectrum and power can be monitored.

Many other important microwave components exist. The above review is intended only as a general introduction to this important area of microwave engineering.

APPENDIX

E

PHYSICAL CONSTANTS AND UNITS

E.1 CONSTANTS

ϵ_0	Permittivity of free space	8.854×10^{-12} F/m
		$\simeq \dfrac{1}{36\pi} \times 10^{-9}$ F/m
$K = 1/4\pi\epsilon_0$		$8.89 \times 10^{9} \simeq 9 \times 10^{9}$ m/F
μ_0	Permeability of free space	$4\pi \times 10^{-7}$ H/m
$c = 1/\sqrt{\mu_0\epsilon_0}$	Speed of light in vacuum	2.998×10^{8} m/sec $\simeq 3 \times 10^{8}$ m/sec
m	Rest mass of electron	9.11×10^{-31} kg
e	Charge of electron	1.602×10^{-19} C
$\eta_0 = \sqrt{\mu_0/\epsilon_0}$	Impedance of free space	$120\pi\ \Omega = 377\ \Omega$

E.2 UNITS (SI)

I	1 ampere = 1 C/sec
ρ	C/m^3
V	1 volt = 1 J/C
J	A/m^2
D	C/m^2
E	V/m
B	1 tesla = 1 Wb/m^2 = 1 N/A-m = 10^4 gauss
H	1 A/m = $4\pi \times 10^{-3}$ oersted
C	1 farad = 1 C/V
P	W/m^2

INDEX